RUBBERY MATERIALS AND THEIR COMPOUNDS

RUBBERY MATERIALS AND THEIR COMPOUNDS

J. A. BRYDSON

Formerly Head, Department of Physical Sciences and Technology, Polytechnic of North London, UK

ELSEVIER APPLIED SCIENCE
LONDON and NEW YORK

ELSEVIER SCIENCE PUBLISHERS LTD
Crown House, Linton Road, Barking, Essex IG11 8JU, England

Sole Distributor in the USA and Canada
ELSEVIER SCIENCE PUBLISHING CO., INC.
52 Vanderbilt Avenue, New York, NY 10017, USA

WITH 82 TABLES AND 102 ILLUSTRATIONS

British Library Cataloguing in Publication Data

Brydson, J. A. (John Andrew), *1932–*
Rubbery materials and their compounds.
1. Polymer composite materials & polymer blends
I. Title
620.1′92

ISBN 1-85166-215-4

Library of Congress Cataloging in Publication Data

Brydson, J. A.
Rubbery materials and their compounds / J. A. Brydson.
p. cm.
Includes index.
ISBN 1-85166-215-4
1. Rubber. 2. Elastomers. I. Title.
TS1890.B83 1988 678′.2--dc19 88-11681
CIP

Printed in Great Britain by Galliard (Printers) Ltd, Great Yarmouth, Norfolk

Preface

Whilst there are a number of texts on the chemistry of rubbers and on a general overview of rubber technology I have been aware for some time of a lack of a single-volume text that surveys rubbers as materials. It is the aim of this book to provide such a text reviewing the polymers available, explaining the differences between grades within a polymer type, additives used in rubber compounds and then indicating with reference to specific products how polymers and additives may be formulated to give suitable compounds.

In some ways this may be considered as being intermediate to two earlier texts of mine, *Plastics Materials* and *Rubber Chemistry*. Although covering the same group of materials as the latter, *Rubbery Materials* has more the format of the former with emphasis on technology and application rather than on theoretical chemistry. However, because of its relevance to the subject matter of this text, Chapter 3 on structure and properties is closely based on Chapter 4 of *Rubber Chemistry*.

The text may be considered in four, unequal, parts. The first three chapters are general, covering the historical development of rubbery materials and the relationship between structure and properties. The next thirteen, and thus the main body of the book, provide a review of the various commercial rubbery polymers including their preparation, structure, properties, compounding, processing and applications. The next five chapters are concerned with additives whilst a final chapter considers compound design in the light of the contents of the previous chapters.

I have taken some topics out of the main body of the book. As a prefix I have listed common abbreviations for elastomeric materials whilst in appendices I have tabulated some specific gravities of additives, presented a note on the use of the Mooney viscometer, results from which are quoted in most chapters, given some references to additional general literature sources and described an ASTM system of classification of rubbers. After some indecision I have located a tabulated summary of the properties of the major rubbers in the last chapter.

For such a wide-ranging survey as this text provides I must acknowledge not only much original research and development work but also many excellent review papers, especially those in *Rubber Chemistry and Technology*. Amongst other important sources are the publications of David Blackley, Ken Lee and Tony Whelan, whom I was privileged to have as members of staff in my department at the Polytechnic of North London for several years. My thanks are also due to the Polytechnic for generously providing me, as an ex-member of staff, with library and other facilities.

I should also like to acknowledge the excellent trade manuals and other publications of polymer manufacturers whose staff were extremely helpful. In particular I would like to thank Bayer, B. F. Goodrich, Bunawerke Hüls, Dow Corning, Du Pont, Polysar, Shell International and Wacker amongst manufacturers and the Malaysian Rubber Producers' Research Association and the International Institute of Synthetic Rubber Producers for their assistance.

A Note on Abbreviations

In the text I have frequently made use of abbreviations. In the case of rubbery polymers I tried to ensure that the first time an abbreviation is used the full name is given alongside. However a list of abbreviations for rubbers is given on pages xix–xxi for ready reference. I have also made wide use of abbreviations for accelerators, in some cases without spelling out the full name the first time the abbreviation is used. The abbreviations for the principal accelerators with their chemical name and main characteristics are given on pages xxi–xxii and in Table 18.1. In the case of other additives I have used the full name alongside the abbreviation on the first occasion but Chapters 19–21 should be used for reference purposes.

J. A. Brydson

Contents

Abbreviations

STANDARD ABBREVIATIONS FOR RUBBERY MATERIALS

(Based on ISO Recommendation R1629 and ASTM D 1418)

ABR Acrylate–butadiene rubber
ACM Copolymer of ethyl or other acrylates and a small amount of a monomer which facilitates vulcanization
AECO Terpolymer of allyl glycidyl ether, ethylene oxide and epichlorhydrin
AEM Copolymer of ethyl or other acrylate and ethylene
AFMU Terpolymer of tetrafluoroethylene, trifluoronitrosomethane and nitrosoperfluorobutyric acid
ANM Copolymer of ethyl or other acrylate and acrylonitrile
AU Polyester urethane
BIIR Bromo-isobutene–isoprene rubber (brominated butyl rubber)
BR Butadiene rubber
CFM Polychlorotrifluoroethylene
CIIR Chloro-isobutene–isoprene rubber (chlorinated butyl rubber)
CM Chlorinated polyethylene
CO Epichlorhydrin rubber
CR Chloroprene rubber
CSM Chlorosulphonated polyethylene
ECO Ethylene oxide and epichlorhydrin copolymer
EAM Ethylene–vinyl acetate copolymer
EPDM Terpolymer of ethylene, propylene and a diene with the residual unsaturated portion of the diene in the side chain
EPM Ethylene–propylene copolymer
EU Polyether urethane
FFKM Perfluoro rubber of the polymethylene type having all substituent groups on the polymer chain either fluoro, perfluoroalkyl or perfluoroalkoxy groups
FKM Fluororubber of the polymethylene type having substituent fluoro and perfluoroalkoxy groups on the main chain

FVMQ Silicone rubber having fluorine, vinyl and methyl substituent groups on the polymer chain
GPO Polypropylene oxide rubber
IIR Isobutene–isoprene rubber (butyl rubber)
IM Polyisobutene
IR Isoprene rubber (synthetic)
MQ Silicone rubber having only methyl substituent groups on the polymer chain
NBR Nitrile–butadiene rubber (nitrile rubber)
NIR Nitrile–isoprene rubber
NR Natural rubber
PBR Pyridine–butadiene rubber
PMQ Silicone rubber having both methyl and phenyl groups on the polymer chain
PSBR Pyridine–styrene–butadiene rubber
PVMQ Silicone rubber having methyl, phenyl and vinyl substituent groups on the polymer chain
Q Rubber having silicon in the polymer chain
SBR Styrene–butadiene rubber
T Rubbers having sulphur in the polymer chain (excluding copolymers based on CR)
VMQ Silicone rubber having both methyl and vinyl substituent groups in the polymer chain
XNBR Carboxylic–nitrile butadiene rubber (carboxynitrile rubber)
XSBR Carboxylic–styrene butadiene rubber
Y Prefix indicating thermoplastic rubber
YBPO Thermoplastic block polyether–polyester rubbers

MISCELLANEOUS ABBREVIATIONS FOR RUBBERY MATERIALS

In addition to the nomenclature based on ISO and ASTM recommendations, several other abbreviations are widely used. Those most likely to be encountered are:

ENR Epoxidized natural rubber
EPR Ethylene–propylene rubber (either EPM or EPDM)
EVA Ethylene–vinyl acetate copolymer (instead of EAM)
OESBR Oil-extended SBR

SBS	Styrene–butadiene–styrene triblock copolymer
SEBS	Hydrogenated SBS
SIR	Standard Indonesian Rubber
SIS	Styrene–isoprene–styrene triblock copolymer
SMR	Standard Malaysian rubber

During the Second World War the United States Government introduced the following systems of nomenclature which continued in use, at least partially, until the 1950s and is used in many publications of the period.

GR–A	Government Rubber–Acrylonitrile (modern equivalent, NBR)
GR–I	Government Rubber—Isobutylene (IIR)
GR–M	Government Rubber—Monovinyl acetylene (CR)
GR–P	Government Rubber—Polysulphide (T)
GR–S	Government Rubber—Styrene (SBR)

ABBREVIATIONS FOR SOME COMMON VULCANIZATION ACCELERATORS

BDTM	2-benzothiazole-dithio-*N*-morpholine
CBS	*N*-cyclohexylbenzothiazole-2-sulphenamide
DCBS	*N*,*N*-dicyclohexylbenzothiazole-2-sulphenamide
DOTG	Di-*o*-tolylguanidine
DPG	Diphenylguanidine
DPTT	Dipentamethylene thiuram tetrasulphide
DTDM	4,4′-dithiodimorpholine
MBS	*N*-morpholinothiobenzothiazole-2-sulphenamide
MBT	2-mercaptobenzothiazole
MBTS	Dibenzothiazole disulphide
NOBS	*N*-oxydiethylbenzothiazole-2-sulphenamide
OTOS	*N*-oxydiethylenethiocarbamyl-*N*-oxydiethylene sulphenamide
PPD	Piperidine pentamethylene dithiocarbamate
SDC	Sodium diethyldithiocarbamate
SIX	Sodium isopropylxanthate
SMBT	Sodium mercaptobenzothiazole
TBBS	*N*-*t*-butylbenzothiazole-2-sulphenamide
TDEDC	Tellurium diethyldithiocarbamate
TETD	Tetraethylthiuram disulphide
TMTD	Tetramethylthiuram disulphide
TMT	Tetramethylthiuram disulphide

TMTM	Tetramethylthiuram monosulphide
TPG	Triphenylguanidine
ZBDP	Zinc dibutyldithiophosphate
ZBX	Zinc butylxanthate
ZDC	Zinc diethyldithiocarbamate
ZEDC	Zinc diethyldithiocarbamate
ZDBC	Zinc dibutyldithiocarbamate
ZDMC	Zinc dimethyldithiocarbamate
ZIX	Zinc isopropylxanthate
ZMBT	Zinc mercaptobenzothiazole

Chapter 1

Introduction

1.1 RUBBER AND ADDITIVES

The ability of a rubber to extend in length several hundred per cent on application of a stress, with virtually immediate and complete recovery on release of the stress, cannot fail to fascinate both the scientist and the layman. Rubbery materials are, however, not simply of academic interest nor toys, as they were when first observed by the sailors of Columbus in the late 15th century when they saw the native Central American Indians using playballs made from natural rubber. Today, rubber is a vital material, particularly in engineering applications, of which the pneumatic tyre is the best known example.

What is perhaps surprising is that rubbery behaviour is shown by many materials, polymeric in nature, of which the first to be exploited was natural rubber. Still an all-important material, the natural product is now complemented by a range of synthetic materials which have found niches in the market because of such characteristics as heat resistance, oil resistance, resistance to burning and low air permeability. The selection of a particular material would require that several factors be taken into consideration, of which the raw material cost is usually one.

Until about 1930, the rubber industry had only natural rubber available but nevertheless it was able to produce a diversity of products such as tyres, elastic bands, erasers, ebonite battery boxes, rubber flooring, cables and shoe soles by judicious use of additives, giving products with a very wide range of properties using formulations which also varied considerably in cost per unit weight.

Today, with an ever-increasing range of polymers and additives available, the rubber technologist has a greater scope than ever before for tailoring formulations to meet specifications and price. It is the purpose of this book to provide the background information to facilitate this task.

1.2 THE HISTORICAL DEVELOPMENT OF RUBBER TECHNOLOGY

The earliest written references to rubber are found in works published early in the 16th century (d'Anghiera, 1516; Valdes, 1535–1537). There is, however, no evidence that the natural polymer itself ever reached Europe prior to the travels of Voltaire's friend, Charles de la Condamine, who travelled widely in South America between 1736 and 1743. In 1770 the Englishman, Joseph Priestley, recommended the material for use as an eraser, or rubber, the latter term then being adopted by the English-speaking world as a generic term for materials of high reversible elasticity. It is interesting to note that in many other parts of the world the original South American Indian name has been adopted with minor variations. Typical examples are caoutchouc (French), caucho (Spanish), kauchuk (Russian) and kautschuk (German). In England the word caoutchouc is still often used to describe the pure rubber polymer hydrocarbon.

Natural rubber remained little more than an academic novelty until the early years of the 19th century, when there was some use for shapes cut from the blocks of raw rubber produced in Brazil. The first notable progress came with the discovery of solvents of the material, originally by French scientists, but later exploited by Charles Macintosh who patented the use of coal tar naphtha in 1823. Macintosh then interposed the resultant solution between two fabrics, which became united as the solvent evaporated, to produce an impermeable laminated material suitable for rainproof outer garments.

The principles of polymer processing have somewhat inelegantly yet economically been summarized by the statement 'Get the shape; set the shape'. Macintosh was able to shape by using rubber in solution form and setting it by evaporation. This approach was of necessity somewhat limited to thin film applications.

At this time attempts to shape rubber by pressing operations were frustrated by the fact that, after pressing, the rubber returned to its original shape. We now know that this is due to the fact that, although the flexible long-chain molecules are capable of being highly extended, entanglements together with tiny crystal structures through which the molecules pass produced a network structure which prevented any permanent large-scale distortion.

About 1830 Thomas Hancock, working in North London, subjected pieces of rubber to intensive shearing and discovered that the material changed from being primarily elastic to a more tractable material, capable of permanent flow. This process, *mastication*, is now known to result from the breaking down of the molecules so that there are fewer entanglements and flow becomes possible. This discovery was the first of two keys that opened the door to the exploitation of conventional rubber technology. This was because it was now possible to compound, in the pharmaceutical rather than the strict chemical sense, the rubber with other materials such as fillers, oils and colours, and then to shape the material, eventually by moulding, calendering and extrusion.

Unfortunately, such masticated rubber had very limited use since the desirable attribute of rubberiness had been largely lost in the process. In 1839 the American, Charles Goodyear, found that heating rubber with sulphur and white lead gave it far superior properties. Hancock exploited this process, termed *vulcanization* by his friend Brockeden, to provide an elastic material less liable to become sticky in hot weather and stiff in cold and which was insoluble in common solvents. Whereas mastication made possible 'getting the shape', vulcanization allowed 'setting the shape'. One hundred and fifty years on, mastication and sulphur-vulcanization remain essential features of most non-latex natural rubber technology processing.

1.2.1 The Advent of Synthetic Rubber

In 1826 Faraday showed that natural rubber was a hydrocarbon of empirical formulae $(C_5H_8)_n$. Later, Greville Williams and others working independently found that destructive distillation of natural rubber yielded a volatile liquid of formula C_5H_8, to which Williams gave the name isoprene:

$$
\begin{array}{c}
\quad\ CH_3 \\
\quad\ | \\
CH_2{=}C{-}CH{=}CH_2
\end{array}
$$

In 1879 F. G. Bouchardat reversed the process by producing a rubbery material from isoprene which had itself been obtained from natural rubber. By 1884 Tilden produced isoprene from turpentine and then obtained a rubber-like material by methods similar to those used by Bouchardat. This may be considered as the first production of a rubber by synthetic methods.

In due course it was established that the polymer molecule in natural rubber was a high-molecular-weight *cis*-1,4-polyisoprene. Whilst it was soon appreciated that the natural product was not obtained in nature by polymerization of isoprene, it was clearly an attractive synthetic approach.

Already in the late 19th century it began to be recognized that conjugated diene hydrocarbons other than isoprene could be converted into elastic substances. Indeed, in 1881 von Hofmann had prepared a rubber from

1,3-pentadiene, and in 1892 Couturier polymerized 2,3-dimethyl-1,3-butadiene. (In passing it may be noted that the first reference to butadiene polymerization appears to be some 17 years later with the issue of a German Patent to Farbenfabriken Bayer, naming E. Hofmann and C. Coutelle as inventors.)

The turn of the century saw the development of the automobile and with it the pneumatic tyre. The consequent demands for natural rubber, then only obtainable from jungle plants rather than plantations, resulted in a severe shortage of the product with substantial increases in price. This led to the rise of the plantation industry and to the first serious research work aimed at producing a rubber synthetically.

In 1908 Strange and Graham Ltd, in London, commenced such a programme of research with a team of highly distinguished chemists, including F. E. Matthews, W. H. Perkin, Jr, and Chaim Weizmann (who later became the first President of Israel). In 1910 Matthews sealed a tube containing isoprene and sodium and set it aside. After a month he noted that the contents had become viscous, and in two months a solid amber-coloured rubber. A British Patent was filed on 25 October 1910. Similar experiments were carried out by C. Harries working for the Bayer Company in Germany, which independently also led to the discovery of sodium-catalysed polymerization of isoprene. In spite of these successes the work was not exploited because of the appearance of plantation rubber. Development work virtually ceased in England, and was severely curtailed in Germany.

However, during 1911 the Bayer Company actually produced a few tonnes of Methyl Rubber (poly-2,3-dimethylbutadiene), but once again the low price of plantation rubber discouraged further development. The situation changed with the outbreak of World War I, and the shortage of raw rubber in Germany. Eventually in 1917 the first commercial production of a man-made rubber commenced. There were two types: Methyl Rubber W (W for *weich*—soft) and Methyl Rubber H (H for *hart*—hard). Type W was prepared by heat polymerization at about 70°C over a period of five months, and Methyl Rubber H, a hard grade suitable for battery boxes and other ebonite-type uses, required some three to four months' polymerization at the lower temperature of 30–35°C. During the war a total of 2350 tons of Methyl Rubber was produced but, as it was inferior to the natural product, production ceased at the end of the war with the renewed availability of natural rubber in Germany. For a second, but not the last, time research work into man-made rubbers went into a severe recession.

1.2.2 Additives for Rubber

Long before the commercial appearance of the first synthetic rubbers, it had been recognized that natural rubber could be modified by the use of additives. Whilst the use of sulphur to vulcanize the raw material must be the most outstanding example, it became the practice during the 19th century to blend rubber with a variety of other additives. Vulcanized unsaturated oils (factices) seem to have been used since the 1850s as softeners and soon became known as *rubber substitutes*, frequently using the French equivalent caoutchouc factice.

By 1880 materials such as chalk were being used to reduce costs, sand as an abrasive for erasers, and zinc white and iron oxide as pigments.

The first patent for what was effectively an antioxidant was taken out by Murphy in 1870, although the age-old practice of South American Indians of smoking drying rubber had already made use of the preservative effect of phenolic bodies. A substantial improvement in antioxidant technology took place with the development by Ostwald and Ostwald in Germany in 1908 of their discovery that a number of neutral or basic nitrogen-containing aromatic or heterocyclic compounds were useful for retarding ageing.

Soon after it was observed that some antioxidants, for example piperylene, had the extra effect of increasing the rate of vulcanization, such materials now being known as accelerators. The first German patents were taken out in 1913; it was later established that Oenslager, working for the Diamond Rubber Company in Akron, Ohio, had discovered the accelerating value of thiocarbanilide in 1906, but that this knowledge was kept as a manufacturing secret.

At about the same time, the beneficial reinforcing effects of carbon black were being recognized. Whilst the reinforcing effect appears to have first been observed in England in about 1910, by 1912 tyres were being manufactured in the United States reinforced with substantial amounts of channel black.

The use of a variety of softening materials, in addition to the factices, was also becoming common. Textbooks published in the early 1920s refer to a bewildering array of materials, including asphaltic bitumens, cryptically known as mineral rubbers, rosin, pitches, tars, mineral oils, such materials acting both as processing aids and to reduce the hardness of the finished product.

There were thus available by the early 1920s most of the classes of rubber additive that we recognize today, although the mechanism by which they functioned was far from adequately understood.

1.2.3 The Return of Synthetic Rubber

In the period immediately following World War I, synthetic rubber research was minimal and largely confined to minor work in the United States and the Soviet Union. However, about 1926 two important stimuli occurred. Firstly, Staudinger produced a series of papers that conclusively demonstrated the long-chain polymeric nature of the natural material. Secondly, the price of natural rubber was rising rapidly. This led the newly-formed IG Farbenindustrie AG (a cartel embracing older firms including Bayer, BASF and Hoechst) to restart synthetic rubber research.

The ultimate success of the programme actually had its roots in two pre-war discoveries. One was the finding by Holt and Stimmig in 1911 that peroxides were polymerization initiators; the other was the observation by Hofmann, Gottlob and their co-workers at Bayer that there were advantages in emulsion polymerization.

Nevertheless, the first commercial products were made by sodium-catalysed polymerization of butadiene and known as Buna Rubbers, which became available in the late 1920s (the term Buna being derived from butadiene and natrium, the New Latin term for sodium). One grade, Buna 85, in fact used potassium instead of sodium. Almost simultaneously with the advent of the Buna Rubbers, German workers prepared poly-butadienes by peroxide-initiated emulsion polymerization, but the properties of the rubber were disappointing. However, in 1929 Tschunker and Boch suggested the use of styrene as a second monomer to produce a butadiene–styrene copolymer, and the following year Conrad and Tschunker mentioned the use of acrylonitrile as comonomer.

The butadiene–styrene rubber, designated Buna S, potentially a general-purpose rubber, could not compete with the natural product and production only commenced in 1937 with the desire of Germany to become self-sufficient in general-purpose rubbers. The butadiene–acrylonitrile copolymers, today's important petrol-resisting nitrile rubbers, were marketed somewhat earlier in 1935 as Buna N. The trade name Perbunan is a clear derivative of the original designation.

The late 1920s also saw the development of other synthetics. Some time between 1920 and 1924 J. C. Patrick and N. M. Mnookin had reacted ethylene dichloride with sodium polysulphide by a condensation reaction to produce a novel rubber. This discovery came as the result of an unsuccessful attempt to utilize refinery olefin streams for making glycols by chlorinating the olefins and then reacting the chlorination products with alkali. Not only did these rubbers exhibit extremely good oil resistance, but they differed from the other rubbers which had been produced before this time in containing no carbon–carbon double bonds, as did the polymers of isoprene, butadiene and dimethyl butadiene. Commercial production commenced in 1929 and, although current commercial materials have a somewhat different structure, they may be considered as being the oldest synthetic rubber-like materials still in production.

The advent of the polysulphide rubbers, marketed under the trade name Thiokol, was soon followed by the second major American contribution to the production of synthetic rubbers. This was polychloroprene, which had its origin in work on acetylene chemistry which was started by Niewland in 1906 and which led in the mid-1920s to the development of a process for making a vinylacetylene from acetylene. The addition of hydrogen chloride to the vinylacetylene yielded 2-chloro-1,3-butadiene, usually known as chloroprene, which could be polymerized to form a rubber with good heat and oil resistance. This was initially marketed as Duprene in 1932 by Du Pont, who shortly afterwards changed the name to Neoprene.

Within the next ten years American chemists were also responsible for the development of another elastomer, butyl rubber. In 1930 the Standard Oil Company of New Jersey had entered into an agreement with IG Farben in Germany for the two companies to assist each other in developing new products and processes based on raw materials from petroleum. This led the German company to disclose to the Americans the fact that isobutylene could be polymerized to a high-molecular-weight polymer by a strong Lewis acid catalyst such as boron trifluoride at the low temperature of −75°C. Although the management of Standard Oil were primarily interested in the use of the material as a fuel additive, two chemists within the company, R. M. Thomas and W. J. Sparks, were intrigued by the unusual elastomeric properties of the material. As a consequence of their work, the American company produced copolymers containing small amounts of isoprene, the latter providing a few double bonds in an otherwise saturated hydrocarbon, to allow cross-linking to occur. Commercial production of this material first commenced in 1942. Butyl rubber remains an important synthetic rubber, particularly because of its low air permeability.

By the time that the United States and Japan had entered the Second World War, there were thus available a number of special-purpose rubbers but only one general-purpose rubber, natural rubber. The occupation in 1942 by the Japanese of Malaya and what was then the Dutch East Indies consequently severely curtailed the availability of the natural product to Britain and to the United States. Fortunately for them, the IG Farben–Standard Oil agreement had provided Esso with technical details for the production of Buna S, the peroxide-initiated emulsion-polymerized copolymer of butadiene and styrene. As a result, the US and Canadian governments built, between 1942 and 1944, a total of 87 factories with a total annual output capacity of about 1 000 000 tons to produce materials based on the Buna S process. It is difficult to overrate the role that this crash programme played in determining the course of the war. The rubber produced, then designated as GR–S, was, however, like Methyl Rubber before it in the First World War, generally inferior to the natural product. At the end of the war production dropped sharply and GR–S was largely regarded as something of the past.

1.2.4 Post-war Developments

For the years immediately following World War II the rubber industry continued to be dominated by natural rubber, although materials such as butyl rubber for inner tubes, nitrile rubber for petrol-resisting applications, and polychloroprene for general heat and oil resistance, had important markets. However, following the pattern established after the First World War, a combination of political and economic events, together with technical developments, provided the stimulus to the production of man-made general-purpose rubbers. The political event was the start of the Korean war, which served as a reminder that supplies of such an important commodity as rubber should not be dependent on a relatively small geographic area. Simultaneously, development of low-temperature polymerization processes led to much better products, which became known as 'cold' rubbers to distinguish them from the older materials which now became known as 'hot' rubbers in reference to their higher polymerization temperatures. For a number of purposes the cold rubbers were indeed preferable to the natural material, particularly when factors such as cost and consistency of quality were taken into account. These factors led to a strong recovery in the use of butadiene–styrene rubbers and, in the mid-1950s, the term GR–S, a term which was never more than a United States government designation, was dropped in favour of the designation SBR, which is now used almost universally. In due course SBR consumption was to exceed that of the natural material.

The success of SBR arose from the fact that the copolymer exhibited better properties than the homopolymer from butadiene. The poor quality of polybutadiene obtained by free-radical-initiated emulsion polymerization was largely due to the lack of control of the polymer structure so that only about 10% of the polymer units were in the desired *cis*-1,4 conformation. The possibility of preparing more regularly structured polybutadienes, as well as a number of other new synthetic rubbers, became possible in the 1950s because of the advent of new organometallic catalyst systems.

In some ways this was a revival of the old Strange and Graham/Bayer approach which in 1910 had led to the sodium-catalysed polymerization of isoprene. In 1912 Hans Labhardt of BASF, in a German Patent, had suggested that butadiene and its homologues could be polymerized with the aid of alkali-metal alkyls. In the 1920s Carl Ziegler and his co-workers, who were familiar with Labhardt's patent, began a programme of research which, by the mid-1950s, had enabled Ziegler to develop stereospecific catalyst systems which were commercially exploited in the production of high-density polyethylenes. Ziegler's discoveries were then further developed by Natta and others to produce what became known as Ziegler–Natta catalyst systems, and this led to the availability of a new generation of rubbers with a much more controlled molecular architecture than had been obtained before. For the first time it could be claimed, with some reason, that a true synthetic 'natural rubber' had been produced. Not only were polyisoprenes and polybutadienes of high steric purity produced, but there also became available the ethylene–propylene rubbers, as well as new forms of SBR produced by solution polymerization. Whilst these new rubbers have in no way replaced the older natural rubber and SBR, they have found important applications in both the tyre and general rubber goods industries.

At the same time, the number of special-purpose and speciality rubbers has also increased. The post-war period has seen the development of silicone and fluoro rubbers of exceptional heat resistance, the acrylic and epichlorhydrin rubbers with good oil and heat resistance, and the ubiquitous polyurethanes.

1.2.5 Markets and Applications of Rubbery Materials

Before World War II rubbery materials, mainly based on natural rubber, were used for a great diversity of applications, many of which did not require the property of reversible high elasticity, i.e. rubberiness. Examples included cables, footwear and flooring. The rapid growth in the availability of thermoplastics, and in particular polyethylene and plasticized PVC, led to the replacement of rubber, at least partially, in many such applications. Whilst the non-rubbery properties of rubbers have often been such that there has been continued use in many of these applications, there has been a change in the overall pattern of usage. Thus the rubber industry has become very dependent on the automotive industry with about 60% of new rubber being consumed in tyre products alone. Not surprisingly those rubbers widely used in tyres (natural rubber, SBR and polybutadiene rubber) comprise about 80% of the world rubber supply in tonnage terms (Table 1.1). Other rubbers such as the ethylene–propylene and chloroprene rubbers which are widely used in the general rubber goods industries have at most only about 5% of the market.

TABLE 1.1
World rubber supply by type—excluding CPEC
(After IISRP, 1988)

Rubber	%			
	1977	*1980*	*1983*	*1986*
Natural rubber (NR)	34·2	35·7	38·6	43·2
Styrene–butadiene rubber (SBR)	38·4	34·6	31·6	27·7
Butadiene rubber (BR)	10·3	10·9	11·2	11·0
Ethylene–propylene rubber (EPDM)	3·3	3·8	4·7	5·2
Butyl rubber (IIR)	4·5	5·0	5·1	4·5[a]
Chloroprene rubber (CR)	3·7	3·7	3·7	3·3
Nitrile rubber (NBR)	2·2	2·5	2·3	2·4
Synthetic isoprene rubber (IR)	2·4	2·6	1·6	1·4
Acrylic rubbers (ACM and AEM)	0·3	0·4	0·4	0·3
Other rubbers	0·7	0·8	0·8	1·0[a]
	Tonnes $\times 10^6$			
SBR supply (solid + latex)	3·33	2·89	2·56	2·56

[a] The 1986 figures for IIR and 'Other rubbers' have been estimated by the author as IIR data is now included with 'Other rubbers' by the IISRP.

For the manufacturer of general-purpose rubbers, both natural and synthetic, the dependence on the tyre industry has been a mixed blessing. The development of the radial tyre, more durable than its cross-ply predecessor, cut down on the rate of replacement of tyres per car. Furthermore the huge rise in the price of petroleum in the 1970s and early 1980s led to many measures to improve economy in the use of petrol. Moves included the trend to lighter and smaller cars and lower maximum road speeds, both of which meant that the tyre rubber market did not grow as fast as any growth in car production. For much of the 1980s there has been little profit in the manufacture of general-purpose rubbers because of their dependence on the tyre industry.

More positively, the automotive engineer has made increasing use of the properties of rubber in modern car and truck design. As specifications have become more stringent there has been particular interest in the speciality rubbers. Examples of elastomers in modern cars include chlorosulphonated polyethylenes in power steering hose, ethylene–propylene rubbers in radiator hose, tail-light gaskets and window seals, ethylene–acrylic rubbers in axle and prop shaft seals and crankshaft dampers, fluoroelastomers for valve stem seals and O-rings, epichlorhydrin rubbers for control system hose, acrylic rubbers for pinion seals, and silicone rubbers for spark plug boots. 'Puncture-proof' rubber fuel tanks are used in racing cars. It has to be emphasized that such a list is by no means comprehensive; it is only intended to give examples, and in many of the applications cited other rubbers may also be used.

Whilst rubbers have lost much of their earlier market as an electrical insulation material they are finding increasing use in specialist electrical insulation applications. The ability to obtain flexible, non-melting, heat-resisting materials which can be made non-burning (often without recourse to chlorine or phosphorus in the base polymer or any additives) is of particular value in aerospace, marine and underground applications. Synthetic rubbers are also of particular value in the high-temperature environments encountered in automotive under-the-bonnet insulation applications. The toughness of rubbers makes them particularly suitable for power cable jacketing.

In civil engineering, natural rubber and chloroprene rubbers are now widely accepted for bridge bearings whilst ethylene–propylene and butyl rubbers are important for reservoir lining. Other important engineering-related uses include conveyor belting, power transmission belting, fire hose liners, vibration insulation pads, dock fenders and chemical plant equipment such as valve diaphragms. Many automotive applications have parallels in aerospace and rail transport. Medical uses include blood transfusion and dialysis tubing, hypodermic syringe seals, condoms and babies' dummies, whilst in sport and recreation activities rubbery materials are used for example for squash and tennis balls and fabric coatings for balloons and inflatable boats. Familiar applications in the home include shoe soles and heels, carpet backing and elastic bands. There are indeed few areas of human endeavour where rubbery materials are not employed.

What is also quite clear is that the technical specifications are now much more demanding than in the past. It is becoming more and more appreciated that the correct choice and employment of an elastomer which may involve only a small part of the cost of a product can have a crucial role in determining the success or otherwise of that product. It is surely no coincidence that the greater reliability of cars in recent years is related to the recognition by car manufacturers of the need to pay particular care to the selection of the elastomer in critical applications.

BIBLIOGRAPHY

Historical Aspects

The following texts provide a good introduction to the history of rubber science and technology.

Pickles, S. S. (1951). *Trans. Inst. Rubber Ind.* **27**, 148 (This is the text of the sixth IRI Foundation Lecture entitled 'The chemical constitution of the rubber molecule').

Schidrowitz, P. & Dawson, T. R. (1952). *History of the Rubber Industry*, published for the Institution of the Rubber Industry by W. Heffer & Sons, Cambridge.

Törnqvist, E. G. M. (1968). In *Polymer Chemistry of Synthetic Elastomers*, Part 1, Chapter 2, ed. J. P. Kennedy & E. G. M. Törnqvist. Interscience, New York.

Also of interest is a number of short articles published in 1985 as 'a tribute to the rubber chemists' in *European Rubber Journal*, **167**(1) (Jan.), 15, and **167**(2) (Feb.), 23.

The earliest references to natural rubber trees are said to be in *De orbe nuovo* (8th 'decade') by Pietro Martire d'Anghiera published in Latin in 1516. *La Historia natural y general de las Indias* by Gonzalo Fernandez De Oviedo y Valdes, published in Seville in the period 1535–1537, has also been credited with the first written reference to rubber itself.

Statistical Data

The most comprehensive sources of data are:

The Rubber Statistical Bulletin, published monthly by the International Rubber Study Group, London.

Worldwide Rubber Statistics, published annually by the International Institute of Synthetic Rubber Producers, Houston.

REFERENCES

BASF (1912). German Patent 255 786.

Farbenfabriken Bayer (1909). German Patent 235 423.

IISRP (1988). *Worldwide Rubber Statistics—1988*. International Institute of Synthetic Rubber Producers, Houston.

Murphy (1870). US Patent 99 935, 15 February 1870.

Strange & Graham Ltd (1910). British Patent 24 790.

Chapter 2

Properties and Structure: Rubbery Properties

2.1 INTRODUCTION

Today rubbers are used largely because of their elastic properties. This may seem like a statement of the obvious, but it should be appreciated that before World War II rubber was very widely used in belting, flooring, cables and proofings where flexibility rather than high elasticity was the main requirement. The advent of thermoplastics such as plasticized PVC has led to large inroads being made into many of these outlets so that the rubbers have become more dependent on their elasticity in order to maintain their commercial importance.

Whilst it is the reversible high elasticity, enabling extensions of more than 1000% elongation in unfilled vulcanizates, that tends to attract the more casual interest, most commercial applications such as tyres, springs, seals and gaskets are concerned with much lower deformation levels. Where rubber products are used under static loads, such as in seals and gaskets, it is necessary to take into account that rubbers are imperfectly elastic. Thus they are liable to creep under a fixed load or subject to stress decay under a fixed strain. Where rubbers are used under dynamic conditions where stress is under continuous variation, the time-dependent nature of visco-elastic materials must also be taken into account.

2.2 THE MOLECULAR NATURE OF HIGH ELASTICITY

As is well known, the natural rubber molecule consists of a large number of carbon atoms linked to form a chain-like structure to which are attached hydrogen atoms and methyl groups in a regular manner. Such a chain is quite flexible at room temperature because there is sufficient energy present to enable one segment or atom to rotate relative to another. If the two chain ends are pulled in opposite directions the chain will tend to straighten out, this being ultimately limited by the C—C—C bond angles (Fig. 2.1).

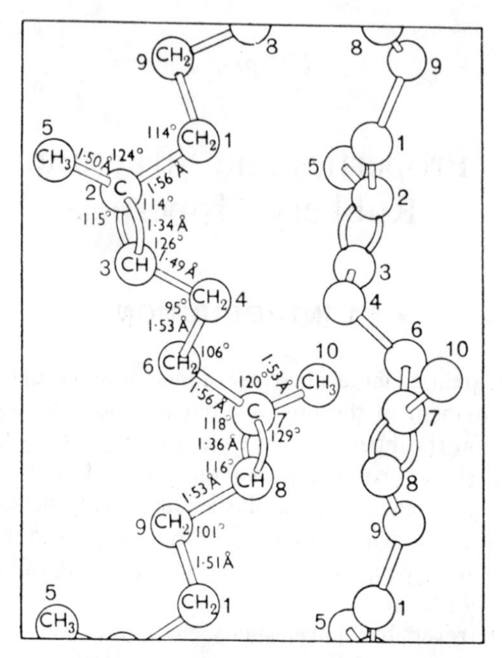

FIG. 2.1. Unit cell structure of the natural rubber molecule. (From Bunn (1942), reproduced by permission of the Royal Society).

If the ends are released the bonds start to rotate with respect to each other and the molecule tends to coil up in a random fashion (Fig. 2.2).

In natural rubber this re-coiling will be substantially complete within a fraction of a second at room temperature. That the rubber returns to its original dimensions before stressing indicates that the molecular chains have not been able to slip past each other during the stressing. In raw natural rubber this is largely due to links between chains through crystal structures or molecular entanglement. With lower-molecular-weight masticated natural rubber and with raw synthetic rubbers such links are less effective and it is necessary to prevent slippage by chemical cross-linking (vulcanization) of the chains to form three-dimensional networks (Fig. 2.3).

Many attempts have been made to derive theoretical quantitative relationships to describe the deformation behaviour of rubbery networks from a consideration of polymer structure and these have been summarized by the author elsewhere (Brydson, 1978) and in greater depth by others (e.g. Smith, 1972; Treloar, 1975). For this reason subsequent sections of this

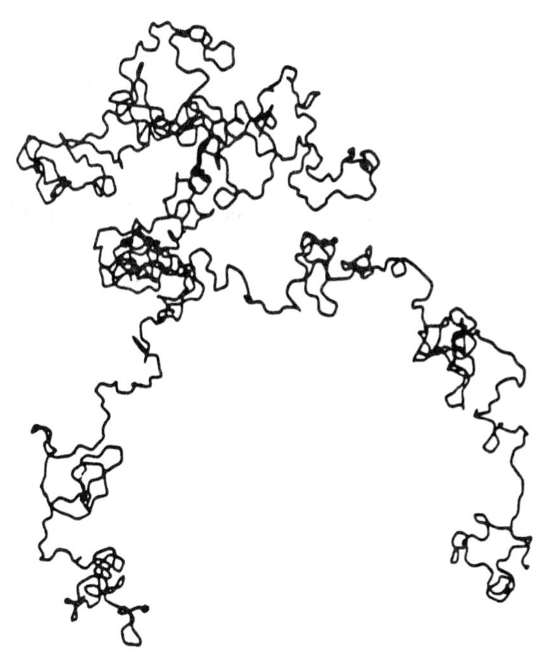

FIG. 2.2. Randomly-coiled rubber molecule. The figure represents a 1000-link wire model of a paraffin (alkane) chain constructed by Treloar. The links were set at the correct valence angle and the angular position of each link in the circle of rotation selected randomly by throwing a die. No two chains would be identical in their randomness, whilst in the rubbery state an individual chain would change with time (From Treloar (1975), reproduced by permission of Oxford University Press).

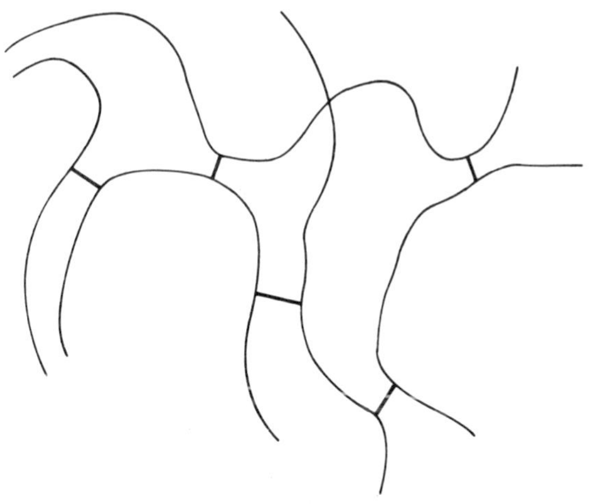

FIG. 2.3. Schematic two-dimensional representation of a cross-linked (vulcanized) structure preventing chain slippage.

chapter will be more concerned with the practical rather than the theoretical aspects of rubber elasticity.

2.3 RUBBER ELASTICITY IN UNIAXIAL TENSION

The Gaussian theory of rubber elasticity leads to the following relationship between tensile stress (f) and extension ratio (λ).

$$f = G\left(\lambda - \frac{1}{\lambda^2}\right) \tag{2.1}$$

where G is the shear modulus. The extension ratio is the extended length divided by the original length.

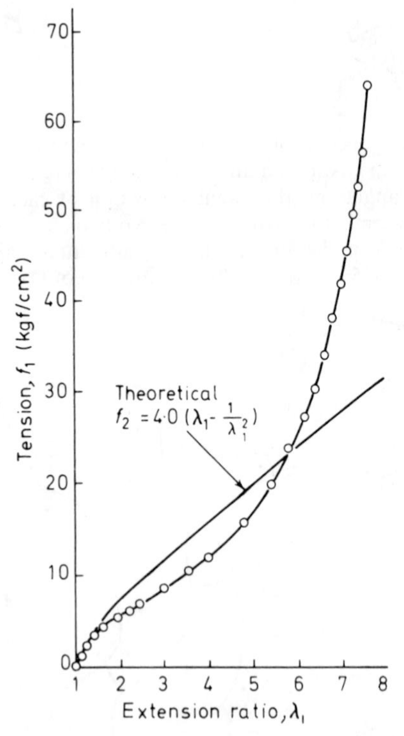

FIG. 2.4. Comparison of theory and experiment for simple extension (f_1, experimental data; f_2, theoretical prediction). (From Treloar (1975), reproduced by permission of Oxford University Press.)

The fit between theory and experiment is generally very good up to about 50% extension, modest up to an extension ratio of about 5 and poor above that value (Fig. 2.4). Since, however, most engineering applications in tension involve strains of less than 50%, the equation is quite useful. The discrepancies at the high strains are because the Gaussian theory does not recognize the finite extensibility of the networks.

The Gaussian analysis also leads to the equation

$$G = \frac{\rho RT}{M_c}\left(1 - \frac{2M_c}{M}\right) \tag{2.2}$$

where ρ is the density, R the gas constant, T the temperature, M_c the molecular weight between cross-links and M the average molecular weight of the polymer before cross-linking. It is also assumed that eqn (2.1) applies to equilibrium stress–strain conditions.

Equation (2.2) is of interest on two counts:

1. The equation enables the molecular weight between cross-links to be obtained using very simple tensile stress–strain experiments. These experiments do not however distinguish between chemical cross-links and physical cross-links due to molecular entanglements so that the results differ somewhat from values of M_c obtained using chemical methods. A calibration between the two approaches has been carried out (Moore & Watson, 1956). Some network 'defects' are illustrated in Fig. 2.5.
2. The term in parentheses in eqn (2.2) is a correction factor for a defect other than entanglements shown in Fig. 2.5, namely free chain ends. These do not form part of a network and are thus not capable of supporting a stress. Where the molecular weight of the polymer chain before cross-linking is high, then the proportion of free chain ends will be small and the correction factor is negligible. If however natural rubber is highly masticated before vulcanization then *the vulcanizate will be much softer for a given degree of cross-linking*. In practice M_c/M is usually of the order of 0·05 so that if the data are not corrected the error is of the order of 10%.

The closed loop, also shown in Fig. 2.5, involves a chemical cross-link without contributing to the load-bearing network. It thus has the opposite effect to an entanglement in that the modulus will be lower than that estimated from chemical measurements of M_c.

A number of theoretical analyses of rubber elasticity have been made which avoid some of the limitations of the Gaussian approach. Whilst they

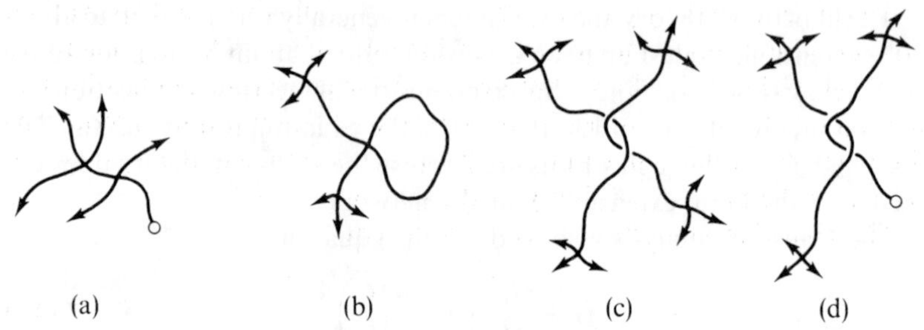

FIG. 2.5. Types of network defect. (a) Free chain end; (b) closed loop; (c) entanglement; (d) entanglement adjacent to free chain end. (From Brydson, 1978.)

provide a better fit to experimental data, the resulting equations have generally been too cumbersome for practical use but the interested reader is referred to the references given in Section 2.2.

Of rather more practical interest are:

1. The Mooney, Rivlin and Saunders equation; and
2. the Martin, Roth and Stiehler equation.

Confusingly both have been referred to as the MRS equation, although most commonly this has been used for the second one.

The first equation, proposed by Mooney (1940) and modified by Rivlin & Saunders (1951) takes the form

$$f = 2\left(C_1 + \frac{C_2}{\lambda}\right)\left(\lambda - \frac{1}{\lambda^2}\right) \tag{2.3}$$

This may be rearranged in the form

$$\frac{f}{2\left(\lambda - \frac{1}{\lambda^2}\right)} = C_2\left(\frac{1}{\lambda}\right) + C_1 \tag{2.4}$$

From this it follows that a plot of $f/2(\lambda - 1/\lambda^2)$ against $1/\lambda$ will give a slope C_2 and an intercept $(C_1 + C_2)$ on the vertical axis at $1/\lambda = 1$.

Figure 2.6 illustrates some results obtained by Gumbrell *et al.* (1953) for various rubbers.

This work, together with a number of other observations, leads to the following conclusions.

1. C_2 is not much affected by the type of rubber (but see Boyer & Miller, 1977.

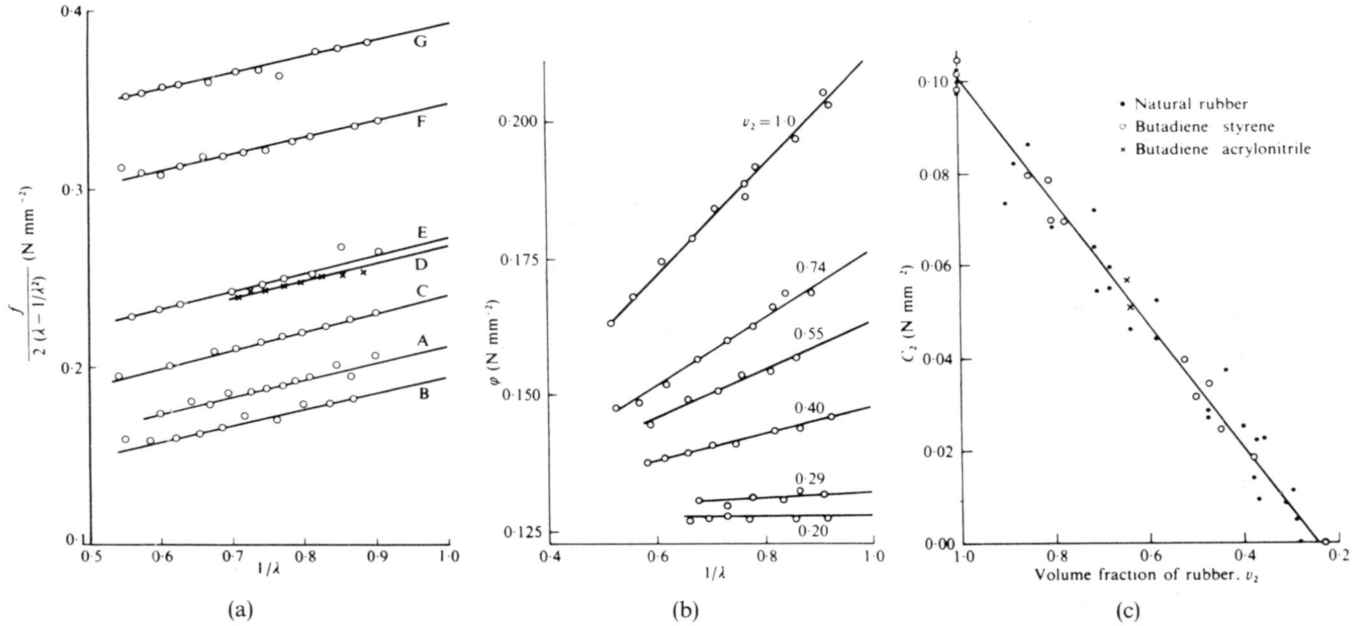

FIG. 2.6. The Mooney C_2 term. (a) Plot of $f/2(\lambda - 1/\lambda^2)$ against $1/\lambda$ for various rubbers A–G. The slope (C_2) is virtually independent of the type of rubber. (b) Plot of ϕ against $1/\lambda$ when $\phi = C_1 + C_2/\lambda$. The slope (C_2) decreases with increased swelling (v_2 is the volume fraction of rubber). (c) Plot of C_2 against v_2. Line is independent of polymer type. (From Gumbrell *et al.* (1953), reproduced by permission of the Royal Society of Chemistry.)

2. C_2 is not much affected by the intensity of cross-linking but does tend to zero as the cross-link density increases in the case of uniform network structures (see Morton, 1977).
3. C_2 tends to zero if the material is highly swollen.
4. C_2 tends to zero if great care is made to allow the rubber to come to equilibrium before measurements are made. It should however be noted that there is not universal agreement on this point.
5. C_2 appears to have the most significance in tensile tests.

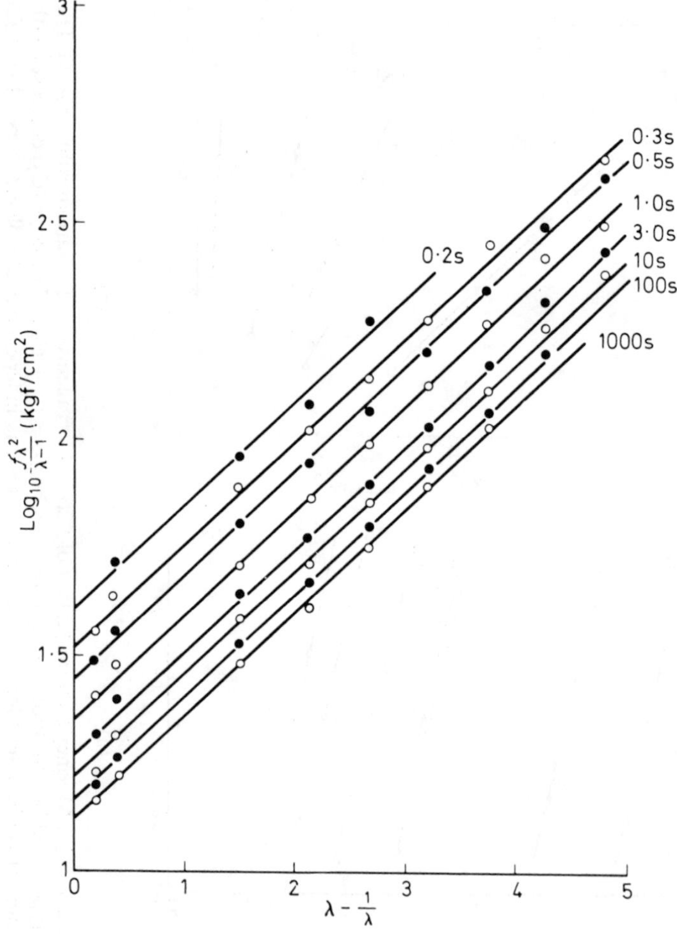

FIG. 2.7. MRS plot for NR at −45°C.

The physical nature of C_1 and C_2 continues to be a matter of academic debate. Inspection of eqn (2.3) shows that the term $2(C_1 + C_2/\lambda)$ is a strain-dependent modulus term. The writer is inclined to the view that $2C_1$ can be equated to the shear modulus G of eqn (2.1) arising from both chemical and permanent (i.e. undisentangleable) physical cross-links. On the other hand, the second component of the modulus $2C_2/\lambda$ could be ascribed to disentangleable entanglements which would explain why the term decreases on stretching and is also consistent with the five conclusions listed above.

The Martin, Roth and Stiehler equation (Martin *et al.*, 1956) states that

$$f = E\left(\frac{1}{\lambda} - \frac{1}{\lambda^2}\right)\exp A\left(\lambda - \frac{1}{\lambda}\right) \tag{2.5}$$

where E is a modulus and A a constant. This may be rearranged in the form

$$\log_{10}\left(\frac{f\lambda^2}{\lambda - 1}\right) = \log_{10} E + 0{\cdot}434A\left(\lambda - \frac{1}{\lambda}\right) \tag{2.6}$$

Thus if $\log_{10}(f\lambda^2/(\lambda - 1))$ is plotted against $(\lambda - 1/\lambda)$ a straight line of slope $0{\cdot}434A$ and intercept $\log_{10} E$ should be obtained. Harwood & Schallamach (1967) found that if stress–strain experiments were carried out at various

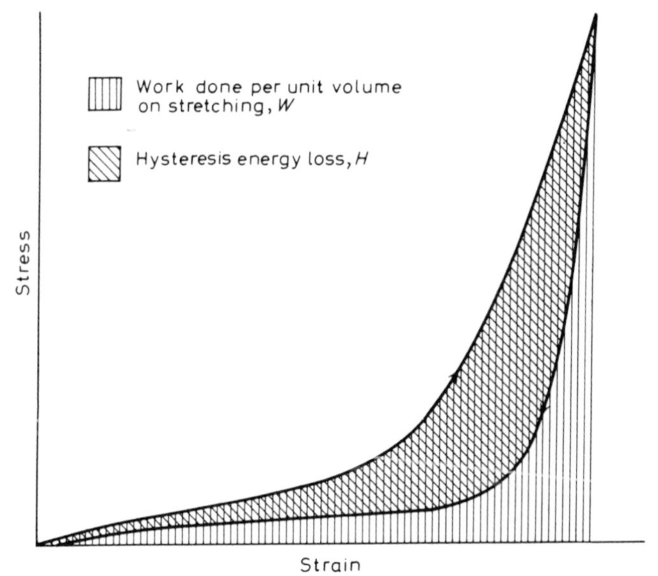

FIG. 2.8. Typical stress–strain extension–retraction cycle.

strain rates and a series of sets of plots made where the data for each set were obtained in the same time from zero strain (i.e. isochronously) the MRS equation was valid over a range of strain rates with the modulus being time-dependent (Fig. 2.7).

It has been further argued that E is useful in predicting the stresses during retraction and hence the hysteresis energy losses during a tensile stress–strain cycle. Such hysteresis losses are shown in Fig. 2.8.

2.4 RUBBER ELASTICITY IN SHEAR AND COMPRESSION

For all practical purposes the relationship between shear stress and shear strain in a rubber is given by the simple linear equation

$$f = Ge \tag{2.7}$$

where e is the shear strain, defined by x/t as in Fig. 2.9.

For a shear mounting of the dimensions indicated in Fig. 2.9 it is also useful to define the shear stiffness of the mounting as

$$\text{Shear stiffness } (K_s) = \frac{\text{Shearing force } (F)}{\text{Shear deflection } (x)} \tag{2.8}$$

Combination of eqns (2.7) and (2.8) leads to the relation for the shear stiffness of a shear mounting K_s as

$$K_s = \frac{GA}{t} \tag{2.9}$$

where A is the area over which the shearing force is applied.

It is to be noted that the shear stiffness of a mounting is not affected by the insertion of metal plates parallel to the direction of shear, providing that the value for the thickness t excludes the thickness of the plates.

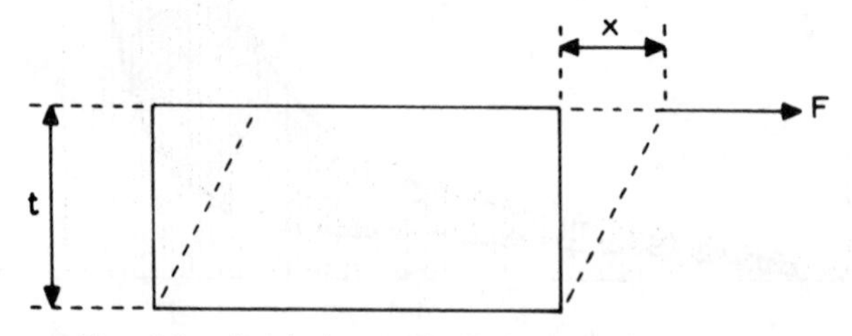

FIG. 2.9. Symbols used in defining shear stiffness.

In compression the stress–strain relationship may also be regarded as linear, this leading to an expression for the compression stiffness K_c:

$$K_c = \frac{AE_c}{t} \tag{2.10}$$

There is however a very important difference. In most compression mountings there is no slip between the rubber and the retaining plates during compression, either because the rubber is bonded to the plates or because of friction. The rubber part, which is effectively acting as a spring, bows out and tensile stresses are set up on the free surface of the rubber. This has the overall effect of increasing the effective stiffness of the mounting. The extra stiffness will depend on the *shape factor*, which is defined as

$$S = \frac{\text{Area of one loaded face}}{\text{Total unloaded surface area}} \tag{2.11}$$

In general it is found that the apparent compression modulus E_c, the shear modulus G and the shape factor S are related by the equation

$$E_c = 3G(1 + 2S^2) \tag{2.12}$$

Thus the squatter the mounting, be it rectangular or cylindrical, the stiffer it will be.

If we insert metal plates between the rubber (Fig. 2.10) we will in effect

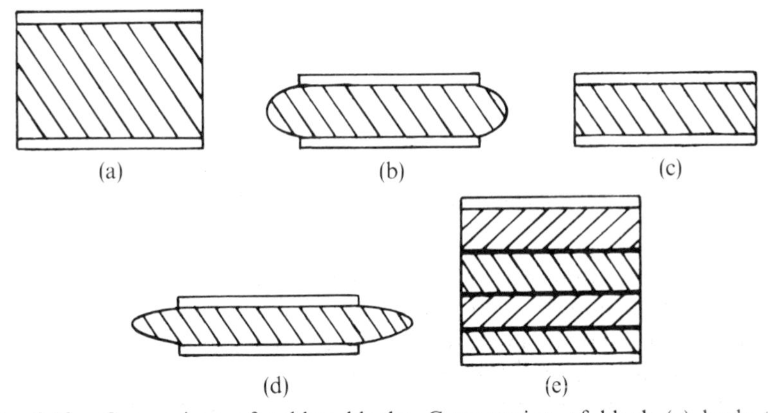

FIG. 2.10. Comparison of rubber blocks. Compression of block (a) leads to tension stresses on the surface of resulting bulge (b). The effect is greater when a squatter block (c) of higher shape factor S is compressed (d). Insertion of metal plates (e) produces a spring of enhanced stiffness.

have an array of rubber springs. If these springs have deflections x_1, x_2, $x_3 \ldots$ the total deflection will be

$$x_1 + x_2 + x_3 + \cdots$$

Since the deflecting force F will be the same for each element, it follows that the total stiffness for the unit (K_{tot}) will be given by

$$\frac{1}{K_{tot}} = \frac{1}{K_1} + \frac{1}{K_2} + \frac{1}{K_3} + \cdots \tag{2.13}$$

If all the rubber 'springs' between the plates are equal and there are n separating metal plates, i.e. $(n+1)$ rubber springs, then

$$K_{tot} = \frac{K}{n+1} \tag{2.14}$$

where K is the compression stiffness of each element.

TABLE 2.1
Effect of the insertion of metal plates on the shear stiffness of rubber springs

No. of plates	*S*	*Increase in stiffness, K/K_0*[a]
1	1	3
2	2	9
3	3	19
5	5	51
7	7	99
10	10	201

[a] K is the stiffness of the composite and K_0 the stiffness of the block with no additional metal plates.

Simple calculations show that insertion of such metal plates markedly increases the compression stiffness. Table 2.1 shows how the insertion of a number of metal plates increases the stiffness of a rectangular mounting of loaded face area L^2 and thickness $t = L/4$.

Thus insertion of rigid plates can increase vertical stiffness but not horizontal stiffness. This has many practical implications.

By inclining the shear mounts as in Fig. 2.11 it is also possible to control the stiffness in each of the horizontal directions. Calculation shows that the

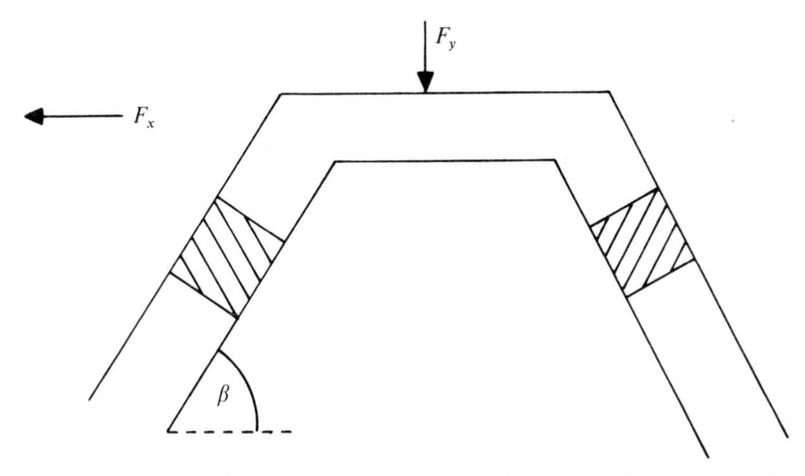

FIG. 2.11. Inclined shear mounts. Stiffness in three mutually perpendicular directions may be controlled independently.

stiffness in the three mutually perpendicular directions x, y and z will be given by

$$K_x = 2(K_c \sin^2 \beta + K_s \cos^2 \beta) \tag{2.15a}$$

$$K_y = 2(K_c \cos^2 \beta + K_s \sin^2 \beta) \tag{2.15b}$$

$$K_z = 2K_s \tag{2.15c}$$

2.5 VISCO-ELASTICITY AND RUBBERS

When a polymer is subjected to a tensile stress the total deformation may be considered as the sum of three components:

1. an instant deformation due to bond bending and stretching similar to that of ordinary elastic deformation in metals (designated as D_{OE});
2. chain uncoiling which is not instantaneous and leads to high elastic deformation (D_{HE});
3. chain slippage leading to viscous deformation (D_{visc}).

Hence

$$D_{total} = D_{OE} + D_{HE} + D_{visc} \tag{2.16}$$

The relative proportions of each component vary according to type of polymer, molecular weight and temperature. For example, the deformation of polystyrene at room temperature will be dominated by D_{OE} but at 230°C, when the material is molten, by D_{visc}. Vulcanized rubber at room temperature will be dominated by D_{HE} whilst highly masticated raw natural rubber will have substantial components of D_{HE} and D_{visc} in any deformation.

It is instructive to consider three idealized materials each of which has only one component of deformation, namely ideal solids, ideal rubbers and ideal liquids, to subject them to a constant load for an interval of time and to note the deformation throughout that time interval. These are illustrated in Fig. 2.12. Figure 2.13 shows the deformation–time curve of a similarly loaded hypothetical material exhibiting significant amounts of each type of deformation.

Two features illustrated in Fig. 2.12 are of particular importance with commercial rubbery polymers. They are:

1. the time dependence of the high elastic deformation;
2. the *permanent set* that occurs due to irreversible chain slippage.

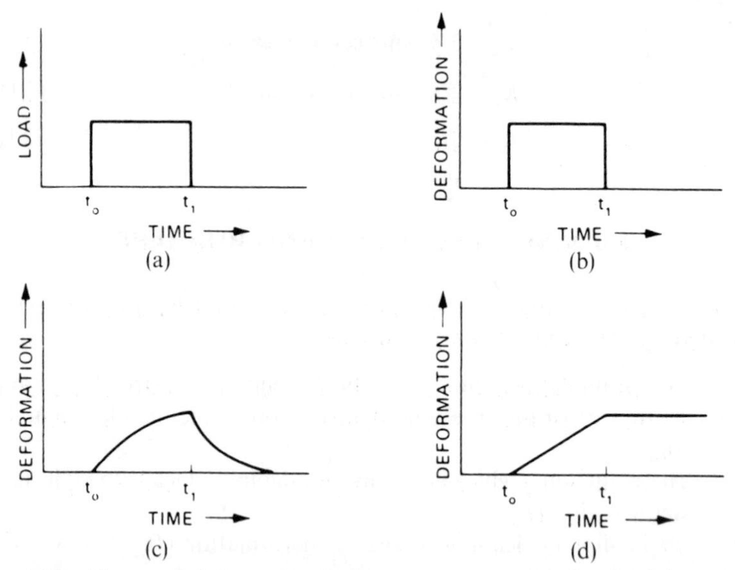

FIG. 2.12. Types of deformational response as a result of a fixed load being imposed between times t_0 and t_1 (a). (b) Ordinary elastic material. (c) Highly elastic material. (d) Viscous material. (From Brydson, 1982).

Mathematical models based on simple mechanical analogies have been developed to mimic visco-elastic behaviour and whilst by their very nature they cannot provide a perfect description of visco-elasticity in polymers, they do provide a useful framework for study. In this so-called phenomenological approach the D_{OE} component of deformation is represented by a spring obeying Hooke's Law and the D_{visc} component by a dashpot containing a viscous oil and obeying Newton's Law that the shear stress is proportional to the rate of strain. The high elastic component, D_{HE},

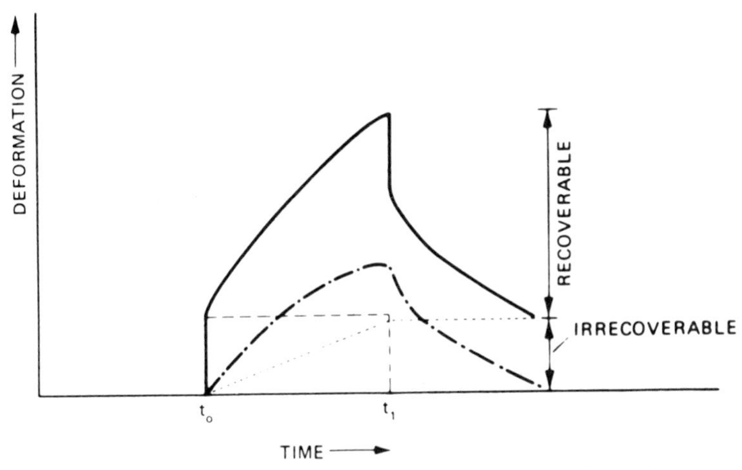

FIG. 2.13. Deformation–time curve for material showing substantial ordinary elastic, high elastic and viscous components of deformation. (From Brydson, 1982.)

is represented by a spring and dashpot in parallel, the dashpot retarding the spring and hence the attainment of equilibrium. Thus a real material could be considered by the model represented in Fig. 2.14.

The combination of a spring and dashpot in series is known as a Maxwell unit and is particularly useful in describing melt flow in polymers and stress decay in rubbers. The corresponding combination in parallel is known as a Voigt unit and describes an 'ideal' rubber (the term ideal being used strictly in the rheological sense only).

The total load (f) on a Voigt unit, or element, will be the sum of the loads on the spring (f_s) and the dashpot (f_d). If the spring component is Hookean, then the load on the spring will be given by

$$f_s = E\gamma \tag{2.17a}$$

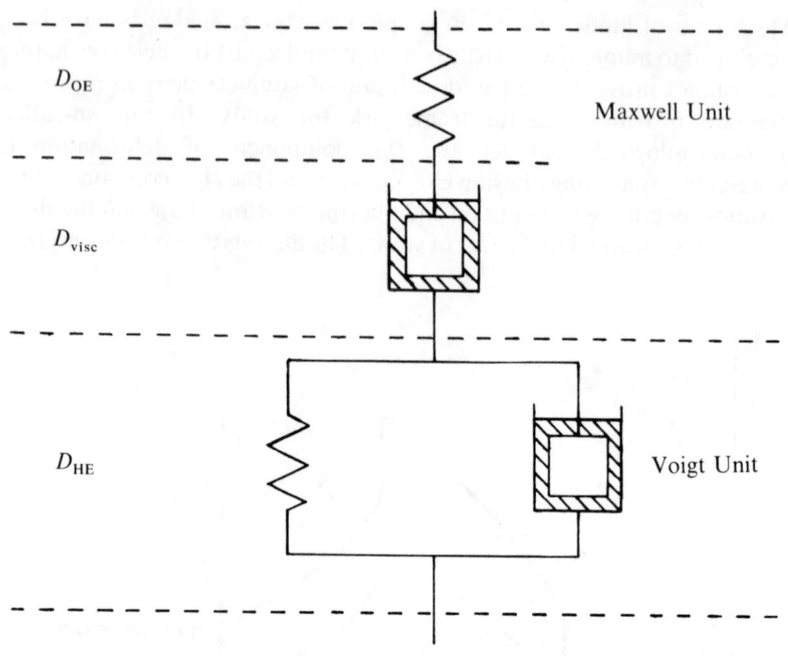

FIG. 2.14. Spring and dashpot model representation of rubber elasticity.

where γ is the strain. If the dashpot is Newtonian the load on the dashpot will be given by

$$f_d = \eta \frac{d\gamma}{dt} \tag{2.17b}$$

from which it follows that

$$f = E\gamma + \eta \frac{d\gamma}{dt} \tag{2.18}$$

For the condition that at $t = 0$ then $\gamma = 0$, this equation has the solution

$$J = \left(\frac{\gamma}{f}\right) = \frac{1}{E}[1 - \exp(-t/\tau)] \tag{2.19}$$

where J is known as the creep compliance in shear and τ (defined by η/E and with the dimensions of time) as the retardation time. In practice different parts of a rubber network will vary in the ease by which a chain can be uncoiled and so each part will have its own retardation time. It is therefore necessary to integrate all of the uncoiling processes and this yields the expression

$$J(t) = \int_{-\infty}^{\infty} L(\tau)[1 - \exp(-t/\tau)]\, \mathrm{d} \ln \tau \qquad (2.20)$$

where $L(\tau)$ (which is equal to τ/E) is known as the retardation spectrum.

If the rubber is subjected to a rapid series of loading–unloading cycles (for example where the load is varied sinusoidally with time) there may not be time for the rubber to reach equilibrium deformation and it will appear to be stiffer. Similarly if the rubber is subjected to an imposed deformation, then under these conditions higher stresses will develop than required for a static deformation. Analyses also show that the strain will tend to lag behind the stress, and arising from this some energy will be dissipated as heat.

Crucial to these considerations is the value of t/τ. When the time scale available for deformation is much greater than the retardation time τ, the material will behave like a rubber but when $t \ll \tau$ then there will be little time for chain uncoiling and the material will appear rigid. There will be transitions and hence a critical situation where $t \simeq \tau$.

In many real rubbers some chain slippage may also occur and this will add a D_{visc} component to the total deformation so that eqn (2.20) will need to be modified to eqn (2.21).

$$J(t) = \int_{-\infty}^{\infty} L(\tau)[1 - \exp(-t/\tau)]\, \mathrm{d} \ln \tau + \frac{t}{\eta} \qquad (2.21)$$

In this case creep will continue indefinitely. This is illustrated schematically in Fig. 2.15, which shows the compliance as a function of time both for an 'ideal' rubber and for one where chain slippage occurs.

Related to creep is *stress decay*. This occurs when a piece of rubber is deformed and them maintained at constant deformation. Because of chain slippage the stress required to maintain the constant strain decays. This situation can be represented by a Maxwell model where

$$\gamma_{total} = \gamma_{OE} + \gamma_{visc} \qquad (2.22)$$

Differentiating with respect to time and putting $\mathrm{d}\gamma_{total}/\mathrm{d}t = 0$,

$$\frac{\mathrm{d}\gamma_{OE}}{\mathrm{d}t} + \frac{\mathrm{d}\gamma_{visc}}{\mathrm{d}t} = 0 \qquad (2.23)$$

and since

$$f = E\gamma_{OE} \qquad \text{and} \qquad f = \eta \frac{\mathrm{d}\gamma_{visc}}{\mathrm{d}t}$$

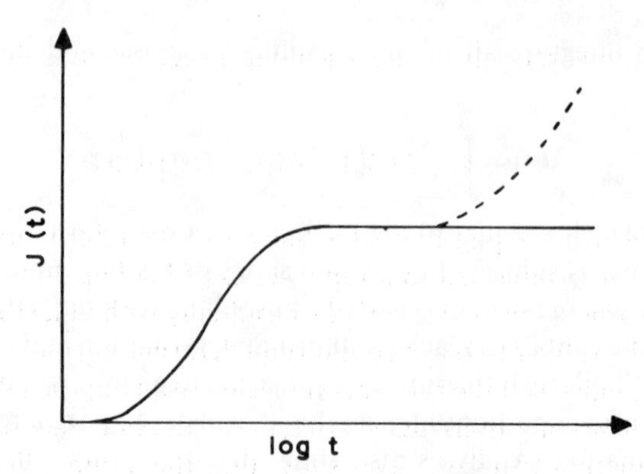

FIG. 2.15. Compliance as a function of time t for 'ideal' cross-linked rubber (——) and one in which viscous creep may occur (----).

then

$$\frac{df}{dt} + E\left(\frac{f}{\eta}\right) = 0 \tag{2.24}$$

The solution of this equation is

$$G(t) = \frac{f}{\gamma_0} = E\exp(-t/\tau) \tag{2.25}$$

where $G(t)$ is known as the retardation modulus and $\tau(=\eta/E)$ the *relaxation time*. As with the case of the treatment of the Voigt units it is more accurate to modify the equation because different parts of a network will have different relaxation times. For a continuous distribution of relaxation times it may be argued that

$$G(t) = \int_{-\infty}^{\infty} H(\tau)\exp(-t/\tau)\,\mathrm{d}\ln\tau \tag{2.26}$$

where $H(\tau)$ is the relaxation spectrum given by τE.

It is important to stress that these equations are based on empirical models. A molecular theory of visco-elasticity has been developed by Rouse (1953) for dilute solutions and subsequently developed by other research workers for less constricting circumstances. Discussion of these theories is however outside the scope of this book and the interested reader is referred to the specialist literature (e.g. Smith, 1972).

2.6 STRESS RELAXATION DURING AGEING

In the previous section it was assumed that no change in the chemical structure of the rubber occurred during the course of a creep or stress decay experiment. In practice oxidative ageing may occur within the time scale of an experiment, or more importantly, within the time scale of desired service of a rubber component. Such long-term ageing can lead to chain scission and breakdown of networks in the case of natural rubber vulcanizates and can be monitored and studied by the use of stress–relaxation techniques. Any such analysis assumes that slippage in the unaged network is negligible. If however chain scission occurs, then the stress required to maintain a constant extension will drop.

If it is assumed that eqn (2.1) holds and also that the shear modulus is inversely proportional to the molecular weight between cross-links (both assumptions are subject to some error), then it can be seen that

$$f=f_0\frac{(M_c)_0}{(M_c)} \tag{2.27}$$

where f_0 and f are the stresses required to maintain a constant strain before and after ageing and $(M_c)_0$ and M_c are the molecular weights between cross-links before and after ageing.

It can thus be seen that the decay in stress at constant extension during ageing provides a direct measure of the degradation of the elastic network. (It should be noted in passing that some rubbers harden on oxidative ageing and this will require an increase in stress to maintain a constant strain.)

Two types of stress–relaxation measurement may be distinguished:

1. Continuous stress–relaxation measurements.
2. Intermittent stress–relaxation measurements.

In the continuous stress–relaxation experiment the extension is maintained throughout the experiment. If, in addition to the chain scission reactions, cross-linking occurs then although the original network will be under strain the 'new network', i.e. the contribution made to the overall network by the new cross-links, will not be under strain and will therefore not be part of the stress-supporting network. It will therefore make no contribution to the magnitude of the stress.

If the sample is strained only in order to make a measurement but for most of the time is held at rest (intermittent stress relaxation), then most cross-link formation will occur when the sample is at rest, and these will contribute to the stress-supporting network. The difference obtained in

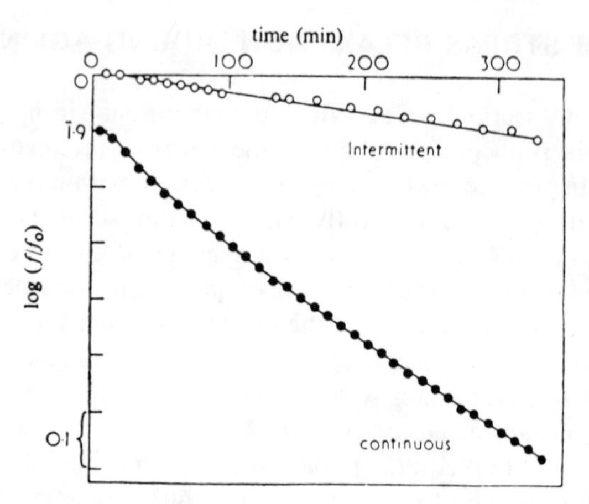

FIG. 2.16. Intermittent and continuous stress–relaxation of an extracted CBS-accelerated vulcanizate of smoked sheet at 100°C in air (f_0 original stress; f, stress at a given point in time). (From Dunn *et al.* (1959), reproduced by permission of The Royal Society of Chemistry.)

continuous and intermittent experiments thus provides a measure of the extent of cross-linking during oxidation (Fig. 2.16).

BIBLIOGRAPHY

Brydson, J. A. (1978). *Rubber Chemistry*. Applied Science, London.

Freakley, P. K. & Payne, A. R. (1978). *Theory and Practice of Engineering with Rubber*. Applied Science, London.

Hepburn, C. & Reynolds, R. J. W. (1979). *Elastomers—Criteria for Engineering Design*. Applied Science, London.

Smith, K. J. (1972). In *Polymer Science*, A. D. Jenkins. North-Holland, Amsterdam, Chapter 5.

Treloar, L. R. G. (1975). *The Physics of Rubber Elasticity*, 3rd edn. Clarendon Press, Oxford.

REFERENCES

Brydson, J. A. (1978). *Rubber Chemistry*. Applied Science, London.

Brydson, J. A. (1982). *Plastics Materials*, 4th edn. Newnes-Butterworth, London.

Boyer, R. F. & Miller, R. L. (1977). *Rubber Chem. Technol.*, **50**, 798.

Bunn, C. W. (1942). *Proc. Roy. Soc.*, **A180**, 40, 67, 82.

Dunn, J. R., Scanlan, J. & Watson, W. F. (1959). *Trans. Faraday Soc.*, **55**, 667.
Gumbrell, S. M., Mullins, L. & Rivlin, R. S. (1953). *Trans. Faraday Soc.*, **49**, 1495.
Harwood, J. A. C. & Schallamach, A. (1967). *J. Appl. Polym. Sci.*, **11**, 1835.
Martin, G. M., Roth, F. L. & Stiehler, R. D. (1956). *Trans. Inst. Rubber Ind.*, **32**, 189.
Mooney, M. J. (1940). *J. Appl. Phys.*, **11** , 582.
Moore, C. G. & Watson, W. F. (1956). *J. Polym. Sci.*, **19**, 237.
Morton, M. (1977). Paper presented at the International Rubber Conference, Brighton, England, May 1977. (Preprints produced by the Plastics and Rubber Institute, London.)
Rivlin, R. S. & Saunders, D. W. (1951). *Phil. Trans. Roy. Soc.*, **A243**, 251.
Rouse, P. E. (1953). *J. Chem. Phys.*, **21**, 1272.
Smith, K. J. (1972). In *Polymer Science*, ed. A. D. Jenkins. North-Holland, Amsterdam, Chapter 5.
Treloar, L. R. G. (1975). *The Physics of Rubber Elasticity*, 3rd edn. Clarendon Press, Oxford.

Chapter 3

Structure–Property Relationships in Rubbers

3.1 INTRODUCTION

In Chapter 2 some characteristic features of rubber elasticity were discussed. Not all polymers, however, are rubbery at normal ambient temperatures and in the first part of this chapter we shall consider the molecular structure requirements of a rubber. Since the technical use of a rubber may also require other properties such as low-temperature flexibility and heat and oil resistance, the second part of the chapter will consider the dependence of these properties on molecular structure. The treatment here will be limited by reasons of space. Somewhat longer qualitative descriptions have been given by the author elsewhere (Brydson, 1966–1982) with more detailed studies on glass transition temperatures (Brydson, 1972) and the swelling of vulcanized rubber (Brydson, 1978).

3.2 STRUCTURE AND RUBBERY PROPERTIES

Whether or not a polymer is likely to be an effective rubber, as opposed to being for example a glass, a leatherlike material, a wax, a grease or an oil, will depend on the molecular and supramolecular structure. In order to exhibit effective rubbery properties there are three primary requirements concerning structure:

1. The chain backbone must be flexible (over the range of temperatures within which it is intended that the rubber should operate).
2. Individual chains should be capable of being linked one with another with a minimum of ineffective free chain ends, i.e. there should be sites available for cross-linking.
3. The main chain backbone should be free of weak links which could be the site of unwanted chain breakage (chain scission).

3.2.1 Chain Flexibility

The most flexible chains are those where the lowest amount of energy is required for one bond to rotate with respect to its neighbour, i.e. there should be low bond rotational energies. Such low rotational energies are usually obtained with chain backbones of aliphatic C—C links, C—O links or Si—O links and with simple side groups attached, such as hydrogen atoms.

On the other hand, the presence of aromatic *p*-phenylene groups and other ring structures in the chain backbone will impart stiffness. The nature of the side groups is also most important. Bulky side groups such as phenyl and *t*-butyl will increase the rotational energy whilst side groups capable of hydrogen bonding with adjacent groups will have a similar effect. In general, it is the aim to keep the side groups simple although this does not necessarily result in a rubber. For example, neither polyethylene $-(CH_2-CH_2)_n-$ nor polyformaldehyde $-(CH_2O)_n-$ is rubbery at room temperature. This is because in each case the structure is so regular that the polymer is capable of crystallization. It is only when the temperature is raised above the crystalline melting point that these polymers can become rubbery.

To say that, for rubberiness, the polymer structure should be free of bulky side groups or other groups that raise the rotational energy, and also be sufficiently irregular to be incapable of crystallization, is also an oversimplification. The world's second most used rubber, SBR, contains phenyl side groups and the world's most used rubber, natural rubber, has a very regular structure capable of crystallization. It is simply a matter of degree. SBR is typically about 25% styrene and 75% butadiene by weight. With the styrene molecule about twice as heavy as the butadiene molecule there are about six butadiene molecules per styrene molecule. If we assume for the moment that there is 1,4-polymerization only with each butadiene molecule contributing four carbon atoms to the main chain and the styrene molecule contributing two, then there will be about one pendant phenyl group per 26 main chain carbon atoms. So few phenyl groups do not seriously affect the flexibility of the overall chain although it is to be noted that polybutadiene, which has no pendant groups, is flexible down to lower temperatures.

The regularity of the natural rubber molecule means that the raw polymer can crystallize, but it has a low melting point with the last trace of crystallinity disappearing at about 50°C. When vulcanized, the cross-links introduce irregularities and reduce the ability of the rubber to crystallize and at room temperature this phenomenon only becomes apparent on stretching the rubber.

Natural rubber, SBR and other diene rubbers contain double bonds in the main chain. Whilst these bonds have very limited flexibility they make the two adjacent single bonds more flexible, with the net effect that diene polymers remain rubbery at lower temperatures than their saturated counterparts even when crystallization effects are eliminated.

3.2.2 Networks

It is no mean feat that the first material to be successfully used to cross-link rubber, sulphur, is still the main vulcanizing agent for the rubber industry. When it is realized that the discovery was made soon after the first-ever synthesis of an organic compound (from inorganic materials) its longevity becomes particularly noteworthy. Its low cost, low toxicity and adaptability will assure its role for some time to come, but at the same time it must be recognized that its very success has had something of a constricting influence in rubber technology. On many occasions the nature of a newly introduced type of synthetic rubber has been changed just to make it sulphur vulcanizable.

It is now well recognized that the cross-linking site does not have to be at or adjacent to main chain carbon–carbon double bonds such as is common with natural rubber, SBR, NBR, polybutadiene and butyl rubber. Such bonds may have the disadvantage that they provide a weak link susceptible, for example, to attack by oxygen and ozone. One development is to build the double bond into a ring structure within the chain so that on rupture the effect is less catastrophic—a practice used with ethylene–propylene terpolymers.

Covalent bonding using non-sulphur vulcanizing agents such as peroxides, polyisocyanates, polyamines and *p*-quinone dioximes is also practised with, in many cases, network formation not requiring a double bond. Some methods of introducing covalent cross-links into polymers are indicated in Fig. 3.1. These methods are, of course, equally applicable to plastics, surface coatings and adhesives.

Network structures can also be produced without covalent bonding taking place. They may be produced by ionic linkages or by hydrogen bonding, either directly from one chain to another, via metal ions or via such diverse materials as plasticizer molecules, carbon black particles or even proteinous material. Small crystalline domains in the rubber can also lead to a form of network.

A more recent approach has been to prepare block copolymers in which a length of a rubbery polymer molecule, such as that of polybutadiene, is linked at one or both ends to a segment of a polymer which is not rubbery at

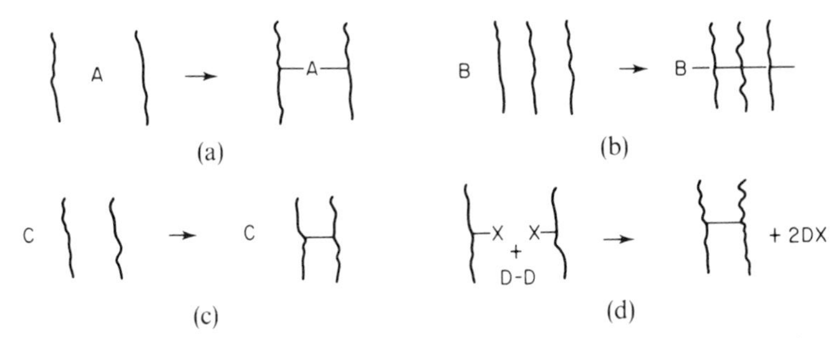

FIG. 3.1. Some general methods of introducing covalent cross-links into a polymeric system. (a) Bridging agents; (b) cross-linking initiators; (c) catalytic cross-linking agents; (d) active site generators. (From Brydson, 1978.)

room temperature, e.g. polystyrene. In suitable proportions the polystyrene groups aggregate into glassy domains effectively linking the rubbery segments at their ends into a network structure. Heating such a copolymer to above the softening temperature (strictly the glass transition temperature) of the polystyrene segment causes the polystyrene domains themselves to soften and the network breaks down. Providing the overall molecular weight is not too high such a heated polymer is thermoplastic, i.e. it is capable of flow (see Chapter 16).

It is possible to raise the maximum useful temperature of the block copolymer by replacing the styrene segments with those which have a higher glass transition temperature or by using as terminal segments highly crystalline moieties as are employed in some of the polyester and polyurethane rubbers.

The network is an essential feature of a rubber and it is not fully effective if there are free chain ends. If the rubber is being cross-linked by a process that does not specifically involve the chain ends but occurs in a more or less random fashion along the chain, then the lower the molecular weight of the rubber immediately before vulcanization the greater the chain end concentration (for equal cross-link intensities). If cross-linking involves all of the chain ends, then this comment does not apply. Some rubbers have been developed with terminally reactive groups but it is usually difficult to guarantee that all such groups will have in fact reacted.

3.2.3 Weak Links

It is only necessary for one link to break per chain for the average chain molecular weight to halve. For this reason weak bonds such as those of oxygen–oxygen and sulphur–sulphur are generally to be avoided in the main chain. In addition, the in-chain carbon–carbon double bond present in diene rubbers, by its proneness to attack by oxygen, ozone and other agents, is, as has already been mentioned, something of a liability although it does have an important role in the mastication process.

3.3 NON-RUBBERY PROPERTIES

From the foregoing it will be apparent that effective rubbery networks may be prepared from a variety of polymers using a range of cross-linking systems. The selection from such a range of rubbers for a particular application will therefore often depend on what might very loosely be termed non-rubbery properties, and on cost. Amongst such properties may be considered heat resistance, low-temperature resistance, chemical properties and electrical insulation properties. Also of importance are properties such as tensile strength, hardness, abrasion resistance and hysteresis characteristics as well as processing behaviour. The rest of this chapter provides an outline of the dependence of these properties on molecular structure.

3.3.1 Heat Resistance

The heat resistance of a rubber should normally be considered under three types of situation:

1. *In vacuo* or in some inert atmosphere;
2. under oxidative conditions;
3. in the presence of other active agents such as those causing hydrolysis.

The second and third of these situations will be dealt with subsequently and comments here will be restricted to in-vacuo heat resistance.

Such heat resistance will depend on three factors:

1. The bond dissociation energies of the various bonds in the rubber, particularly of those in the main chain;
2. weak links;
3. chain reactions involving degradation of the polymer chain.

Table 3.1 gives some typical bond dissociation energies obtained as a result of making many observations on the energetics of formation and dissociation of many molecules.

It is of interest to note that the main chain bonds in most common use (Si—O, C—C and C—O) have very similar bond energies. It is also reasonable to conclude that those rubbers containing only C—C, C—O, C—H and/or C—F bonds, even when cross-linked, would have good heat resistance under anaerobic conditions. Such a conclusion is borne out by the ethylene–propylene rubbers, the acrylic rubbers, fluorine-containing rubbers and the polyepichlorhydrins.

TABLE 3.1
Typical dissociation energies for some selected primary bonds

Bond	*Mean bond energy* ($kJ\,mol^{-1}$)	*Bond*	*Mean bond energy* ($kJ\,mol^{-1}$)
O—O	146	Si—O	368
Si—Si	178	N—H	389
S—S	270	C—H	430–510
C—S	272	O—H	464
Si—C	301	C—F	485
C—N	305	C=C	611
C—Cl	327	C=O	*c.* 740
C—C	346	(aldehydes and ketones)	
C—O	358	C≡N	890

Whilst the bond energy of the silicon–carbon bond is comparatively low it is clearly adequate for many purposes. For example, the vapour of tetramethylsilane is stable to 600°C. This indicates that in polymers factors other than bond energy are of importance.

One such factor is the presence of weak links which can seriously affect the heat resistance. One polymer that is particularly affected in this way is polyvinyl chloride where stepwise dehydrochlorination can occur by means of an unzippering reaction initiated from a weak link. In rubbers the weak link, as has been mentioned, is rather more important in causing degradation under the more common oxidative conditions to be dealt with later.

Many polymers degrade by chain reactions, some of which, as with polyformaldehyde and polymethyl methacrylate, lead to depolymerization to monomer. It has been suggested that the good heat resistance of the silicones is due not to high bond strengths but to the absence of a favourable reaction path by which the polymer system may degrade. Chain reactions may often be reduced in number by the use of one or more of the following approaches.

1. By preventing the initial formation of weak links. This may involve, amongst other things, the use of rigorously purified monomer.
2. By deactivating any active weak link. For example, some commercial polyformaldehyde (polyacetal) plastics have their chain ends capped by a stable grouping. (This will be of little use where the initiation of chain degradation is not at the terminal group.)
3. By copolymerizing with a small amount of a second monomer which acts as an obstruction to the unzipping reaction, in the event of this being allowed to start.
4. By the use of certain additives which divert or moderate the degradation reaction. A wide range of antioxidants and stabilizers function by this mechanism.

3.3.2 Low-temperature Resistance

Progressive cooling of a rubber eventually leads to the loss of rubberiness as the polymer passes through its glass transition temperature (T_g), or by crystallization.

A number of interpretations of the nature of the T_g are possible but for the purposes of this chapter it may be treated as the temperature below which main chain bonds no longer have sufficient energy for rotation and hence wriggling of the molecular chains. In practice the T_g is not a sharp transition and there is a finite temperature range through which bond rotation becomes more and more difficult and the deformational response to a stress takes longer and longer. The T_g thus becomes somewhat dependent on the method of test. Slow dilatometric experiments will give lower values than mechanical and electrical tests carried out at high frequencies. It is even possible for a polymer to appear rubbery in one test at room temperature and glassy in another.

The molecular factors controlling the T_g have been discussed at length elsewhere (Brydson, 1972) but a brief summary of these is given below.

The main factors determining the T_g are:

1. Chain stiffness;
2. interchain attraction;
3. molecular symmetry;
4. copolymerization;
5. branching and cross-linking;
6. solvents and plasticizers.

Chain stiffness must be considered as one of the most dominating factors and some theories of the glass transition have assumed this to be the only factor. Suffice it to repeat the statement in Section 3.2.1 that if a material is to be rubbery then the bulk of the backbone bonds should be flexible and have low rotational energy requirements such as occur with C—C, C—O, Si—O and C—S bonds, and where the bulk of the groups attached directly to the backbone do not substantially increase the bond rotational energy. Such an increase can occur, for example, by steric hindrance of phenyl groups on adjacent main chain carbons, by hydrogen bonding between carboxylic groups attached to the main chain and in close proximity, or by the stiffening effect of higher branched hydrocarbon moieties such as a *t*-butyl group.

The glass transition temperature may also be increased by raising the interchain attraction. Amorphous non-polar molecules will be held together primarily by dispersion forces. However, if there are polar groups in the structure or groupings that allow hydrogen bonding across the chains, then the T_g will be raised. Examples are PVC (T_g 80°C) and *cis*-polychloroprene (T_g − 20°C) when compared with their non-halogen-containing equivalents, polyethylene (T_g − 25°C) and *cis*-polybutadiene (T_g − 107°C). A further example of this rule appears to be with *trans*-polychloroprene and *trans*-polybutadiene which have respective reported T_gs of −45°C and −106°C.

The type of side chain is also important. It is commonly observed that for an aliphatic homologous series (e.g. polymethyl acrylate, polyethyl acrylate, poly-*n*-propyl acrylate, etc.) increasing the length of the side chain decreases the T_g, presumably by the chain separation effect. However, when the number of carbon atoms in the side chain is eight or more, the T_g frequently increases with an increase in the length of the side chain. This has been ascribed to a number of causes, including side chain crystallization.

Molecular symmetry plays a role, not yet completely understood, some aspects of which are particularly important with rubbers. These include:

1. Symmetrical disubstitution on a single chain;
2. *cis*/*trans* isomers;
3. tactic isomers.

It is well known that polyisobutylene (**I**) has a lower T_g than polypropylene (**II**) whilst that of polyvinylidene chloride (**III**) is lower than that of polyvinyl chloride (**IV**):

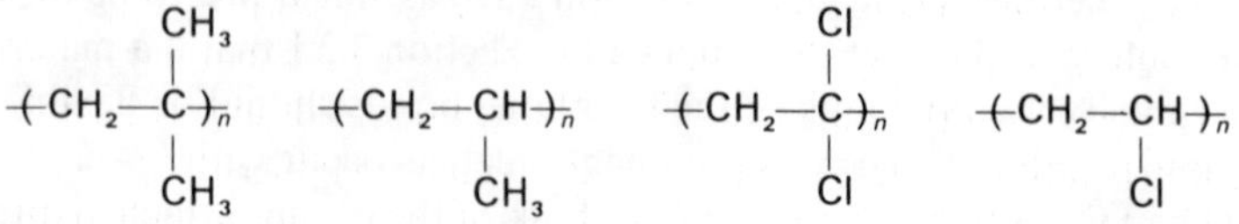

$$\mathrm{-\!(CH_2{-}\overset{\overset{\large CH_3}{|}}{\underset{\underset{\large CH_3}{|}}{C}}\!)_{\!n}-} \quad \mathrm{-\!(CH_2{-}\underset{\underset{\large CH_3}{|}}{CH}\!)_{\!n}-} \quad \mathrm{-\!(CH_2{-}\overset{\overset{\large Cl}{|}}{\underset{\underset{\large Cl}{|}}{C}}\!)_{\!n}-} \quad \mathrm{-\!(CH_2{-}\underset{\underset{\large Cl}{|}}{CH}\!)_{\!n}-}$$

(I) (II) (III) (IV)

This has been described variously as being due to chain symmetry (Boyer, 1963), to the fact that disubstitution lowers the value of the difference between the lowest potential energy minima even though it raises the actual rotational energy (Gibbs & di Marzio, 1958), and to the reduction in dipole moment due to symmetry of substitution (Würstlin, 1950).

Of greater significance to rubber chemistry is the difference in T_g between *cis*- and *trans*-diene polymers (Table 3.2).

TABLE 3.2
Glass transition temperatures and crystalline melting points of *cis*- and *trans*-diene polymers

	T_g (°C)	T_m (°C)		T_g (°C)	T_m (°C)
cis-Polybutadiene	−107	+3	*trans*-Polybutadiene	−106	+145
cis-Polyisoprene	−73	+25	*trans*-Polyisoprene	−60	+65
cis-Polychloroprene	−20	—	*trans*-Polychloroprene	−45	+75

Data now available indicate that the T_gs of *cis*- and *trans*-polybutadienes are very similar. This would seem reasonable and it had indeed been difficult to explain earlier reported differences. These results now give renewed support to the early explanations (Bunn, 1955) of the difference in T_gs of the *cis*- and *trans*-polyisoprenes as being due to interference between methyl groups. In addition, it would seem reasonable that chlorine groups would act in the opposite sense to explain the differences between the two polychloroprenes.

Comparisons of isotactic, syndiotactic and atactic polymers also cause problems of explanation. For many polymers the type of tacticity appears to have little effect on T_g. Exceptions are provided by the methacrylate polymers and, in contrast to other acrylates, polyisopropyl acrylate.

The effects of copolymerization vary from one system to another but a large number of copolymer rubbers obey the Gordon–Taylor relation quite closely. In effect this proposes that the T_g of a copolymer can be linearly interpolated from the T_gs of the appropriate homopolymers. Butadiene–styrene random copolymers provide a good fit but in general interactions between the monomer residues can produce such a variety of effects that a single simple relationship is not of general applicability.

The influence of molecular weight on T_g is more clearly understood. Chain ends have greater freedom than in-chain bonds and have lower effective rotational energies. As a consequence the molecular weight goes down linearly with an increase in chain end concentration, i.e. with reciprocal molecular weight. It would seem that the significance to the rubber technologist is that for optimum low-temperature properties (i.e. a low T_g), networks should contain a significant proportion of chain ends and that network formation via terminal groups will be disadvantageous, all else being equal. This writer is not aware of any experimental verification of this prediction.

Cross-linking reduces chain flexibility and hence raises the T_g. The best-known apparent demonstration of this is the fact that when natural rubber is reacted with large amounts of sulphur the product, ebonite, is a rigid material. In fact ebonite is not a particularly good example since in addition to network formation there is also a considerable amount of intramolecular reaction. Nevertheless, the following data (Ueberreiter & Jenckel, 1939) in Table 3.3 show a steady increase of T_g with combined sulphur.

It may be reasoned that each cross-link causes an additional restriction to the flexibility of a polymer molecule. This leads to a relation of the form

$$T_{gx} = T_g + K_x\rho \qquad (3.1)$$

where ρ is the number of cross-links per unit volume or some other expression of cross-link density and K_x is a constant, T_g is the glass transition of the uncross-linked polymer and T_{gx} the corresponding transition of the network. A peroxide cross-linked polyisoprene increases

TABLE 3.3
Effect of combined sulphur on the glass transition temperature of natural rubber

Sulphur (%)	T_g (°C)	*Sulphur* (%)	T_g (°C)
0	−64	10	−40
0·25	−65	20	−24

in T_{gx} by about 6°C for every 10^{20} cross-links per gram. Where the cross-links are different in character from the monomer units which create the main chains, for example sulphur in diene rubbers, the situation is more complex.

The effect of chain branching on T_g is also somewhat complex. Branching increases the chain ends, which tends to depress the T_g, whilst the junction points tend to increase the value of the transition. Providing there is no overlap of effects, the influence of chain ends is usually greater than that of the junction points and the T_g tends to be depressed.

In addition to the molecular structure of the polymer, the T_g may be affected by the presence of monomers, solvents and plasticizers. These may be considered as providing the system with a high concentration of chain ends and such groups of chemicals commonly depress the T_g and improve the low-temperature properties. Fillers usually have only a small effect, increasing the T_g slightly.

Low-temperature properties will also be affected by another polymer characteristic, the ability or otherwise of a polymer to crystallize: such crystallization should not be confused with the common and popular concept of a crystal which normally implies a single crystal—a particle that has grown without interruption from a single nucleus, has a definite external shape and is relatively free from defects. Many polymers exhibit polycrystallinity (a property which may also be seen in non-polymeric materials). In such substances aggregates of tiny crystals have been developed, often interlocked with amorphous material. Regions of high order are discernible and unit cell structures may be elucidated.

Crystallization can usually, but not always, occur if the polymer structure is regular. Except in those cases where isomorphous replacement is possible, irregular structures cannot crystallize. If it is desired to avoid crystallization, irregularities may be introduced in the following ways:

1. By random copolymerization;
2. by introduction of groups in an irregular manner after polymerization;
3. by extensive chain branching and cross-linking;
4. by polymerizing to give a lack of stereoregularity;
5. by control of geometrical isomerism.

All of these techniques are significant in rubber chemistry and examples will be given in the chapters dealing with the various polymer types.

As with the glass transition, the crystalline melting point (T_m) of polymers

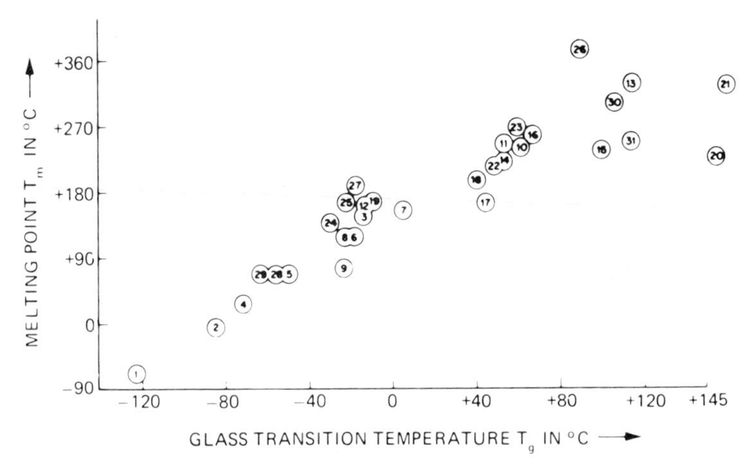

FIG. 3.2. Relationship between glass transition temperature and melting point. (1) Poly(dimethyl siloxane); (2) *cis*-1,4-polybutadiene; (3) *trans*-1,4-polybutadiene; (4) *cis*-1,4-polyisoprene; (5) *trans*-1,4-polyisoprene; (6) polyethylene; (7) polypropylene; (8) polybutene-1; (9) polypentene-1; (10) poly-3-methylbutene-1; (11) poly-4-methylpentene-1; (12) polyoxymethylene; (13) polytetrafluoroethylene; (14) polychlorotrifluoroethylene; (15) isotactic polystyrene; (16) poly(ethylene terephthalate); (17) isotactic poly(methyl methacrylate); (18) poly(*t*-butyl acrylate); (19) isotactic poly(isopropyl acrylate); (20) bisphenol polycarbonate; (21) poly(vinyl carbazole); (22) nylon 6; (23) nylon 66; (24) poly(vinyl methyl ether); (25) poly(vinyl isobutyl ether); (26) poly(vinyl cyclohexanone); (27) poly(vinylidene chloride); (28) poly(ethylene oxide); (29) poly(propylene oxide); (30) cellulose triacetate; (31) poly(methyl isopropenyl ketone). (From Brydson, 1972.)

is not particularly sharp and it is usual to quote as the T_m the temperature at which the last crystalline traces disappear.

As a general rule those factors that tend to raise the T_g also tend to raise the crystalline melting point T_m. Therefore, although there is no unique relation between T_g and T_m, the correlation between the two is quite good (Fig. 3.2). A useful rule-of-thumb (Brydson, 1972) is given by

$$T_g/T_m = 2/3 \quad \text{(when temperatures are in K)} \tag{3.2}$$

One glaring exception to this rule is given by *trans*-1,4-polybutadiene (Table 3.2).

It is interesting to note that data on a large number of non-polymeric substances, although somewhat more scattered, give a similar average value for T_g/T_m.

3.3.3 Chemical Reactivity

There are two opposing requirements concerning the chemical reactivity of a rubber. On one hand it should be sufficiently reactive to allow desirable reactions such as cross-linking, whilst on the other hand it should not react during service in any adverse manner. Natural rubber and indeed all the hydrocarbon diene rubbers provide excellent examples of this. In such polymers the double bond provides a site for vulcanization. At the same time it is a vulnerable point for attack by oxygen, ozone and other agents which can, in some circumstances, have a catastrophic effect.

Modern strategy is to develop a rubber which for the most part has a very low chemical reactivity but to incorporate a few reactive sites for cross-linking. First successfully applied to butyl rubber in 1940, this approach has been extensively used since with, for example, the ethylene–propylene terpolymers, the acrylic rubbers and the most recent types of fluoro-elastomer.

The chemical reactivity of a commercial rubber compound will depend on the nature of the initial polymer, the nature of the cross-links introduced during polymerization and, to some extent, the additives that may be present. It is stating the obvious to say that the first two are dependent on the nature of the chemical bonds present. Nevertheless, it is useful to make some general observations about the various groups and bonds that occur in rubbers by reference to the following list of examples:

1. Olefin (alkene) polymers such as polyisobutylene and ethylene–propylene copolymers contain only C—C and C—H bonds and may be considered as high-molecular-weight paraffins (alkanes). Like the simpler paraffins they are somewhat inert and those chemical reactions that occur do so by substitution, e.g. halogenation. They may be attacked by peroxides and by high-energy irradiation. In some cases this leads to cross-linking, in others degradation.
2. The fluoro-olefins contain, in addition, C—F bonds. These are exceptionally stable and polymers with C—C and C—F bonds only are very inert. Polymers containing such groups only are virtually impossible to cross-link and so an additional 'reactive site monomer' is incorporated to facilitate such cross-linking.
3. The diene rubbers and a number of other polymers contain double bonds. These react with many agents such as oxygen, ozone, hydrogen halides and halogens. If the double bond is part of the backbone chain then the reaction with ozone, and sometimes

oxygen, will cause scission of the main chain and the rupture of one such bond per chain will halve the number-average molecular weight. The activity of a double bond, and the adjacent α-methylene group, is influenced by the presence of substituent groups. Thus a methyl group, as with natural rubber, will activate the double bond whilst the electron-withdrawing chlorine atom, as in polychloroprene rubbers, has the reverse effect. These differences may have profound effects both on processing and on applications.

4. Ester, amide and carbonate groups are susceptible to hydrolysis. When such groups are in the backbone their hydrolysis will also result in a molecular weight reduction. Where hydrolysis occurs in a side chain then the effect on molecular weight is usually insignificant. The presence of aromatic ring structures adjacent to these groups may offer some protection against hydrolysis except where organophilic hydrolysing agents are present.
5. Hydroxyl groups, of particular importance with the polyurethanes, are very reactive and can be involved in a wide range of reactions.
6. Benzene rings in either the backbone (seldom the case in rubbers) or in side groups can be subjected to substitution reactions. Such changes do not normally cause great changes in the fundamental nature of the polymer since they seldom lead to cross-linking or scission. Perhaps the most well-known examples are the modified polystyrenes used as ion-exchange resins. Presumably, if the demand existed, an ion-exchange rubber, possibly in an open cell cellular form, could be made from SBR.

Polymer reactivity differs from the reactivity of simple molecules in two important respects. The first is the considerable influence of groups present in only trace amounts. Some of these are 'weak links', being either thermally unstable or undesirably reactive, and lead to premature degradation. Others, such as the trace aldehydic groups noted in natural rubber, can cause undesirable cross-linking.

The other special feature of polymer reactivity is the possibility of one group reacting within an adjacent group on the chain. A good example also occurs in natural rubber where, under appropriate conditions, the rubber is cyclized to give a ring–chain structure.

3.3.4 Solution Properties

The solution properties of polymers in general are of interest for two reasons. First, they influence the choice of processing method and of polymer for a specific application, whilst secondly, being very different from those of simple molecules, they arouse academic curiosity. In the particular case of rubbers the swelling behaviour of networks is of interest in characterizing network structure.

When simple substances are immersed in their solvents the outside of the substance dissolves away, first generating fresh surfaces which in turn dissolve, the process repeating until the whole mass has dissolved. As an example, one may cite a cube of sugar in water. In the case of high-molecular-weight linear polymers the solvent first diffuses into the polymer mass, which swells continuously until the polymer molecules become so diluted that they disentangle, the solvent becomes the continuous phase and a solution has been formed.

In practice one is concerned both with rate of dissolution and with equilibrium properties. The rate of dissolution is very dependent on the initial rate of diffusion of solvent into the polymer. This in turn, being governed by the laws of diffusion, will depend on the size of the diffusing solvent molecules, by the 'holes' (free volume) in the polymer mass (generally higher in rubbers than in glassy and crystalline rigid polymers), by the initial interfacial area, and by the temperature. A spectacular example of rapid swelling will occur if a piece of a rubber foam is immersed in an appropriate solvent; in this instance the holes are of macro-dimensions.

Amongst the equilibrium characteristics of interest is whether or not a given liquid is a good solvent for a given polymer. One commonly used rule-of-thumb is that 'like dissolves like', e.g. polar polymers dissolve in polar solvents and non-polar polymers in non-polar solvents. Attempts to express this quantitatively have been made via several approaches, of which the solubility parameter concept has been the most widely used.

This is based on the argument that if a polymer and a solvent are to be compatible in the liquid phase, the force of attraction between polymer segments should be similar to the force of attraction between solvent molecules. One measure of the force of attraction is the latent heat of evaporation (L) less the work done in evaporation (RT) per unit molecular weight (M). If expressed per unit volume this expression, known as the cohesive energy density, becomes

$$\frac{L - RT}{M/D}$$

where R is the gas constant, T the absolute temperature and D the density.

The square root of this expression is encountered sufficiently often to be given a name of its own, the solubility parameter (δ), i.e.

$$\delta = \sqrt{\frac{L - RT}{M/D}} \tag{3.3}$$

It has been shown that the heat of mixing between two substances A and B is given by

$$\Delta H = V_{\mathrm{m}}(\delta_{\mathrm{A}} - \delta_{\mathrm{B}})a_{\mathrm{A}}a_{\mathrm{B}} \tag{3.4}$$

in the absence of specific interaction (Hildebrand & Scott, 1950), where V_{m} is the total volume of the mixture and a is the volume fraction of each component.

Since the (Gibbs) free energy is given by

$$\Delta G = \Delta H - T\Delta S \tag{3.5}$$

solution will proceed if $T\Delta S > \Delta H$. Because entropy increases as a polymer dissolves, solution will occur where ΔH is zero, or negative, or slightly positive.

Thus, where Hildebrand's equation is valid, if a polymer and a solvent have similar solubility parameters then solution will occur. Such validity covers amorphous hydrocarbon rubbers and can also be used qualitatively with caution with the more polar rubbers. The solubility parameter of natural rubber, in $(\mathrm{MJ\,m^{-3}})^{1/2}$, is 16·5, which suggests that a not-too-entangled mass of rubber molecules will dissolve in turpentine (16·5), carbon tetrachloride (17·6) and toluene (18·2) but not in acetone (20·4) or ethanol (26·0). (All figures are expressed in units of $(\mathrm{MJ\,m^{-3}})^{1/2}$.) This expectation is realized in practice.

Table 3.4 gives some solubility parameters of polymers including rubbers and Table 3.5 some data on common solvents. More comprehensive lists have been published elsewhere (e.g. Gardon, 1965; Burrell, 1955), whilst a method of predicting solubility parameters from consideration of molecular structure has been proposed and widely used (Small, 1953).

Experimentally the solubility parameter of a network-forming rubber is most readily obtained by immersing identical vulcanized samples into solvents with a range of solubility parameters and observing the equilibrium swelling. The solubility parameter of the solvent giving the greatest swelling is taken as the solubility parameter of the rubber.

Useful information can be obtained by studying the thermodynamics of network swelling. A polymer may be expected to dissolve eventually in a

TABLE 3.4
Solubility parameters of polymers

Polymer	δ	
	$(MJ\,m^{-3})^{1/2\,a}$	$(cal\,cm^{-3})^{1/2}$
Polytetrafluoroethylene	12·6	6·2
Polychlorotrifluoroethylene	14·7	7·2
Polydimethyl siloxane	14·9	7·3
Ethylene–propylene rubber	16·1	7·9
Polyisobutylene	16·1	7·9
Polyethylene	16·3	8·0
Polypropylene	16·3	8·0
Polyisoprene (natural rubber)	16·5	8·1
Polybutadiene	17·1	8·4
Styrene–butadiene rubber	17·1	8·4
Poly(*t*-butyl methacrylate)	16·9	8·3
Poly(*n*-hexyl methacrylate)	17·6	8·6
Poly(*n*-butyl methacrylate)	17·8	8·7
Poly(butyl acrylate)	18·0	8·8
Poly(ethyl methacrylate)	18·3	9·0
Polymethylphenyl siloxane	18·3	9·0
Poly(ethyl acrylate)	18·7	9·2
Polysulphide rubber	18·3–19·2	9·0–9·4
Polystyrene	18·7	9·2
Polychloroprene rubber	18·7–19·2	9·2–9·4
Poly(methyl methacrylate)	18·7	9·2
Poly(vinyl acetate)	19·2	9·4
Poly(vinyl chloride)	19·4	9·5
Bisphenol A polycarbonate	19·4	9·5
Poly(vinylidene chloride)	20·0–25·0	9·8–12·2
Ethyl cellulose	17·3–21·0	8·5–10·3
Cellulose di(nitrate)	21·6	10·55
Poly(ethylene terephthalate)	21·8	10·7
Acetal resins	22·6	11·1
Cellulose diacetate	23·2	11·35
Nylon 66	27·8	13·6
Poly(methyl α-cyanoacrylate)	28·7	14·1
Polyacrylonitrile	28·7	14·1

Because of difficulties in their measurement, published figures for a given polymer can range up to 3% on either side of the average figure quoted.

[a] $(MJ\,m^{-3})^{1/2} \equiv MPa^{1/2}$.

TABLE 3.5(a)
Solubility parameters of some common solvents

Solvent	δ	
	$(MJ\,m^{-3})^{1/2}$[a]	$(cal\,cm^{-3})^{1/2}$
Neopentane	12·8	6·3
Isobutene	13·7	6·7
n-Hexane	14·9	7·3
Diethyl ether	15·1	7·4
n-Octane	15·5	7·6
Methyl cyclohexane	15·9	7·8
Ethyl isobutyrate	16·1	7·9
Di-isopropyl ketone	16·3	8·0
Methyl amyl acetate	16·3	8·0
Turpentine	16·5	8·1
Cyclohexane	16·7	8·2
2,2-Dichloropropane	16·7	8·2
sec-Amyl acetate	16·9	8·3
Dipentene	17·3	8·5
Amyl acetate	17·3	8·5
Methyl *n*-butyl ketone	17·6	8·6
Pine oil	17·6	8·6
Carbon tetrachloride	17·6	8·6
Methyl *n*-propyl ketone	17·8	8·7
Piperidine	17·8	8·7
Xylene	18·0	8·8
Dimethyl ether	18·0	8·8
Toluene	18·2	8·9
Butyl cellosolve	18·2	8·9
1,2-Dichloropropane	18·3	9·0
Mesityl oxide	18·3	9·0
Isophorone	18·6	9·1
Ethyl acetate	18·6	9·1
Benzene	18·7	9·2
Diacetone alcohol	18·7	9·2
Chloroform	19·0	9·3
Trichloroethylene	19·0	9·3
Tetrachloroethylene	19·2	9·4
Tetralin	19·4	9·5
Carbitol	19·6	9·6
Methyl chloride	19·8	9·7
Methylene chloride	19·8	9·7
Ethylene dichloride	20·0	9·8
Cyclohexanone	20·2	9·9
Cellosolve	20·2	9·9

(*continued*)

TABLE 3.5(a)—*contd.*

Solvent	δ	
	$(MJ\,m^{-3})^{1/2\,a}$	$(cal\,cm^{-3})^{1/2}$
Dioxane	20·2	9·9
Carbon disulphide	20·4	10·0
Acetone	20·4	10·0
n-Octanol	21·0	10·3
Butyronitrile	21·4	10·5
n-Hexanol	21·8	10·7
sec-Butanol	22·0	10·8
Pyridine	22·2	10·9
Nitroethane	22·6	11·1
n-Butanol	23·2	11·4
Cyclohexanol	23·2	11·4
Isopropanol	23·4	11·5
n-Propanol	24·2	11·9
Dimethylformamide	24·7	12·1
Hydrogen cyanide	24·7	12·1
Acetic acid	25·7	12·6
Ethanol	26·0	12·7
Cresol	27·1	13·3
Formic acid	27·6	13·5
Methanol	29·6	14·5
Phenol	29·6	14·5
Glycerol	33·6	16·5
Water	47·7	23·4

[a] $(MJ\,m^{-3})^{1/2} \equiv MPa^{1/2}$.

solvent if such a process causes a decrease in the Gibbs free energy of dilution, this being defined by eqn (3.5). Experimental studies have shown that in typical systems of rubber-effective solvent (e.g. natural rubber–benzene) ΔH is usually small but positive and $T\Delta S$ several-fold greater.

It is possible to estimate the increase in entropy on mixing from statistical considerations. The heat of dilution is postulated as being in proportion to the square of the volume fraction of polymer, a proposition with some theoretical basis and found to apply to many simple liquid mixtures. From such estimates Flory (1942) and, separately, Huggins (1942) derived similar equations for ΔG which when slightly simplified both yielded the expression

$$\Delta G = RT\{\ln(1 - v_r) + v_r + \chi v_r^2\} \tag{3.6}$$

TABLE 3.5(b)
Solubility parameters for some common plasticizers

Plasticizer	δ	
	$(MJ\,m^{-3})^{1/2\,a}$	$(cal\,cm^{-3})^{1/2}$
Paraffinic oils	15·3 approx.	7·5 approx.
Aromatic oils	16·4 approx.	8·0 approx.
Camphor	15·3	7·5
Di-iso-octyl adipate	17·8	8·7
Dioctyl sebacate	17·8	8·7
Di-isodecyl phthalate	18·0	8·8
Dibutyl sebacate	18·2	8·9
Diethyl hexyl phthalate	18·2	8·9
Di-iso-octyl phthalate	18·2	8·9
Di-2-butoxyethyl phthalate	18·9	9·3
Dibutyl phthalate	19·2	9·4
Triphenyl phosphate	20·0	9·8
Tritolyl phosphate	20·0	9·8
Trixylyl phosphate	20·2	9·9
Dibenzyl ether	20·4	10·0
Triacetin	20·4	10·0
Dimethyl phthalate	21·4	10·5

[a] $(MJ\,m^{-3})^{1/2} \equiv MPa^{1/2}$.

where v_r is the volume fraction of polymer in the mixture and χ is an adjustable parameter, a function of polymer–solvent interaction and considered to be made up of two components, one entropic and the other energetic in origin. Equation (3.6) is generally referred to as the Flory–Huggins equation. Typical values for χ are given in Table 3.6. Shvarts (1958) has related χ and δ, the solubility parameter, by the equation

$$\delta_1 = \delta_2 \pm \{(RT/V_1)(\chi - \chi_0)\}^{1/2} \tag{3.7}$$

where δ_1 and δ_2 are the solubility parameters of solvent and polymer, V_1 is the molar volume of the solvent and χ_0 the entropic component of χ. Experimental data give a moderate fit to this equation.

In a cross-linked rubber the network prevents total solution of the polymer, the polymer swelling instead to an equilibrium value. In this case the total energy of dilution, ΔG, will be the sum of two terms, ΔG_m, the free energy of dilution in the state prior to cross-linking, and ΔG_e, which takes into account the entropy changes caused by expansion of the polymer network.

TABLE 3.6
Values of Flory–Huggins parameter χ (the third decimal figure is probably not significant)
(After Sheehan & Bisio, 1966)

Liquid	*Natural rubber*	*Poly-chloroprene (Neoprene)*	*Butyl rubber*	*Butadiene–styrene rubber (71·5% styrene)*	*Butadiene–acrylonitrile rubber (18% acrylo-nitrile)*	*Silicone rubber (dimethyl siloxane)*
Benzene	0·421	0·263	0·578	0·398		0·52
Toluene	0·393		0·557			0·465
Hexane	0·480	0·891	0·516	0·656	0·990	0·40
Decane	0·444	1·147	0·519	0·671	1·175	
Dichloromethane	0·494	0·533	0·579	0·474	0·394	
Carbon tetrachloride	0·307		0·466	0·362	0·478	0·45
n-Propyl acetate	0·649					
Methyl ethyl ketone	0·856					
Acetone	1·36					

It has been shown (e.g. Brydson, 1978) that

$$\Delta G_e = \frac{\rho RTV_1}{M_c}(A\eta v_r^{1/3} - Bv_r) \tag{3.8}$$

By combining eqns (3.6) and (3.8) and assuming that at equilibrium $\Delta G = 0$,

$$-[\ln(1 - v_r) + v_r + \chi v_r^2] = \frac{\rho V_1}{M_c}(A\eta v_r^{1/3} - Bv_r) \tag{3.9}$$

There is some disagreement about the values of A, B and η.

Three special cases of eqn (3.9) are of interest:

1. Where $A = 1$, $\eta = 1$ and $B = 0$. In this case the equation simplifies to the original Flory–Rehner equation of 1943:

$$-[\ln(1 - v_r) + v_r + \chi v_r^2] = \frac{\rho V_1}{M_c} v_r^{1/3} \tag{3.10}$$

2. Where $A = 1$, $\eta = 1$ and $B = \frac{1}{2}$. This equation simplifies to the modified Flory–Rehner equation of 1950:

$$-[\ln(1 - v_r) + v_r + \chi v_r^2] = \frac{\rho V_1}{M_c}(v_r^{1/3} - v_r/2) \tag{3.11}$$

3. Where $A=1$, $B=1$ and $\eta \neq 1$. The simplified equation obtained in this case is not known to have been used but does take into account increasingly accepted views of the magnitude of B and η:

$$-[\ln(1-v_r)+v_r+\chi v_r^2]=\frac{\rho V_1}{M_c}(\eta v_r^{1/3}-v_r) \tag{3.12}$$

Equations (3.10)–(3.12) do not take into account the fact that chain ends present do not contribute to the modulus of the system. This may be corrected by multiplying the right-hand side of the equations by the factor $(1-xM_c/M)$, where $x=2$ according to Flory (1944) but is also, arguably, equal to 1 or to some figure in the range 1–2.

Further deviations from eqn (3.9) are to be expected where segments are highly extended and Gaussian statistics no longer apply, because of neglect of the experimentally observed C_2 term and by all the other imperfections of the Gaussian theory of high elasticity. In spite of this, eqn (3.9) (or rather its simplifications (3.10) and (3.11)) has been of importance in the elucidation of the structure of vulcanized rubbers.

The above expressions are of particular interest in that they relate the equilibrium swelling ($1/v_r$) to the average molecular weight between cross-links (M_c), which in turn according to the elementary Gaussian theory of rubber elasticity is related to the shear modulus by the expression

$$G=NkT=\rho RT/M_c \tag{3.13}$$

3.3.5 Electrical Properties

The electrical properties of polymers have been widely studied and the results subjected to a number of excellent reviews (e.g. Link, 1972; Parker, 1972; Seanor, 1972). The writer has discussed, in qualitative terms, the effect of molecular structure on some electrical properties such as dielectric constant and power factor (Brydson, 1966–1982). The rubbers show no particular distinctive features in respect of electrical properties over other polymers so this topic will be dealt with very briefly here. Perhaps the main point to be made is that the electrical properties of rubber compounds are often more dependent on the additives than on the rubbery polymer itself.

If a polymer mass containing only truly covalent bonds free from dipoles is placed in an electric field there is an instantaneous electron shift (electron polarization) but no actual movement of the molecules themselves. If this field is between two charged plates the polymer acts as a dielectric in a capacitor and the capacity of the charged plates is increased. The factor by which the capacity is increased is known as the dielectric constant or permittivity. Because it depends on virtually instantaneous electron movement, the value of the dielectric constant is not dependent on temperature nor on the frequency (if the field is subject to alternation). The dielectric constant of such materials is frequently equal to the square of the refractive index and both properties may be calculated from a knowledge of the chemical bonds present. The method of making such a computation is given in most standard texts on physical chemistry.

When dipoles are present but not frozen-in, as in water or in polychloroprene, the dipoles will tend to align with the field and increase the capacity of a dielectric over and above that due to electron polarization, the effect being known as dipole polarization. In a low-frequency, alternating-current field it is usually possible for the dipole movement to keep in phase with changes in the alignment of the electric field. As the frequency is increased the molecules do not have time to complete their alignment before the field changes direction. This reduces the amount of dipole polarization and there is a reduction in the dielectric constant. Eventually a frequency is reached where there is no time for any dipole movement and the dielectric constant is due to electron polarization only.

If the temperature is raised the reduction in the viscosity of the system makes dipole movement easier and there is a shift in the dielectric constant–frequency curve in the direction of higher frequencies. Hence both temperature and frequency affect the dielectric constant of polar polymers. Although of little importance with rubbers, it is to be noted that dipole movement is very restricted below the T_g unless the dipoles are present on side chains which have lower transitions than that of the main chain.

In the range of frequencies where dipole movement is finite but incomplete there is some internal friction as the movement of dipoles fails to keep in step with the change in field. This causes loss of electric power and some building-up of heat in the dielectric. This can be characterized by various parameters such as power factor and loss factor. The peak in the power factor–frequency curves coincides with the point of inflection in the dielectric constant–frequency curves and is also shifted by raising the temperature (Fig. 3.3).

The large-tonnage elastomers (i.e. the polyisoprenes, polybutadienes, SBR, butyl and EP rubbers) are non-polar in themselves but may contain polar additives, polymerization residues or, in the case of natural rubber, natural products such as proteins. Many of the speciality rubbers, particularly those of importance because of their oil resistance (i.e. polychloroprenes, nitrile rubbers, fluorine-containing rubbers, acrylic rubbers, epichlorhydrin rubbers, polysulphides) are polar.

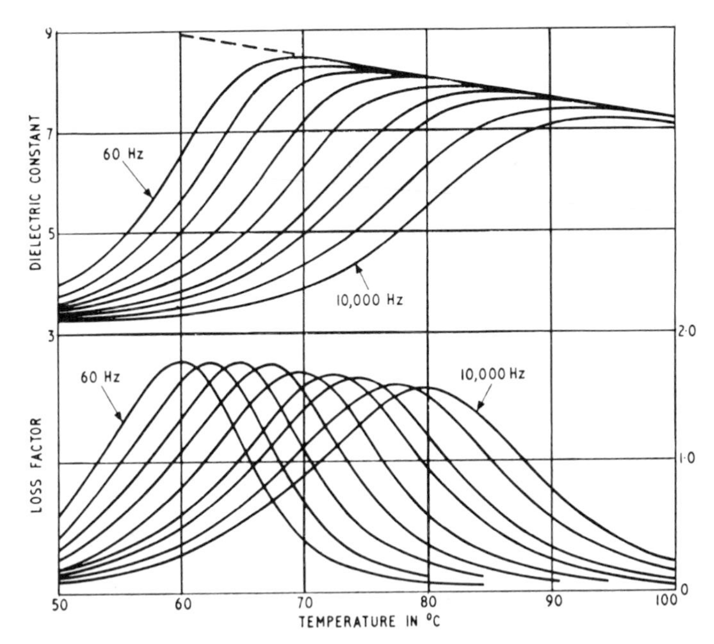

FIG. 3.3. Electrical properties of poly(vinyl acetate) (Gelva 60) at 60, 120, 240, 500, 1000, 2000, 3000, 6000 and 10 000 Hz. (Reproduced by permission from *J. Am. Chem. Soc.*, **63** (1941) 2832. Copyright American Chemical Society.)

3.4 STRUCTURE AND PROCESSING PROPERTIES

Processing properties that are affected by molecular structure include:

1. Viscosity;
2. die swell;
3. extrudate quality;
4. milling characteristics including ease of banding;
5. cold flow and green strength;
6. tack.

For simple liquids, such as water, the coefficient of viscosity is defined as the ratio of shear stress to shear rate. When the viscosity is independent of both shear stress and shear rate the liquid is said to be Newtonian. Polymer melts are almost invariably non-Newtonian in that their viscosity decreases with increasing shear and the term apparent viscosity is used instead of the coefficient of viscosity. As a general rule, the broader the molecular weight distribution the more non-Newtonian the polymer becomes, a phenomenon observed with both rubbery and non-rubbery polymers.

The viscosity of an unfilled polymer depends primarily on three structural variables:

1. Weight-average molecular weight;
2. chain flexibility and other factors that affect the glass transition temperature T_g;
3. long chain branching.

For linear polymers it is commonly observed that the melt viscosity extrapolated to zero shear rate (η_0) is related to the weight-average molecular weight by the relation

$$\eta_0 = K\bar{M}_w^{3\cdot5} \quad \text{where} \quad \bar{M}_w > \bar{M}_{w,c} \tag{3.14}$$

The experimentally observed power-law exponent of about 3·5 has been given some theoretical justification by Bueche (1952). Below the critical value, $\bar{M}_{w,c}$, the molecular weight–viscosity relationship is roughly linear, the greater dependence above this point usually being ascribed to entanglement effects.

As has already been noted the T_g depends on chain flexibility and other molecular factors. It may also be shown that the melt viscosity at some temperature T may be related to the T_g by the following form of the well-known WLF equation:

$$\log_{10}\left(\frac{\eta_T}{\eta_{T_g}}\right) = \frac{-17\cdot44(T-T_g)}{51\cdot6+(T-T_g)} \tag{3.15}$$

This equation indicates that the viscosity, η_T, is decreased by increasing the difference between T and T_g. Thus at a fixed processing temperature T, say 100°C, and for two rubbers of equal $\bar{M}_w$, that with the highest T_g will have the highest viscosity. Since T_g has been related to molecular structure so η_T may be similarly related.

Many polymers have long branches in their molecular structure (Fig. 3.4). Such branched molecules when sheared will be more compact and entangle less with neighbouring molecules than will a linear polymer of similar molecular weight. For this reason branched polymers have lower melt viscosities.

Melt viscosity can also be influenced according to whether or not the polymer can crystallize. If a crystallizable polymer is subject to very high shear at a temperature little above its T_m, shear-induced crystallization can

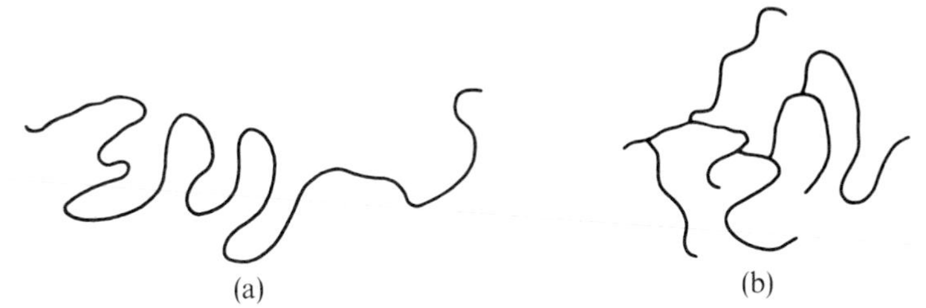

FIG. 3.4. Linear (a) and long-branched molecule (b) of the same size. The branched molecule is more compact and in bulk has lower melt viscosity at equivalent shear rate and temperature. (From Brydson, 1978.)

occur and the shape of the shear stress–shear rate curve shows a pronounced upward sweep (Fig. 3.5).

This can cause difficulties in injection moulding and other high-shear processing operations. For this reason synthetic *cis*-polyisoprene is often easier to mould than the natural polymer.

When polymers are extruded or calendered the cross-sectional area of the extrudate or the calendered sheet is usually greater than the cross-sectional area of the extruder die or the space between the calender rolls. Known as die swell and calender swell these effects are due to relaxation of

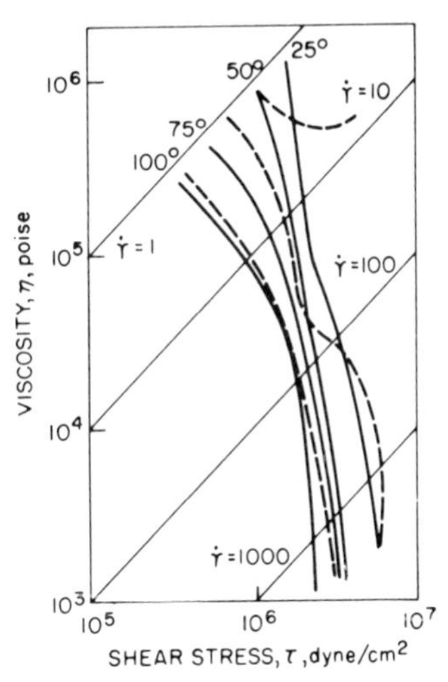

FIG. 3.5. Apparent viscosities of natural rubber (– – –) and synthetic polyisoprene rubber IR-305 (——) at different temperatures. (From Van der Vegt & Smit (1967), reproduced by permission of the Society of Chemical Industry.)

molecules that had been oriented during shear. As a general rule die swell is found, like viscosity, to increase with molecular weight and to decrease with a rise in temperature.

In addition to die swell the rubber processor is also concerned with extrudate quality. Much work has been carried out with thermoplastics on the phenomena of melt fracture and sharkskin. The former, melt fracture, is characterized by an extrudate exhibiting a helical form of distortion. This occurs above a critical shear stress τ_c and at a given temperature it is found that

$$\tau_c \bar{M}_w = K \tag{3.16}$$

for a given polymer.

The value of K appears to be related to the polymer solubility parameter (δ), high values of the latter going together with high values of K. The critical shear stress is only slightly affected by temperature but the corresponding critical shear rate increases considerably with increase in temperature. For the usually desired high values of τ_c the aim must be to have a high δ and a low $\bar{M}_w$ where these are compatible with other requirements.

Sometimes confused with melt fracture (also known as elastic turbulence, bambooing and distortion) is sharkskin. In thermoplastics this occurs as tiny transverse ridges on the surface of an extrudate. It is found to occur above a critical linear output rate. The writer has suggested that this must be due to tearing of weak elastic melts as the surface of the extrudate accelerates in velocity, relative to the centre, as it leaves the die. Most formal studies have been made on thermoplastics but the phenomenon, or something very much like it, is observed with rubbers.

In the case of polyethylene sharkskin appears to be at its worst with polymers of a very narrow molecular weight distribution. What appear to be parallel results have been reported for *cis*-polybutadiene (Short, 1969). In this case the rubbers were extruded through a Garvey die, which yields an extrudate with a wedge-shaped cross-section. There are grounds for believing that the poor extrudates obtained with narrow molecular-weight distribution polybutadienes is associated with a sharkskin-type effect. It is certainly common practice when preparing synthetic rubbers to try to broaden the molecular weight distribution in order to improve processing behaviour. One possible explanation is that the low-molecular-weight molecules prevent tearing by allowing viscous flow whilst the high-molecular-weight component provides a degree of elastic strength. Long chain branching has a similar effect but since this is often associated with a broad molecular-weight distribution it is not clear which molecular feature is operative.

The ability of a rubber compound to form a tight band on a two-roll mill—a useful processing attribute—seems to be related to the above behaviour. A systematic study of polybutadienes and SBR (Short, 1969) has indicated that improved milling behaviour (as characterized by the ability to produce a good tight band round the roll) may be obtained:

1. By incorporating some vinyl groups into the chain;
2. by broadening the molecular weight distribution;
3. by long chain branching;
4. by increasing the styrene content;
5. by blending high-molecular-weight polymer with oil.

Most of these indications again indicate that the high-molecular-weight polymer helps to provide tenacity whilst the low-molecular-weight polymer and the oils provide a viscous component that allows the polymer mass to be sheared without undue tearing.

One problem when handling unvulcanized rubber compounds is that they deform under their own weight—a phenomenon known as cold flow—and they may also have little strength ('green strength'). Cold flow may be reduced and green strength improved by allowing some sort of network to build up in the rubber. This can be brought about by such techniques as:

1. Controlling molecular weight to such a degree that adequate physical entanglements occur;
2. incorporating a small amount of covalent cross-links into the rubber, for example by polymerizing in the presence of a small quantity (0·1%) of divinylbenzene;
3. providing sites for some form of heat-fugitive cross-linking, for example by ionic bonding, by the use of block copolymers containing blocks with a T_g above the storage temperature, by some incipient crystallization, or by incorporating a weak covalent bond which is readily broken down during processing. As an example of the latter, one commercialized rubber (Solprene 250) contains long chain branches linked to the main chain by carbon–tin bonds, which break down during subsequent processing.

More recently Buckler *et al.* (1977) have reported some success in the preparation of modified SBR materials with a high green strength. In this case a small quantity of a third monomer is incorporated into the polymer chain in order to provide a cure site. This cure site is then reacted with an agent to provide cross-links whose bonds are weaker than the chain backbone bonds and which break down during such processing operations as mixing and extruding but which are able to re-form on standing at room temperature. Cross-linking systems quoted include those across aldehyde groups via a primary diamine and across tertiary amine groups via a dihalide (R_3X_2).

$$\sim\mathrm{CH{=}O} \;+\; \mathrm{H_2N{-}R{-}NH_2} \;+\; \mathrm{O{=}CH}\sim \;\rightleftharpoons\; \sim\mathrm{CH{=}N{-}R{-}N{=}CH}\sim \;+\; 2\mathrm{H_2O}$$

$$\sim\mathrm{NR_1R_2} \;+\; \mathrm{X{-}R_3{-}X} \;+\; \mathrm{R_1R_2N}\sim \;\rightleftharpoons\; \sim\mathrm{N^+R_1R_2\,X^-}{-}\mathrm{R_3}{-}\mathrm{N^+R_1R_2\,X^-}\sim$$

The good green strength of natural rubber compounds is normally explained as being due to crystallization which produces a form of physical cross-link. Whilst this is probably the correct explanation the possibility that it may be due to cross-linking across naturally occurring carbonyl groups does not appear to have been investigated.

Many rubber manufacturing operations require that the rubber compound has good 'tack'—in this sense used to mean the ability of unvulcanized rubber to stick to itself (without necessarily being sticky to other surfaces). This useful property of rubbers arises from the mobility of molecules in a mass of rubber. When two such masses are brought into contact the wriggling chain ends in one mass are able to diffuse into the other, the rate decreasing with increasing molecular weight. For equal amounts of diffusion, tack strength will increase with increasing molecular

TABLE 3.7
Effect of reinforcement on mechanical properties of amorphous and crystallizable vulcanizates

	Natural rubber	*SBR*
Tensile strength, MN/m^2		
Gum stock	20	3
Reinforced	30	20
Elongation at break, %		
Gum stock	700	800
Reinforced	600	500
Tensile stress at 400% elongation		
Gum stock	3	1·5
Reinforced	18	15

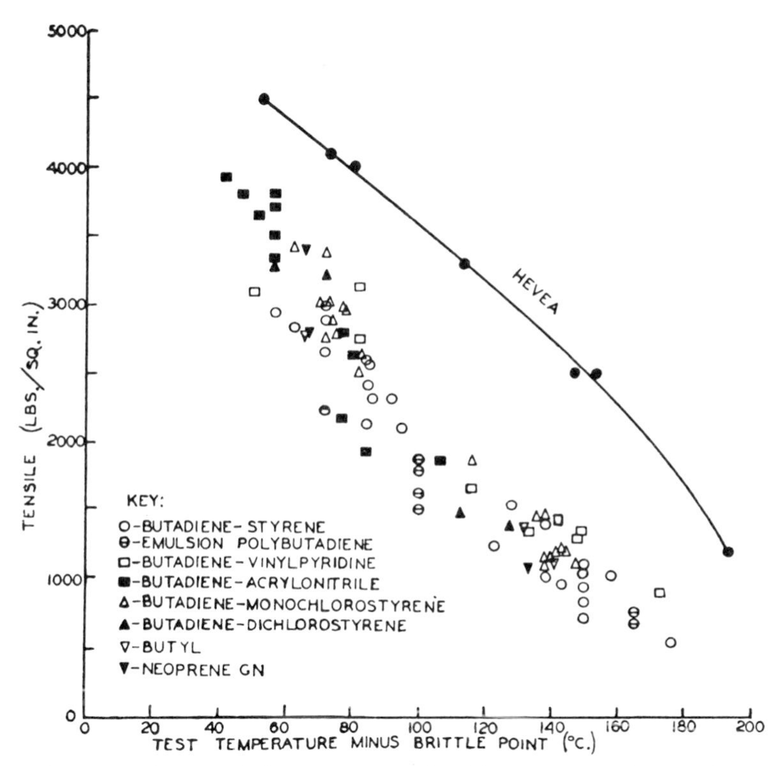

FIG. 3.6. Tensile strength–brittle point relationships of tread stock vulcanizates of natural (Hevea) rubber compared with those of synthetic rubbers of less regular structure. (Reproduced by permission from *Ind. Eng. Chem.*, **38** (1946) 1061. Copyright American Chemical Society.)

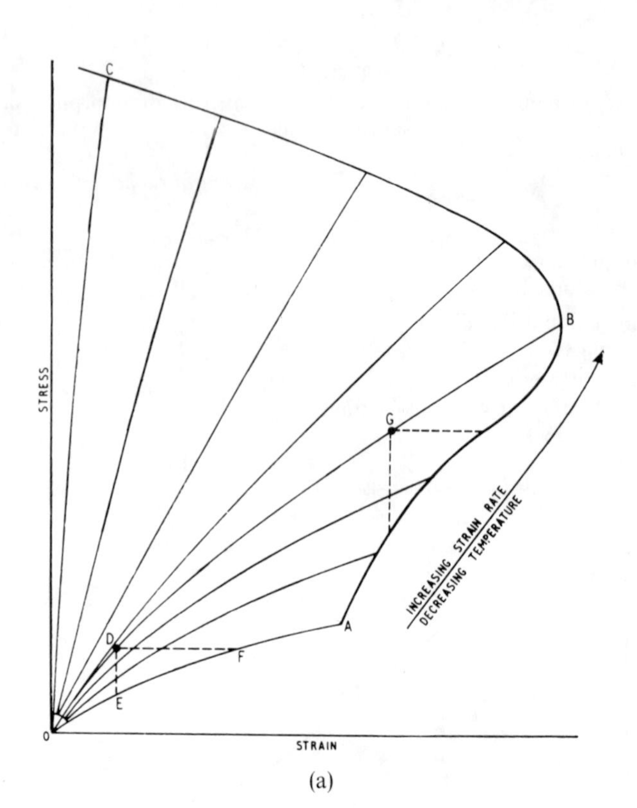

C
B
G
A
D
F
E
O
STRESS
STRAIN
INCREASING STRAIN RATE
DECREASING TEMPERATURE

(a)

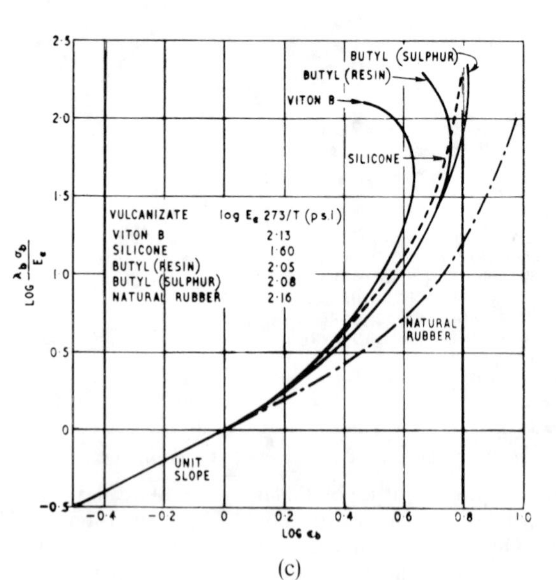

BUTYL (SULPHUR)
BUTYL (RESIN)
VITON B
SILICONE
VULCANIZATE log Eₑ 273/T (p.s.i)
VITON B 2·13
SILICONE 1·60
BUTYL (RESIN) 2·05
BUTYL (SULPHUR) 2·08
NATURAL RUBBER 2·16
NATURAL RUBBER
UNIT SLOPE
LOG λb σb / Ee
LOG εb

(c)

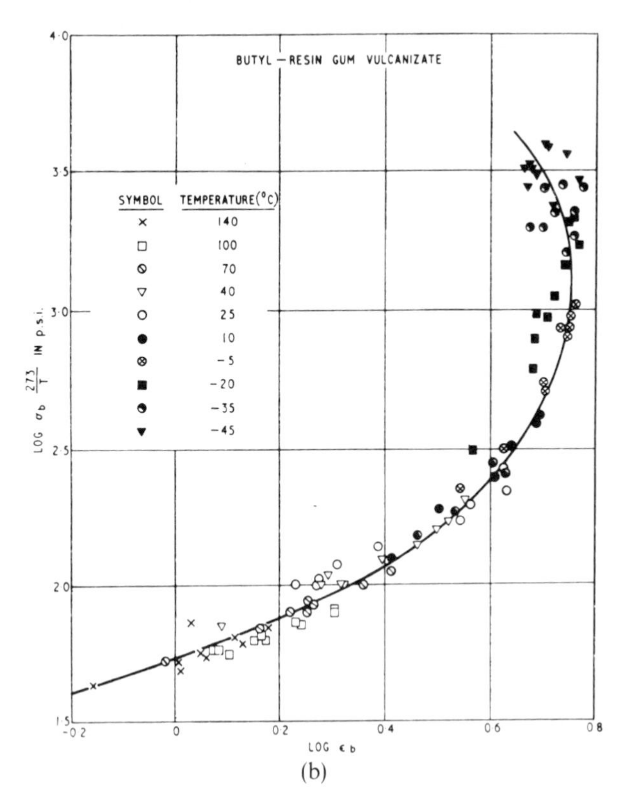

(b)

FIG. 3.7. Failure envelopes from rubber vulcanizates. (a) Schematic representation of the dependence of tensile stress–strain curves for an amorphous rubber on strain rate and temperature. (b) Failure points for a resin-cured butyl rubber gum vulcanizate tested at various rates and temperatures: σ_b = stress at break, ε_b = elongation at break. (c) Comparison of vulcanizates (Viton B is a fluoroelastomer): λ_b = relative extension at break, E_e = equilibrium modulus. (From Smith, 1962.)

weight so that in practice there is an optimum molecular weight to give the best tack.

In the case of amorphous rubbers, providing they are in a state well above the T_g, the degree of tack is not, to a first approximation, very dependent on molecular structure. If, however, the rubber is crystallizable, crystal structures, formed at the interface from molecular segments originally in different masses, will enhance the degree of tack. Many synthetic rubbers, by lacking the ability to crystallize, require incorporation of additives in order to provide adequate building tack.

3.5 STRENGTH OF RUBBERS

It has long been recognized that the strength of a rubber may be increased by:

1. facilitating crystallization during stretching;
2. incorporation of finely divided fillers, in particular carbon blacks.

Hence unfilled SBR compounds are much weaker than corresponding vulcanizates from natural rubber but both are considerably strengthened (reinforced) by the incorporation of carbon black (Table 3.7).

Less well understood are the molecular factors that control strength. Many years ago (Borders & Juve, 1946) (Fig. 3.6) it was shown for amorphous rubbers that the tensile strength was proportional to the difference between the test temperature, T, and the glass transition temperature, T_g. This suggests, with hindsight, that viscosity has some role in determining strength and that a WLF-type relationship, widely used in many branches of polymer science to equate the effects of changing temperature to a change in the rate of strain, might be of some use here.

Tensile stress–strain tests with amorphous rubbers over a range of strain rates and temperatures have shown that for a given rubber the failure point lay along an envelope of the stress–strain diagram (Smith, 1962) (Fig. 3.7) and that the data could be superimposed by a WLF-type shift operation (Fig. 3.8).

Studies with a series of amorphous rubbers gave failure envelopes which almost superimposed when plotted on a common diagram. This suggests that differences reported in the strengths of unfilled amorphous rubbers are largely due to the different extents by which viscous forces retard movement of the chain at different values of $(T - T_g)$.

It is outside the scope of this book to discuss the relationship to tensile

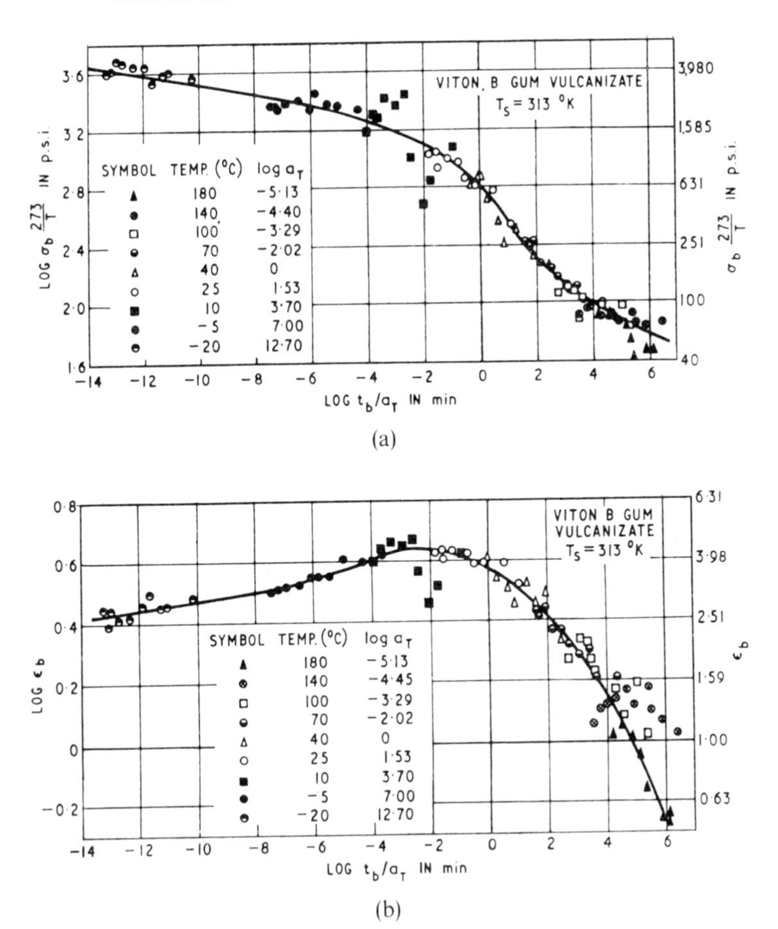

FIG. 3.8. Superposition of ultimate tensile data. Plots of (a) $\log \sigma_b(273/T)$ against $\log (t_b/a_T)$; (b) $\log \varepsilon_b$ against $\log (t_b/a_T)$, for a fluoroelastomer vulcanizate (Viton B): t_b (time to break) $= \varepsilon_b/\dot{\varepsilon}$, where $\dot{\varepsilon}$ is the elongation rate. $\log a_T = \log (t/t_0) = [-8{\cdot}86(T-T_s)/101{\cdot}6 + T - T_s]$ which is a form of the WLF equation, where T_s is an arbitrary reference temperature (313 K for the data above). (From Smith, 1962.)

strength of other mechanical properties involving failure, such as tear strength and abrasion resistance. It is not, however, surprising to find that in many cases the magnitude of many of these properties, when measured at some temperature T, depends on the value of $(T - T_g)$.

Such a dependence on $(T - T_g)$ is not restricted to properties so obviously related to tensile strength. The rate of cut growth of diene rubbers due to

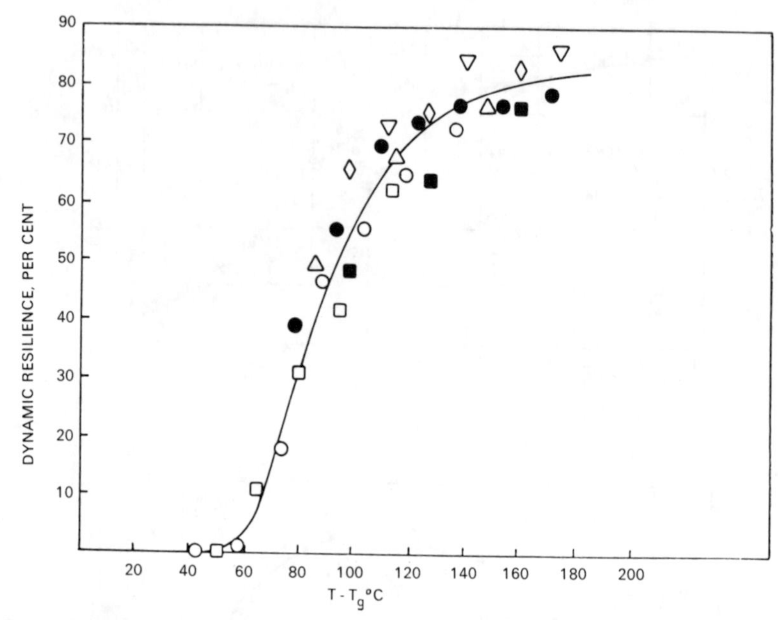

FIG. 3.9. Relationship between dynamic resilience and $(T-T_g)$ for poly(vinyl alkyl ether) vulcanizates: ○, ethyl; △, *n*-butyl; □, isobutyl; ◇, *n*-pentyl; ●, *n*-hexyl; ▽, *n*-octyl; ■, 2-ethylhexyl; T = test temperature, °C. (From Lal *et al.* (1965), reproduced by permission of John Wiley & Sons Inc.)

ozone attack appears to be similarly dependent on T_g whilst a surprisingly unique relationship between dynamic resilience and $(T-T_g)$ appears to hold for a homologous series of poly(vinyl alkyl ether)s (Fig. 3.9) (Lal *et al.*, 1965).

REFERENCES

Borders, A. M. & Juve, R. D. (1946). *Ind. Eng. Chem.*, **38**, 1066.

Boyer, R. F. (1963). *Rubber Chem. Technol.* **36**, 1303.

Brydson, J. A. (1966). *Plastics Materials*, 1st edn. Newnes–Butterworth, London.

Brydson, J. A. (1972). In *Polymer Science*, ed. A. D. Jenkins. North-Holland, Amsterdam, London, Chapter 3.

Brydson, J. A. (1978). *Rubber Chemistry*. Applied Science, London.

Brydson, J. A. (1982). *Plastics Materials*, 4th edn. Newnes–Butterworth, London.

Buckler, E. J., Briggs, G. J., Henderson, J. F. & Lasis, E. (1977). *European Rubber J.*, **159**(7/8), 21.

Bueche, F. (1952). *J. Chem. Phys.* **20**, 1959.

Bunn, C. W. (1955). *J. Polymer Sci.*, **16**, 323.
Burrell, H. (1955). *Interchem. Rev.*, **14**, 3.
Flory, P. J. (1942). *J. Chem. Phys.*, **10**, 51.
Flory, P. J. (1944). *Chem. Rev.*, **35**, 51.
Gardon, J. L. (1965). Cohesive energy density. In *Encyclopaedia of Polymer Technology*, Vol. 3. Interscience, New York, p. 833.
Gibbs, J. H. & Di Marzio, E. A. (1958). *J. Chem. Phys.*, **28**, 373.
Hildebrand, J. H. & Scott, R. L. (1950). *The Solubility of Non-electrolytes.* Reinhold, New York.
Huggins, M. L. (1942). *Ann. N.Y. Acad. Sci.*, **43**, 1.
Lal, J., McGrath, J. E. & Scott, K. W. (1965). *J. Appl. Polymer Sci.*, **9**, 3471.
Link, G. L. (1972). In *Polymer Science*, ed. A. D. Jenkins. North-Holland, Amsterdam, London, Chapter 18.
Mead, D. J. & Fuoss, R. M. (1941). *J. Am. Chem. Soc.*, **63**, 2832.
Parker, T. G. (1972). In *Polymer Science*, ed. A. D. Jenkins. North-Holland, Amsterdam, London, Chapter 19.
Seanor, D. A. (1972). In *Polymer Science*, ed. A. D. Jenkins. North-Holland, Amsterdam, London, Chapter 17.
Sheehan, C. J. & Bisio, A. L. (1966). *Rubber Chem. Technol.*, **39**, 149.
Short, J. N. (1969). *Proc. 4th Int. Synthetic Rubber Symposium*, p. 22.
Shvarts, A. G. (1958). *Rubber Chem. Technol.*, **31**, 691.
Small, P. A. (1953). *J. Appl. Chem.*, **3**, 71.
Smith, T. L. (1962). Technical Documentary Report No. ASD-TDR-62-572, June 1962: *Proc. 7th Joint Army–Navy–Air Force Conference on Elastomers Research and Development, October 22–24*, Vol. 2.
Ueberreiter, K. & Jenckel, E. (1939). *Zeit. Physikal. Chem.*, **A182**, 361.
Van der Vegt, A. K. & Smit, P. P. A. (1967). SCI Monograph No. 26, p. 313.
Würstlin, F. (1950). *Z. Angew. Phys.*, **2**, 131.

Chapter 4

Natural Rubber

4.1 INTRODUCTION

That natural rubber has been able to retain a substantial share of the rubber market is a tribute both to the intrinsic nature of the polymer and to the research and development effort which has helped to ensure a steadily improving quality of the commercial product. Much of this work has been carried out by the research associations of the producer countries, particularly noteworthy being that of Malaysia.

It has been estimated that some 2000 different plant species yield aqueous dispersions or latices of polymers akin to natural rubber and that rubbers of sorts have been obtained from some 500 of them. These include dandelions, the well-known decorative house plant *Ficus elastica*, and the guayule bush of Arizona and Mexico. The commercial market, however, is totally dominated by one plant, *Hevea brasiliensis*.

As indicated by the name, this tree is a native of South America. Because of the high price of rubber in the 1870s some 2000 seeds were smuggled out of the Amazon basin and planted by Sir Joseph Hooker in Kew Gardens, London, in 1873. The seedlings did not survive but a further consignment of 70 000 seeds was brought from Brazil by Sir Henry Wickham and 2000 plants were raised at Kew. Some of these plants were then sent to what was then Ceylon, and it is from the survivors that the rubber plantations of the Far East in Malaysia, Indonesia, Thailand, Sri Lanka and elsewhere have developed.

Today these four Far East producers account for about 80% of the natural rubber marketed, with Malaysia alone responsible for over one-third. There are also substantial plantations in tropical America (particularly Brazil), tropical Africa (such as in Liberia, Nigeria and the Congo basin) and the Indian sub-continent.

The number 80 figures prominently in a consideration of the growing conditions required for *Hevea*. The tree grows up to 80 feet in height with an average temperature requirement of 80°F and an average annual rainfall of

80 inches. Good drainage of soil at somewhat low altitudes, generally below 2000 feet, and a slightly acid soil are further desirable requirements.

The rubber latex occurs in tiny vessels embedded in the inner cortex of bark of the tree which occurs below the corky surface layers or outer cortex. Incisions into the bark cause the latex to exude as a result of osmotic pressure until such time as the latex coagulates, plugging the severed vessels in a manner somewhat reminiscent of the clotting of blood. Typical practice is to cut a sloping half-circumferential incision on alternate days, successive incisions being made immediately below the previous cut so that a tapping panel results in the shape of a parallelogram. Both the yield and character of the latex are affected by the frequency of tapping, the height of the tapping panel and the time of tapping.

The exudate is collected in cups whose contents are subsequently transferred to large containers. The choice of subsequent treatment of the latex can lead to two quite distinct technologies. These alternatives are:

1. Coagulation or coacervation of the latex to produce a solid rubber processed by the conventional technique of dry rubber technology;
2. concentration of the latex to about 60% dry rubber content.

The concentrated latex is then shipped in liquid form to processors around the world who specialize in shaping the rubber directly from the latex or for use in such products as adhesives.

4.2 PREPARATION OF DRY RUBBER

Early descriptions by travellers and explorers to the Amazon basin described the techniques used by the native Indian tribes to produce a dry rubber. In a typical process the latex collector used a bottomless earthen pot in which a smoky fire was made by burning some dried twigs and to which some locally abundant pine nuts were added. The blade of a long wooden paddle was smeared with a wet clay, to act as a release agent, and then dipped into the rubber latex and held above the smoke. This produced a thin layer of coagulated and partly dried rubber. By repeatedly dipping the paddle in the latex and rotating it over the fire, successive layers of rubber were deposited on the paddle until a somewhat ellipsoidal ball of solid rubber was produced. When of the desired size it was removed from the paddle by means of a moistened knife. The smoking is now known to have an antiseptic or anti-putrefying effect which helps to preserve the rubber. In spite of the advent of plantation rubber the writer was somewhat surprised to see in a television travel programme screened in 1985 that this process is still in use in the Amazon basin.

In spite of its crudity the process invoked the process of coagulation, the needs for thin films and short diffusion paths for reducing drying times, the use of release agents, the need for preservatives and also the efficacy of cutting rubber with knives wetted with water. For large-scale production, though, the process was far too slow, and the product was far too variable and subject to high levels of contamination. Thus with the development of plantation rubber new techniques were introduced which themselves have only recently begun to be replaced.

The traditional grades of sheets and crepes are classified into a number of internationally recognized grades (Table 4.1). Probably the most typical of these is *ribbed smoked sheet*. This is produced by diluting the latex to about 12–15% dry rubber content and adding to large aluminium or aluminium-lined tanks, typically of 250 gallons capacity. Dilute glacial acetic or formic acid is added to destabilize the latex. Partitions are inserted vertically into the tanks, commonly in such a way that the gel is formed into a long serpentine sheet which is then fed to the first of four to six two-roll washing mills. The rolls of the final mill are grooved and this confers a diamond patterned rib to the surface.

The surface pattern helps both to reduce adhesion between sheets in a rubber bale and also to accelerate drying by increasing the surface area. The washed sheet is then dried by subjecting it to wood smoke in a long tunnel-like building. A typical drying time is four days. As with the rubber produced by the original Amazonian Indian method, the smoking has a beneficial antiseptic effect on the rubber. The smoked sheet is then baled in blocks up to 250 lb in weight and coated with chalk or talc to stop the bales from sticking to each other during transportation and storage.

Ribbed smoked sheets are dark brown in colour but the better grades (i.e. those with lower numbers) give vulcanizates of generally better mechanical properties than other traditional grades. Processing characteristics do tend to be variable, properties such as cure rate and ageing behaviour depending on the type and amount of non-rubber constituents present. Virtually all of the smoked sheet produced is classified solely on the basis of appearance.

Where a light-coloured rubber is required it is usual to use a *pale crepe*. This is made by first of all treating the latex with a chemical such as sodium bisulphite to prevent enzymes present from darkening the rubber. The rubber is then coacervated in tanks similar to those used for smoked sheet but with fewer partitions in order to obtain a thicker product. After treatment with the coacervating agent, usually acetic acid, the resulting gel

TABLE 4.1
Classification of natural rubber by visual grading in accordance with guidelines laid down in the Green Book—International Standards of Quality and Packing in Natural Rubber

Section	*Grades*
1	Ribbed smoked sheets: Made entirely from coagulated rubber sheets properly dried and smoked 1XRSS Superior quality 1RSS Standard quality 2RSS Good fair average quality 3RSS Fair average quality 4RSS Low fair average quality 5RSS Inferior fair average quality
2	White and pale crepes: Prepared from fresh coagula of NR latex under controlled conditions (classified as either thin or thick) 1X Superior quality pale crepe 1 Standard quality pale crepe 2 Good fair average quality palish crepe 3 Fair average off-colour palish crepe
3	Estate brown crepes: From cup lump, other high-grade scraps and pre-cleaned tree bark scrap (classified as thick or thin) 1X Clean light brown crepe 2X, 3X } Inferior grades
4	Compo crepes: From lumps, tree scraps and smoked sheet cuttings. Power wash mills used to clean away impurities Grades 1, 2, 3
5	Remills—thin brown crepes Grades 1, 2, 3, 4
6	Thick blanket crepes—ambers Grades 2, 3, 4
7	Flat bark crepes: Any scrap material power washed, including earth scraps Grades standard, hard

is thoroughly washed in a crepeing battery, typically consisting of three pairs of washing mills. Pale crepe is air-dried. Extra-white crepes may be produced by removing the β-carotene present either by treatment with xylyl mercaptan or by a fractional coagulation process. As with normal pale crepes, sodium bisulphite is also used to prevent enzyme action. Shoe sole crepes are produced by laminating thin layers of pale crepe. As with the ribbed smoked sheets, pale crepes are classified by visual inspection including comparison with standard samples. These are set using the Green Book: *International Standards of Quality and Packing of Natural Rubber Grades*.

The Green Book also specifies a number of inferior grades. The estate brown crepes are generally obtained from scrap material generated on estates such as the coagulated material which accumulates in the tapping cups. Some pre-cleaned tree bark scrap may also be present. Other grades include the compo crepes obtained from lumps, tree scraps and smoked sheet cuttings and the flat bark crepes which are produced on wash mills from all types of uncompounded natural rubber including both scrap obtained from the tree bark and also earth scraps taken from the ground at the base of the tree.

4.2.1 Technically Specified Rubbers

The development of the synthetic rubber industry after World War II exposed some serious shortcomings in the traditional forms of the natural product. In the first place the bales were of an inconvenient shape and size to handle. Secondly, and more seriously, the natural product, assessed only by visual inspection, was a very variable material.

For example, it was found that rate of cure when vulcanizing with sulphur was dependent on the dilution of the latex prior to coagulation and on the pH at which coagulation occurs. By controlling these variables it was found possible to maintain grades within fairly narrow bands of cure characteristics and this led to the introduction of technically classified rubbers in Malaysia about 1960.

The development of new processes for natural rubber enabled more fully technically specified rubbers to be introduced. The first of these were the Standard Malaysian Rubbers (SMR) which by the mid-1980s accounted for about 40% of total Malaysian production. These were followed by the introduction of Standard Indonesian Rubbers, and also by standard rubbers of other manufacturing countries.

For SMR classified materials standards are laid down (Table 4.2) for maximum dirt content, maximum ash content, maximum nitrogen content,

TABLE 4.2
Standard Malaysian Rubber Specifications as from January 1979
(After Elliot, 1979)

Parameter[a]	*Latex*				*Sheet material, SMR-5*	*Blend, viscosity-stabilized, SMR-GP*	*Field grade material*		
	Viscosity-stabilized		*Not stabilized*						
	SMR-CV	*SMR-LV*[b]	*SMR-L*	*SMR-WF*			*SMR-10*	*SMR-20*	*SMR-50*
Dirt retained on 44-μm aperture, max. % weight	0·03	0·03	0·03	0·03	0·05	0·10	0·10	0·20	0·50
Ash content, max. % weight	0·50	0·50	0·50	0·50	0·60	0·75	0·75	1·00	1·50
Nitrogen content, max. % weight	0·60	0·60	0·60	0·60	0·60	0·60	0·60	0·60	0·60
Volatile matter, max. % weight	0·80	0·80	0·80	0·80	0·80	0·80	0·80	0·80	0·80
Wallace rapid plasticity—minimum initial value, P_0	—	—	30	30	30	—	30	30	30
Plasticity retention index, PRI, min. %	60	60	60	60	60	50	50	40	30
Colour limit, Lovibond Scale, max.	—	—	6·0	—	—	—	—	—	—
Mooney viscosity, ML 1+4, 100°C	—[c]	—[d]	—	—	—	—[e]	—	—	—
Cure[f]	R	R	R	R	—	R	—	—	—
Colour coding marker[g]	Black	Black	Light green	Light green	Light green	Blue	Brown	Red	Yellow
Plastic wrap colour	Trans-parent	Trans-parent	Trans-parent	Trans-parent	Trans-parent	Trans-parent	Trans-parent	Trans-parent	Trans-parent
Plastic strip colour	Orange	Magenta	Trans-parent	Opaque white	Opaque white	Opaque white	Opaque white	Opaque white	Opaque white

[a] Testing for compliance shall follow ISO test methods.
[b] Contains 4 phr light, non-staining mineral oil. Additional producer control parameter: acetone extract 6–8% by weight.
[c] Three sub-grades, viz. SMR-CV50, SMR-CV60 and SMR-CV70 with producer viscosity limits at 45–55, 55–65 and 65–75 units respectively.
[d] One grade designated SMR-LV50 with producer viscosity limits at 45–55 units.
[e] Producer viscosity limits are imposed at 58–72 units.
[f] Cure information is provided in the form of a rheograph (R).
[g] The colour of printing on the bale identification strip.

volatile matter, Wallace rapid plasticity, plasticity retention index (a measure of ageing resistance) and with some grades colour and Mooney plasticity. For some grades cure information for the batch must also be provided.

The new processes may be classified into two basic types:

1. Heveacrumb processes based on patents issued to the Rubber Research Institute of Malaya (RRIM) which use castor oil to promote crumbling of coagulum during passage through crepeing rolls;
2. comminution processes in which the wet coagulum is broken up mechanically by such devices as dicing machines, rotary cutters and hammer mills.

Both crumb or comminuted rubber may then be dried by passing through drying tunnels, typically for 3 h at 105°C, and the dried product then cooled and compressed into convenient-sized flat bales typically about 33 kg in weight and of dimensions 70 × 35 × 17 cm. The bales are then usually wrapped in polyethylene sheet.

In the Heveacrumb process up to 0·7 pphr of castor oil is added to rubber latex or on to wet crumb, promoting crumbling of the coagulum as it is passed through the crepeing rolls. Whilst the residual castor oil appears to have no significant adverse effect on the properties of the rubber, the amount originally added may be reduced by the inclusion of zinc stearate.

Comminution processes vary but are typified by the Dynat process developed by the Guthrie Corporation. When used with prime quality field latex, the first stage of the process consists of blending and bulking of the latex which is then coagulated with dilute formic acid, a process typically taking about 16 h. The resulting coagulum is passed through a turbo mill for granulating and dewatering, and the granules then fed to an extruder to further homogenize the product. The rubber emerges from the extruder in spaghetti-like form and is collected in boxes which are then passed through the drying tunnel. Modified processes exist to handle material which has coagulated in the field, such as cut lump from the collecting cups.

By the mid-1980s about twice as much SMR was prepared by Heveacrumb processes as by comminution processes.

Whilst many producers market the new-type materials under trade names (e.g. Dynat, Harub) their gradings usually follow the SMR and SIR (Standard Indonesian Rubber) classification schemes.

The highest-quality grades are the Latex Grades (also known as Latex Rubber Grades and Whole Field Latex Grades) which are obtained entirely from latex derived from Hevea rubber, diluted but not fractionated, using an approved coagulant system and under the control of the operators of the scheme (RRIM in the case of SMR). Excluding the Constant Viscosity grades (dealt with below) these grades only account for about 10% of SMR rubbers. Their light colour and low impurity level lead to applications in pharmaceutical and surgical goods in particular and also for gum stock compounds (e.g. stationers' bands) and light-coloured compounds.

Field grade materials are those which consist largely of rubber that has been produced by spontaneous coagulation in the collecting cup or elsewhere without deliberate addition of coagulant. These grades account for about 75% of the total market and are widely used in general rubber goods as well as for tyre and automotive applications.

4.2.2 Constant-Viscosity and Peptized Rubbers

When a *Hevea* tree is tapped for the first time, or after a prolonged rest period, it is found that the hydrocarbon in the latex may contain up to 70%

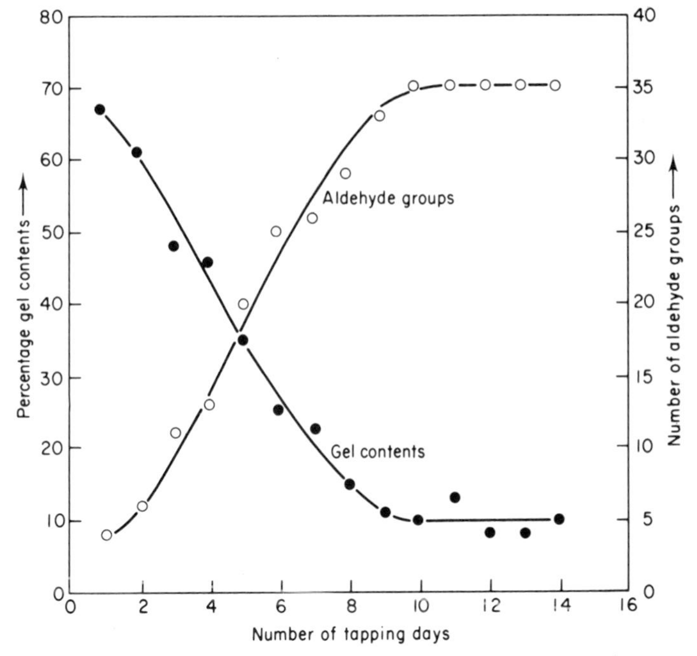

FIG. 4.1. Effect of continuous tapping of a set of newly opened trees on the number of aldehyde groups and gel contents in rubber. (From Sekhar (1962), reproduced by permission of The Plastics and Rubber Institute.)

of *microgel* cross-linked material within individual latex particles. If, however, the tree is tapped for several consecutive days (Fig. 4.1) the microgel content drops, reaching about 10% after two weeks. At the same time it is found that the number of carbonyl groups per molecule increase from 5 to 35. (Sekhar's original paper refers specifically to aldehyde groups but later work by Macey and Gregg suggested that lactones were the active groups. The less specific term 'carbonyl group' is used here although the term 'aldehyde' has been retained in Fig. 4.1 taken from Sekhar's paper.)

Sekhar (1962) suggested that, on storage within the tree cells, carbonyl groups attached to the rubber molecules within the latex reacted with another unidentified group which he called an 'aldehyde-condensing agent' to produce a cross-link. If, however, the tree is regularly tapped there is less time for cross-linking and the tapped latex exhibits lower gel content but a higher carbonyl content. If the latex is then stored under anaerobic conditions outside the tree, the carbonyl group concentration decreases whilst the percentage gel content increases.

This implied that cross-linking between the carbonyl group and the aldehyde-condensing group not only causes microgel but also the hardening of bulk dry rubber that occurs on storage (leading to macrogel). In turn this suggested the possibility that such unwanted hardening reactions may be prevented by either:

1. Reacting the carbonyl groups in the rubber with an added aldehyde-condensing agent that reacts preferentially over groups of similar function in the rubber; or
2. reacting the aldehyde-condensing groups in the rubber with a carbonyl-containing compound of greater reactivity than the trace carbonyl groups in the natural rubber molecule.

This has led to the development of the so-called *constant-viscosity rubbers* which are prepared by treating freshly tapped latex with about 0·15% by weight of either hydroxylamine hydrochloride or, now more commonly, hydroxylamine neutral sulphate as the aldehyde-condensing agent. The term 'CV rubber' is often used for such products.

Because the viscosity of the rubber varies from clone to clone (a clone being a group of plants resulting from the vegetative propagation of one individual), with tapping history and with the seasons (because of the influence of these factors on molecular weight and molecular weight distribution) and with microgel, judicious blending of latex is necessary to produce material in bulk within reproducible limits of viscosity.

At the time of writing the SMR scheme recognizes three sub-grades of

CV rubber, CV50, CV60 and CV70, which have Mooney ML 1+4 (100°C) viscosities of 50 ± 5, 60 ± 5 and 70 ± 5 respectively and of which the CV60 grade dominates the market. Because of these low viscosities and the fact that hardening on storage is largely repressed with these materials, the traditional natural rubber mastication process can be largely eliminated with considerable savings in time and energy. Most CV rubbers are made using a Heveacrumb process.

Low Mooney viscosity rubbers may also be obtained if a small amount of a non-staining naphthenic mineral oil is added to a viscosity-stabilized rubber. The LV50 grade, for which there is a small market, consists of a blend of a moderate-Mooney CV rubber with 4% of oil to give a rubber with a Mooney ML 1+4 (100°C) value in the range 50 ± 5.

A viscosity-stabilized GP-type rubber is also marketed which is a blend of latex grade rubber or sheet material with field coagula material and is intended for general-purpose use.

An alternative to the CV process is to treat the rubber with a so-called *peptizing agent*, usually 0·010–0·016 pphr of pentachlorothiophenol, either added to the latex or to the crumb. The peptizing agent facilitates breakdown, reducing the mastication times necessary. In some grades the peptizer is used in CV rubber which has also been treated with hydroxylamine sulphate.

4.2.3 Speciality Natural Rubbers

A number of speciality grades of natural rubber are available, introduced to provide modified performance in processing (SP rubber), or in service (deproteinized rubber) or in order to use by-product material (skim rubber).

4.2.3.1 Purified and Deproteinized Rubbers

Compared with synthetic *cis*-1,4-polyisoprenes, natural rubber shows a number of differences which cannot be explained in terms of the greater structural regularity of the natural material. These include a faster vulcanization rate, greater tendency to premature vulcanization (scorch) during processing operations and a lower heat build-up during flexing. This has been ascribed to the non-rubber constituents such as denatured protein together with some carbonyl-containing low-molecular-weight poly-isoprene. Removal of such materials from the natural product or addition to the synthetic material bring the properties of the two polymers much closer together. Gregg & Macey (1973) have suggested that there is some bonding between polyisoprene and protein, and that the protein molecule can act like a cross-link for the natural rubber.

Removal of proteinous material also has the important effect of reducing the water absorption which in turn improves the electrical resistivity and also reduces creep. If protein removal is accompanied by the removal of other unwanted non-rubbery material, a generally purer rubber will be obtained that is more acceptable for critical medical applications (although for some of these a post-vulcanization extraction operation will be required).

Partially deproteinized rubber may be obtained simply by centrifuging prior to coagulation, this process removing up to 50% of the protein and mineral matter. Rather more may be removed by the treatment of the latex with soaps or, in particular, by the treatment of fresh field latex with a proteolytic enzyme (Smith, 1974). This process also helps to remove other naturally occurring hydrophilic substances such as traces of inorganic salts.

4.2.3.2 Skim Rubber

In contrast, skim rubber has a high protein content and is a by-product of the centrifuging process used to produce concentrated latex. Surfactant-assisted biological coagulation processes are usually used to give easy-processing, fast-curing rubbers used in cellular products and industrial tapes, and as a processing aid for other grades of rubber.

4.2.3.3 'Superior Processing Rubber' (SP Rubber)

Attempts to produce extrudates and calendered sheet from unfilled or lightly filled rubber compounds are fraught with difficulty. The extrudates have a very rough (rugose) finish and there is considerable die swell. Hollow sections tend to collapse on cure. Related problems occur with calendered sheet.

Greatly improved results may be obtained by blending ordinary latex with vulcanized latex. The latter is a form of latex where rubber molecules *within* the particles are cross-linked before coagulation. Typically blends are in an 80:20 ratio (although 60:40 blends are also available for subsequent blending with unmodified material). Oil-modified grades have also been available.

It is interesting to note that similar effects may also be obtained with synthetic rubbers and special cross-linked grades of SBR, NBR and other rubbers have been made available for similar purposes but, as with the natural product, usage is not large.

4.2.3.4 Oil-Extended Natural Rubber (OENR)

In response to the success obtained commercially with OESBR (see

Chapter 5), oil-extended natural rubbers are supplied by some major rubber producers. Use has not been great since there is little advantage in using such material compared with incorporating oil at the rubber compounding stage.

4.2.3.5 Powdered Rubber

In principle there are many advantages in using rubber in powdered or granule form but this has proved difficult with natural rubber because of the tendency of clean material to stick to itself (natural tack). Powdered rubber is now available in which the particles are partitioned with a fine-particle-size silica. Whilst extensive use of powdered natural rubber has been predicted with extrusion and injection moulding processes, its greatest use to date has been as an additive to bituminous road compounds and for preparing adhesive formulations because of the ease of handling in the first case and faster dissolution rates in the second.

4.3 STRUCTURE

Natural rubber (Hevea) latex varies in its composition but the following may be considered as typical.

	%
Total solids content	36 (including a dry rubber content of 33%)
Proteinous substances	1–1·5
Resinous substances	1–1·5
Ash	<1
Sugars	1
Water	*c.* 60

In addition to their natural biological function, the non-rubber constituents also influence the methods of coagulation, the techniques of rubber latex technology and, by no means least, the vulcanization and ageing characteristics of rubber compounds.

The rubber molecules consist virtually entirely of *cis*-1,4-polyisoprene:

$$\left[\!-CH_2-C(CH_3)=CH-CH_2-\right]_n$$

There is no evidence for any *trans* material or for any 1,2- or 3,4-isoprene polymer in the natural product, in contrast to the synthetic polyisoprenes (see Chapter 6).

However, as indicated in this chapter, there is some evidence for a few carbonyl groups attached to the chain and which may be a site for cross-linking and lead to the formation of gel. There is some dispute as to the nature of these groups, both aldehyde and lactone groups having been suggested. The presence of the occasional epoxy group in the chain has also been postulated.

The molecular weight of the natural polymer is very high but varies between latices from different clones. In one study (Nair, 1970) the number-average molecular weight ($\bar{M}_n$) was found to range from $0{\cdot}25 \times 10^6$ to $2{\cdot}71 \times 10^6$ and the weight-average molecular weight ($\bar{M}_w$) from $3{\cdot}4 \times 10^6$ to $10{\cdot}17 \times 10^6$. Using the data from this paper the average values for $\bar{M}_w/\bar{M}_n$ for 12 different clones is 6·72 with values for individual clones ranging from 3·63 to 10·94.

4.4 PROPERTIES

As seen in the previous section, the natural rubber molecule is an unsaturated aliphatic hydrocarbon polymer of very regular structure.

Whilst the molecular weight of the raw polymer is so high that entanglements make solution difficult, masticated rubber (see Section 4.4.1) will dissolve in aliphatic hydrocarbons and other liquids of similar solubility parameter. It will also react with chemicals known to be reactive to carbon–carbon double bonds, will be a good electrical insulator when pure and will burn in air as may be predicted from the structure.

The very flexible chain backbone leads to a very low T_g of about −73°C. Because of the stiffening effect of the methyl group on the chain backbone this value is somewhat higher than that for *cis*-polybutadiene. On the other hand, the presence of the double bond so reduces the energy of rotation about the adjacent carbon–carbon single bonds that the T_g is somewhat lower than for a 50:50 ethylene–propylene copolymer (*c.* −56°C), although this does not explain the higher figure for the *trans* polymer:

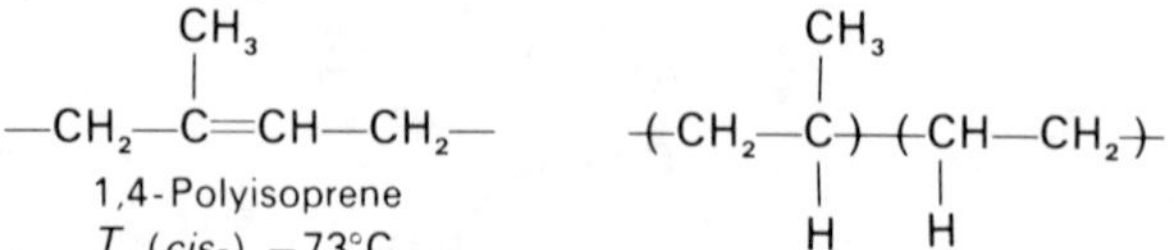

$$-CH_2-\overset{\displaystyle CH_3}{\overset{|}{C}}=CH-CH_2-$$

1,4-Polyisoprene
T_g (*cis*-) −73°C
(*trans*-) −53°C

$$\left(CH_2-\underset{\displaystyle H}{\underset{|}{\overset{\displaystyle CH_3}{\overset{|}{C}}}}\right)\left(\underset{\displaystyle H}{\underset{|}{CH}}-CH_2\right)$$

Ethylene–propylene copolymer
T_g *c.* −56°C

Because of the regular structure, the natural rubber molecule is capable of crystallization and the unstretched raw polymer has a crystalline melting point (T_m) of about +25°C. Crystallization may also be induced by stretching samples such as in a tensile test. As a result of this stress-induced crystallization, the tiny crystal structures formed act rather like reinforcing particles and, unlike SBR which does not crystallize, enable gum and lightly loaded stocks to exhibit high vulcanizate strengths.

4.4.1 Mastication Behaviour

That the properties of natural rubber changed if the material was subjected to intensive working was first demonstrated by Hancock about 1830. Coagulated and dried raw polymer is quite elastic at common ambient temperatures. Attempts to shape the material are frustrated by elastic recovery after a deforming stress. For similar reasons it is difficult, if not impossible, to blend in additives unless the rubber is first dissolved in a suitable solvent. If, however, natural rubber is subjected to intensive working, the process known as *mastication*, then the polymer becomes progressively more plastic rather than elastic, is capable of flow and may therefore be shaped.

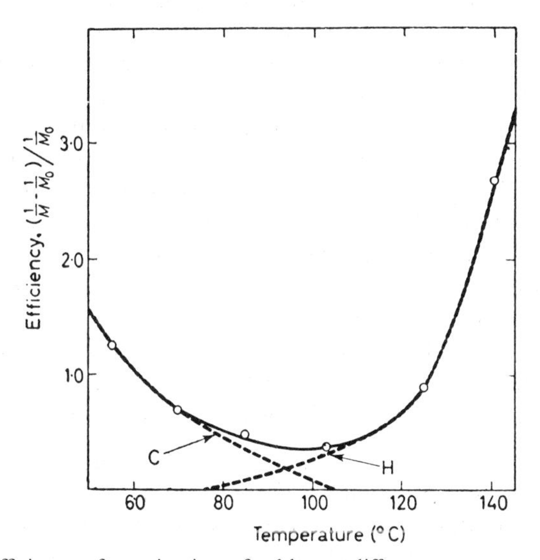

FIG. 4.2. Efficiency of mastication of rubber at different temperatures. Molecular weights (M) measured after 30-min mastication of 200 g natural rubber in a size B laboratory Banbury mixer. (From Pike & Watson (1952), reproduced by permission of John Wiley & Sons, Inc.)

For many years it was recognized that this effect was associated with a reduction in molecular weight. In 1931 Cotton, and independently Busse (1932), found that oxygen was necessary for the 'plasticization' to occur. It was also observed that the rate of breakdown varied with temperature in a complex way, passing through a minimum at about 100°C (Fig. 4.2). At about the same time Staudinger & Bondy (1931, 1932) suggested that mastication involved the rupture of molecules by shearing forces imposed during bulk deformation of the rubber. This was a particularly daring suggestion at a time when Staudinger was still having difficulty persuading chemists of the existence of high-molecular-weight polymers. In 1940 Kauzman & Eyring specifically hypothesized that cold mastication involved the direct breaking of C—C bonds in the polymer backbone in order to generate free radicals.

Experimental verification was eventually provided by Pike & Watson (1952) for what became known as a mechanochemical mechanism. Such evidence may be summarized as follows.

1. The negative temperature coefficient of mastication up to 100°C is consistent with a decrease in viscosity reducing the shearing stresses, at a given shear rate, available to rupture the chains.
2. It is difficult to reduce the molecular weight below about 70 000 for the same reason as above, the lower viscosity allowing slippage and preventing the build-up of high internal stresses.
3. In the absence of oxygen the radicals may recombine or be involved in chain transfer reactions without much change in molecular weight. If, however, the radicals could combine with oxygen to form $RO^{\cdot}$ and $RO^{\cdot\cdot}$ radicals which could not then recombine, there would be a reduction in molecular weight.
4. The confirming demonstration that radicals were obtained from broken polymer molecules was achieved by the use of highly sensitive radical acceptors such as 1,1′-diphenyl-2-picrylhydrazyl (DPPH) (Ayrey *et al.*, 1956) which were found attached to ruptured polyisoprene chains containing several thousand isoprene units.

It was also subsequently demonstrated that the radicals generated could initiate polymerization of monomers such as styrene and methyl methacrylate, and thus allow polymeric chains of these monomers to be grafted on to the natural rubber molecule backbone. The ability to form block copolymers by combination of radicals produced from dissimilar polymers was also demonstrated.

The positive temperature coefficient of mastication rate above 100°C is often explained as being due to a conventional oxidation process. However, it has been demonstrated (Kuz'minskii, 1971) that this oxidation is influenced by mechanochemical reactions. Natural rubber masticated at 100°C oxidized more rapidly (as manifested by the formation of carbonyl groups) than rubber oxidized without working at 150°C. A number of chemicals are known which can induce the rate of breakdown at these high temperatures, some being effective at temperatures as low as 65°C. They are known as peptizing agents, the term presumably originating from the superficial resemblance of this process to the conversion of insoluble protein into soluble peptones, the process known as peptization.

In commercial practice mastication is usually carried out in internal mixers or, for laboratory-scale work, on two-roll mills. The process has changed little in principle for well over a century. The development of the constant-viscosity rubbers (see Section 4.2.2) has provided an important way of substantially reducing and in some cases eliminating the energy-expensive mastication process.

4.4.2 General Chemical Reactivity

Natural rubber may be considered to be an alkene with its chemical reactivity dominated by the presence of the double bonds. The possibility of alkane-type reactions away from the double bonds must, however, not be excluded.

Double bonds consist of a strong σ-bond and a weak π-bond. The π-bond electrons, which lie in clouds above and below the plane of the atoms, are less tightly held than the σ-bond electrons and may be considered as a source of electrons. In effect the double bond acts like a base and tends to react with substances deficient in electrons (electrophilic agents). Free radicals are also attracted to electrons.

The presence of the methyl group in 1,4-polyisoprenes enhances the activity of the double bond, compared with 1,4-polybutadienes, and for this reason natural rubber is generally rather more reactive than SBR or polybutadiene to a wide variety of chemicals, including vulcanization systems. It should also be noted that important reactions also occur at the carbon atoms adjacent to the double bonds, the hydrogen atoms at this point being generally easier to abstract than at the double bond.

One respect in which polymer reactivity is distinct from that of small molecules is that a considerable influence on properties can be brought about by a quantitatively small number of reactions. For example, if there are just two reactions on a chain with (say) a molecular weight of 100 000 which lead to that chain reacting with two neighbouring chains, then this reaction repeated throughout a mass of rubber would lead to the production of a cross-linked network. Similarly a single reaction in a chain that led to chain rupture (chain scission) would lead to a halving of the average molecular weight.

The all-important vulcanization and ageing reactions are dealt with separately. Amongst other reactions of natural rubber the following should be noted, all of which have been discussed at length by the author elsewhere (Brydson, 1978):

1. Hydrogenation. This process is not of commercial significance, the product having a structure effectively that of an alternating ethylene–propylene copolymer.
2. Hydrohalogenation. In the case of HCl the hydrohalide adds onto the double bond according to Markownikov's rule (that the hydrogen adds onto the carbon with the most hydrogen atoms):

$$\sim CH_2{-}\overset{\displaystyle CH_3}{\overset{|}{C}}{=}CH{-}CH_2\sim \xrightarrow{HCl} \sim CH_2{-}\underset{\displaystyle Cl}{\underset{|}{\overset{\displaystyle CH_3}{\overset{|}{C}}}}{-}CH_2{-}CH_2\sim$$

   The product known as rubber hydrochloride is not rubbery but is used as a packaging film (Pliofilm) and as a rubber-to-metal adhesive.
3. Halogenation. Chlorination of natural rubber is a highly complex process involving substitution, addition and cyclization reactions. The product is of considerable importance for chemical- and heat-resistant coatings. Bromination and iodination reactions appear less complex.
4. *cis/trans* Isomerization. If butadiene sulphone is added to natural rubber on a two-roll mill and the resulting mix heated, sulphur dioxide is formed *in situ*. Reaction with the natural rubber leads to isomerization of some of the double bonds to the *trans* form. Thus the polymer has a less regular structure and is less liable to crystallization, particularly in very cold weather.
5. Cyclization. Treatment of natural rubber with certain acids such as concentrated sulphuric acid, or by Lewis acids such as stannic chloride, leads to an isomerization reaction which results in the formation of a series of ring structures along the polymer chain. Cross-linking does not occur and the polymer remains soluble but loses its rubbery nature and becomes stiff. Cyclized rubber is

occasionally used as a means of stiffening a rubber compound without increasing the density.

6. Ene-reactions. These are molecular reactions of the general type

H Y X Δ → H Y X

and which do not require active species such as ions or free radicals as intermediates. In the case of ethyl *N*-phenylcarbamoyl-azoformate (ENPCAF), reaction can occur simply by addition to the rubber on a two-roll mill:

C_6H_5—NH·CO·N=N·CO·O·C_2H_5 + NR → C_2H_5OOC—N(H)—N(CONH—C_6H_5)—(rubber chain)

Whilst this material is of interest in that it reduces storage hardening, ENPCAF-like structures containing trialkoxysilyl groups are especially interesting. If these molecules are attached to a rubber molecule via an ene-reaction it is possible to couple the rubber to a silicaceous filler via the silyl groups. The method has not, however, proved commercially successful. The ene-reaction has also been used to produce network-bound antioxidants and for introducing photosensitive groups, enabling the rubber to be cross-linked by photodimerization.

7. Epoxidation. The natural rubber double bond reacts rapidly with peracids, the double bonds being converted to oxirane rings. Because of extensive recent interest in epoxized natural rubber this material is treated separately in Section 4.9.

4.4.3 Solubility and Swelling

A hydrocarbon with a solubility parameter of about 16·5 $MPa^{1/2}$, masticated natural rubber dissolves in a wide variety of solvents of similar solubility parameter. Dissolution of raw material is more difficult because of the entanglement of the high-molecular-weight molecules, the presence of some microcrystalline structures which at room temperature act rather like cross-links, and also through the possibility of a slight amount of cross-linking via carbonyl groups attached to the chain as discussed in Section 4.2.2.

Cross-linked material, as with other covalently cross-linked rubbers, only swells in solvents of similar solubility parameter, the swelling being at a maximum when the solubility parameters of rubber and solvent coincide.

4.4.4 Burning Behaviour

Natural rubber, as a hydrocarbon, burns quite readily with a sooty flame. Fire retardants can reduce this propensity, but where fire is a serious potential problem alternative rubbers are usually preferred.

4.5 VULCANIZATION

The process of lightly cross-linking, or vulcanizing, natural rubber was first discovered some 150 years ago by Goodyear (see Chapter 1). The vulcanizing agent used by Goodyear was sulphur and today it remains the basis of nearly all commercial vulcanizing practice, alternative systems having very limited use. Today, however, the sulphur is used in conjunction with a number of other agents and the mechanism is very complex. It is only possible to give an outline of what is now known as *accelerated sulphur vulcanization* and the interested reader should refer to more detailed treatments elsewhere (Bateman, 1963; Porter, 1968; Brydson, 1978).

4.5.1 Accelerated Sulphur Vulcanization

Typical accelerated sulphur vulcanization of natural rubber involves four components: sulphur, accelerator, activator and fatty acid, which in a conventional system are used in the quantities indicated in column A of Table 4.3.

The vulcanization system is chosen so that premature vulcanization (scorching) does not occur during shaping operations but as rapidly as convenient for the process once shaping is complete. As an aid to prevent scorching prevulcanization inhibitors may also be used. Accelerators vary enormously in their effect but a typical system used in dry rubber technology (as opposed to latex technology where scorching problems do not exist) will require about 15–20 min cure at 140°C with an approximate doubling of the ratê with every 10°C rise in temperature. Thick section parts require longer times because of the low thermal conductivity of the rubber which prolongs the time before the innermost sections reach

TABLE 4.3
Typical conventional and 'efficient' vulcanization systems for use in natural rubber

	A *Conventional system* (*pphr*)	*B* *'Efficient' vulcanization system* (*pphr*)
Natural rubber	100	100
Sulphur	2·0–3·5	0·4–0·8
Accelerator	1·2–0·4	5·0–2·0
Activator (ZnO)	3–5	3–5
Fatty acid (e.g. stearic acid)	1–1·5	1–1·5

vulcanization temperatures. With such products it is important that the formulation is not sensitive to *reversion* since the outer layers will be subject to longer periods at the vulcanization temperatures.

Extensive studies, particularly by chemists at the Malaysian Rubber Producers' Research Association, have demonstrated that in the case of natural rubber there is a wide variety of structures in a typical vulcanizate, as schematically illustrated in Fig. 4.3.

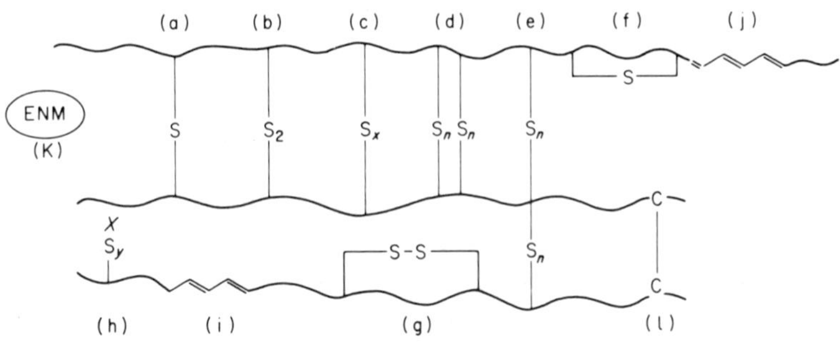

FIG. 4.3. Typical chemical groupings present in a sulphur-vulcanized natural rubber network. (a) Monosulphide cross-link; (b) disulphide cross-link; (c) polysulphide cross-link ($x = 3$–6); (d) parallel vicinal cross-links ($n = 1$–6) attached to adjacent main chain atoms and which have the same influence as a single cross-link; (e) cross-links attached to common or adjacent carbon atoms; (f) intra-chain cyclic monosulphide; (g) intra-chain cyclic disulphide; (h) pendant sulphidic group terminated by moiety X derived from accelerator; (i) conjugated diene; (j) conjugated triene; (k) extra-network material; (l) carbon–carbon cross-links (probably absent). (From Brydson, 1978.)

Some of these groups, such as the cyclic monosulphide and disulphide (f and g), do not contribute usefully to the network. The ratio of monosulphidic to disulphidic and polysulphidic groups has been shown to have an important effect on the properties of the vulcanizate, the proportion of monosulphidic being higher with longer vulcanization times and with efficient vulcanization systems being used (see column B of Table 4.3).

In very general terms the efficient vulcanization (EV) systems confer improved heat ageing and reversion resistance with low compression set at 70°C. On the other hand, they tend to have lower tensile strength, fatigue resistance and resilience but higher physical creep rates than conventional vulcanizates. Compromise systems are available and discussed more fully in Chapter 18.

Work at MRPRA and its predecessor organizations has also demonstrated that the sulphur is involved in a great diversity of reactions with the natural polymer. It has been clearly demonstrated that, contrary to a one-time widely held belief, the reaction with sulphur is not simply a reaction across double bonds but one more likely to occur at an adjacent methylene group. Thus one typical reaction would be

$$
\begin{array}{c}
\sim CH_2-\overset{\displaystyle CH_3}{\overset{|}{C}}=CH-CH_2\sim \\[2em]
\sim CH_2-\underset{\displaystyle CH_3}{\underset{|}{C}}=CH-CH_2\sim
\end{array}
\xrightarrow[\text{accelerator}]{S+}
\begin{array}{c}
\sim CH_2-\overset{\displaystyle CH_3}{\overset{|}{C}}=CH-CH\sim \\
\qquad\qquad\quad | \\
\qquad\qquad\quad S_x \\
\qquad\qquad\quad | \\
\sim CH_2-\underset{\displaystyle CH_3}{\underset{|}{C}}=CH-CH\sim
\end{array}
$$

The structure on the right is particularly typical of those formed at an early stage of cure. Continued heating has been found not only to reduce the value of x down to 2 and 1 but also leads to rearrangement reactions which, amongst other changes, may effectively shift the position of the double bond.

4.5.2 Non-sulphur Vulcanization

Variations in the detailed formulation have enabled the rubber technologist to have a considerable level of control over the vulcanization process, particularly with respect to scorching tendencies, cure rate and reversion but also to control of mechanical properties. This is an advantage not enjoyed by the many non-sulphur systems that have been developed.

Whilst they are discussed in more detail in Chapter 18 brief mention will be made here of systems that have been, or are being, used with natural rubber.

1. Sulphur chloride. This material was found by Alexander Parkes in 1846 to be capable of curing natural rubber at room temperature and this 'cold cure' process continues to be used for curing proofed and solution-dipped goods.
2. Selenium and tellurium were used at one time to give enhanced heat resistance but are seldom, if ever, used today because of their odour and toxicity.
3. Peroxides were first used in rubber by Ostrosmislenskii in 1915 but it is only very recently that the tendency to premature curing in natural rubber has begun to be controlled successfully. They have never been popular with the natural material although quite widely used with some synthetics.
4. Radiation curing is potentially attractive because of the absence of toxic chemicals but is of only limited effectiveness when anti-oxidants are present and only give weak vulcanizates.
5. The so-called urethane cross-linking agents were developed by MRPRA because of their conferring good resistance to reversion; this is attractive when moulding thick sections but they have not become widely used.
6. Phenolic resins, like the peroxides, have found greater use with the synthetics in order to improve heat resistance. Their value in natural rubber is limited by the tendency of the resin to impede crystallinity and thus adversely affect many mechanical properties, such as tensile strength.

4.6 NATURAL RUBBER EBONITE

Very soon after the discovery of the sulphur vulcanization process for natural rubber it was observed that if rubber was heated with substantial quantities of sulphur a hard black ebony-like material was obtained: this is known variously as ebonite, hard rubber and, today, rarely as vulcanite. The first patents were taken out by Nelson Goodyear in 1851 and the material may truly be considered to be the first man-modified product to be used as a plastics material, predating Parkesine, the forerunner of Celluloid and Xylonite, by some 10 years. Widely used for 100 years, it was particularly important for battery boxes, for smokers' pipe stems and for rubber-to-metal bonding. Today it has been largely replaced by thermoplastics materials.

In common practice the rubber:sulphur ratios in ebonite compositions were about 68:32; this corresponds to the empirical formula $(C_5H_8S)_x$. For many years this gave rise to the erroneous belief that ebonite was produced by a simple reaction of a single sulphur atom across the double bond of the polymer. In due course it was established that up to 43% of the product could be comprised of sulphur and that polysulphidic cross-links and intramolecular cyclic sulphides were present. Most of the evidence is that the cross-link efficiency is quite low. Natural rubber ebonites undergo a softening at about 80°C; however, this is not a melting process as with a thermoplastics material, but more akin to a glass transition (T_g) effect. There is indeed some evidence showing how the T_g of natural rubber is related to the combined sulphur content.

Long cure times are required for curing ebonite although some accelerators can be partially effective. However, the curing is associated with a high exotherm and therefore care must be taken to prevent the reaction getting out of hand. The more extensive cross-linking also leads to a higher moulding shrinkage, which can be up to 6% unless non-reactive fillers such as ebonite dust are used. Normal reinforcing fillers such as carbon black commonly have an adverse effect on strength.

4.7 THE STATUS OF NATURAL RUBBER

In spite of considerable scientific and technical developments in polymerization chemistry, natural rubber retains a major role amongst rubbery materials and is currently, in tonnage terms, the world's leading rubber. This has been achieved in spite of the fact that NR is not oil-resistant, nor particularly heat-resistant, it burns, is attacked by oxygen and ozone and is affected by ultraviolet light.

Its virtue lies in that most important property of a rubber: high reversible elasticity coupled with high strength and toughness. It is comparatively easy to formulate compounds with a high resilience, which is a desirable feature of rubber springs and tyre sidewalls. It is also possible to prepare unfilled compounds with good tensile strength and filled compounds with good non-skid and abrasion resistance appropriate, for example, for tyre treads.

As a tyre rubber, it recovered earlier lost ground with the changeover to radial tyres. It has further enhanced its major role with truck and lorry tyres whose thicker sections are more susceptible to heat build-up than the smaller passenger-car tyres. This heat build-up occurs through dissipation of energy on deformation of a less than perfectly resilient material. The lower the percentage resilience of the material, the higher is the heat build-up. In general rubber goods applications, NR has lost some markets to the ethylene–propylene rubbers in recent years for uses where a high resilience is less important than good weathering behaviour. For lightly loaded applications requiring good strength, for example medical goods, it continues to be widely used.

In contrast to many major synthetic rubbers, an increase in annual production has generally been maintained in recent years. In part this is inevitable as it is not easy to effect a short-term control on production as is possible with synthetic rubber production. The natural rubber produced has to be sold eventually and this tends to lead to rather wider price variations than for the synthetics.

4.8 GUTTA PERCHA, BALATA AND RELATED MATERIALS

In addition to natural rubber, many other materials occur in nature that may be considered as polymers of isoprene or their derivatives. Many are of low molecular weight; they include the terpenes and carotenes. High-molecular-weight polymers also occur and some of these have found commercial application.

Gutta percha is based on a *trans*-1,4-polyisoprene structure. It is obtained from the trees of the genera *Palaquium* and *Payena*, of which *Pal. oblongifolium* has been the most important. This plant occurs in Malaysia and Indonesia, and has been cultivated mainly in Sumatra. The latex is more viscous than that of natural rubber and is most commonly extracted from the leaves rather than by making incisions in the bark. The coagulated latex gives a typical analysis of 70% hydrocarbon, 11% resin, 3% dirt and 16% moisture.

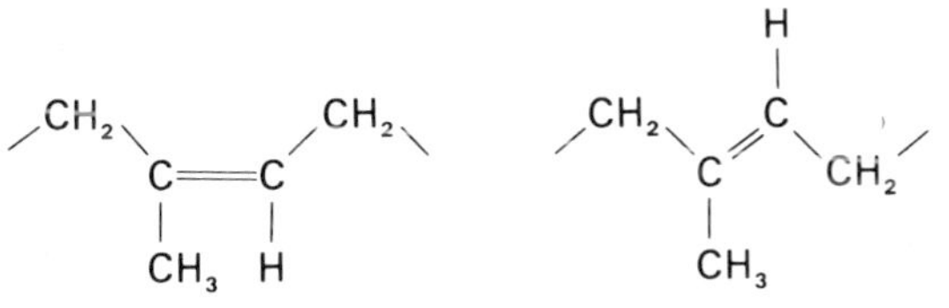

$$-CH_2-C(CH_3){=}C(H)-CH_2-$$

cis-1,4 Natural rubber

$$-CH_2-C(CH_3){=}C(H)-CH_2-$$

trans-1,4 Gutta percha

The *trans* polymer has a more compact structure than the *cis* and crystallizes more readily. X-ray evidence indicates that when first produced from the tree the crystals exist in a form (known as the α-form) which melts at 65°C. Whilst on slow cooling the α-form is produced, rapid cooling produces a β-form melting at 56°C and which slowly reverts to the α-form. At room temperature gutta percha is about 60% crystalline and has the characteristics of a rigid solid somewhat reminiscent of high-density polyethylene.

Like natural rubber the polymer may be hydrogenated, hydrochlorinated and vulcanized with sulphur. Ozone causes rapid degradation and the material is so badly affected by air and light that it is often stored under water. Deterioration when working under dry conditions is also rapid so that it is common practice to process with a minimum water content of 1%.

The solubility of gutta percha is similar to natural rubber; it is dissolved by carbon disulphide, chloroform and benzene. It is chemically resistant to alkaline solutions and dilute acids. One important use before the advent of polyethylene was for hydrofluoric acid storage bottles.

In Victorian times gutta percha was widely used for making highly decorative and elaborate mouldings such as ink stands. The earliest polymer extruders were developed about 1850 in order to process gutta percha and the material was widely used until after World War II for undersea cable insulation. It is rarely encountered today.

Balata is another natural material based on *trans*-polyisoprene; it is obtained from the *Mimosups balata* occurring in Venezuela, Barbados and Guyana. In this case the latex is thin and may be tapped in the same way as natural rubber latex. The coagulated material contains only about 50% of the *trans* polymer, the bulk of the residue consisting of resinous material. At one time it was widely used in belting applications; today its main use is for the covers of high-quality golf balls. In cheaper golf balls it has largely been replaced by the ionomer resins.

The latex of the *Sapota achras* (also variously named *Achras zapota* and *Saprota achros*), which occurs in Central America, contains a mixture of *cis*- and *trans*-polyisoprenes in the approximate ratio of 1:3. Known as *chicle*, the main use of the product is in the manufacture of chewing gum.

An alternative chewing-gum base is obtained from *jelutong*, a mixture of polyisoprene and resin obtained from the *Dyera costulata*. Although it is found in many countries, Borneo is the principal source. Before the development of plantation rubber it was an important rubber substitute but is not known to be used for this purpose any longer.

4.9 EPOXIDIZED NATURAL RUBBER (ENR)

Natural rubber was first epoxidized more than 50 years ago (Bloomfield & Farmer, 1934) but it has only been in the 1980s that interest has become of more than academic interest. As a result of work initiated by the Malaysian Rubber Producers' Research Association, commercial production of up to 1500 tonnes per annum (tpa) of ENR was announced by Guthrie in 1987, the product being marketed as Dynaprene.

The material is of interest in that although capable of crystallizing like natural rubber and thus giving non-black stocks a high strength, it also exhibits a good level of oil resistance and low air permeability similar to that of nitrile rubber (see Chapter 9).

Early attempts to produce epoxidized natural rubber led to products that not only contained the desired epoxy (oxirane) ring but also the products of secondary epoxide ring-opening reactions because of the pronounced tendency of the highly strained three-membered ring to open. Careful control of the reaction conditions (Baker *et al.*, 1985; Gelling, 1985) has led to these secondary reactions being avoided. Products are now available in which the epoxidation is random, the products have little gel and a viscosity of about 80 Mooney units. These rubbers retain their stereoregularity and may be represented by the formula

```
       CH3                    CH3
         \                      \
          C—CH                   C=CH
  ~~~~/   \ /   \              /      \~~~~
           O     CH2—CH2
```

Epoxidation of natural rubber results in a systematic increase in polarity. This will not only enhance hydrocarbon oil resistance but also increase the glass transition temperature (Fig. 4.4).

In turn the glass transition temperature will influence such properties as abrasion resistance, resilience and road grip.

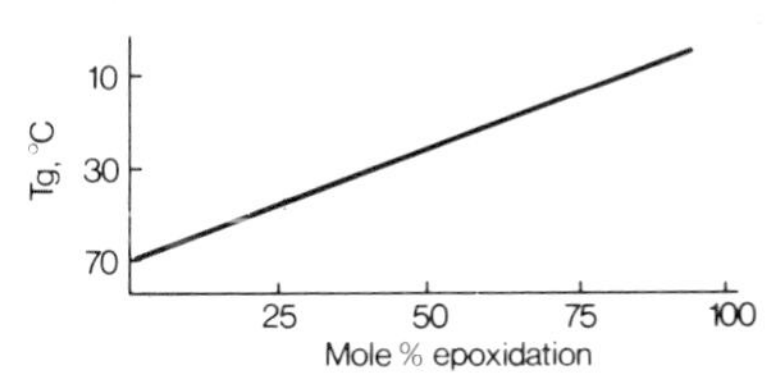

FIG. 4.4. Variation of T_g with level of epoxidation. (From Gelling (1987), reproduced by permission of The Malaysian Rubber Producers' Research Association.)

TABLE 4.4
Properties of gum vulcanizates of NR, ENR and NBR

	Natural rubber	*ENR-25*	*ENR-50*	*NBR (34% ACN)*
Modulus at 100% elongation, MPa	0·93	0·89	0·90	0·94
Modulus at 300% elongation, MPa	2·3	2·1	2·2	1·3
Tensile strength, MPa	29·0	25·0	31·0	15·1
Elongation at break, %	660	670	650	770
Tension fatigue, kc to failure				
0–100% extension	144	113	134	69
50–150% extension	1 350	910	813	24
Crystallinity (400% elongation), %	11	11	10	0

Initial work on ENR has concentrated on two grades, ENR-50 with 50 mol% epoxidation and ENR-25 with 25 mol% epoxidation. These materials have glass transitions of −23 and −48°C respectively.

Vulcanization may be carried out by any of the systems used to cross-link unsaturated rubbers but semi-EV accelerated sulphur systems are preferred (conventional systems exhibiting inferior ageing characteristics). Some typical physical properties of a gum vulcanizate are given in Table 4.4 and

TABLE 4.5
Properties of black-filled ENR vulcanizates[a]
(After Gelling, 1987)

	ENR-25	*ENR-50*
IRHD hardness	70	82
Modulus at 100% elongation, MPa	2·93	4·10
Modulus at 300% elongation, MPa	12·9	14·2
Tensile strength, MPa	25·1	22·0
Elongation at break, %	540	455
Dunlop resilience (23°C), %	38	15
DIN abrasion index	75	82
Oil resistance (70 h/100°C), %vol. swell		
ASTM No. 1 oil	16	−0·5
ASTM No. 2 oil	86	14·1
ASTM No. 3 oil	167	36·9

[a] ENR, 100; aromatic oil, 5; ZnO, 5; stearic acid, 2; quinoline-type antioxidant, 2; sulphur, 1·5; MBS, 1·5; *N*-(cyclohexylthio)phthalimide, 0·2; calcium stearate, 5 pphr.

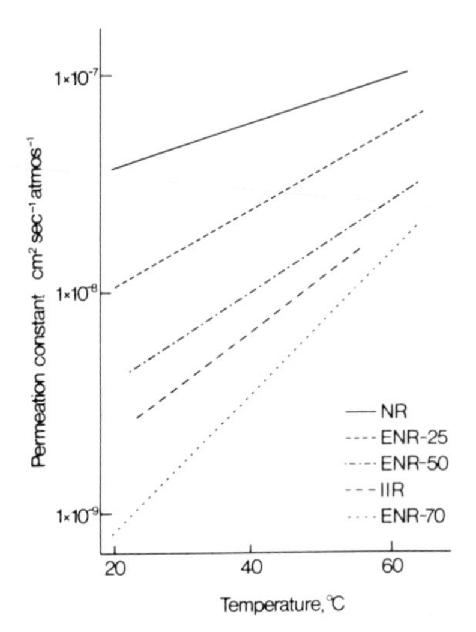

FIG. 4.5. Comparative air permeabilities of natural rubber, IIR and ENR. (From Gelling (1987), reproduced by permission of The Malaysian Rubber Producers' Research Association.)

for a black-filled (50 pphr N-220 (ISAF) black) vulcanizate are given in Table 4.5.

An increase in level of epoxidation also increases the air impermeability (see Fig. 4.5), with ENR-50 being close to butyl rubber (IIR) in its permeability behaviour.

The high hysteresis of ENR gives good tyre road grip together with low rolling resistance (a sought-after combination) but abrasion resistance is not as good as with conventional SBR-based tyre tread compounds.

At the time of writing ENR is a small-tonnage development product. Interest in the material at present arises largely from the ability to produce strong non-black rubbers that are oil-resistant, of low air permeability and with good damping properties. Also of interest are the compatibility with polar materials such as PVC as well as natural rubber and the high road grip which may be of interest in speciality tyre applications.

REFERENCES

Anon. (1969). *International Standards of Quality and Packing for Natural Rubber Grades (The Green Book)*. Rubber Manufacturers' Association Inc., New York.

Ayrey, G., Moore, C. G. & Watson, W. F. (1956). *J. Polym. Sci.*, **19**, 1.
Baker, C. S. L., Gelling, I. R. & Newell, R. (1985). *Rubber Chem. Technol.*, **58**, 67.
Bateman, L. (ed.) (1963). *The Chemistry and Physics of Rubber-Like Substances.* Applied Science, London.
Bloomfield, G. F. & Farmer, E. H. (1934). *J. Soc. Chem. Ind.*, **53**, 121T.
Brydson, J. A. (1978). *Rubber Chemistry.* Applied Science, London.
Busse, W. F. (1932). *Ind. Eng. Chem.*, **24**, 140.
Cotton, F. H. (1931). *Trans. Inst. Rubber Ind.*, **6**, 487.
Elliot, D. J. (1979). In *Developments in Rubber Technology—1*, ed. A. Whelan & K. S. Lee. Applied Science, London, Chapter 1.
Gelling, I. R. (1985). *Rubber Chem. Technol.*, **58**, 86.
Gelling, I. R. (1987). *NR Technology*, **18**, 21.
Gregg, E. C. & Macey, J. H. (1973). *Rubber Chem. Technol.*, **46**, 47.
Kauzman, W. & Eyring, H. (1940). *J. Am. Chem. Soc.*, **62**, 3113.
Kuz'minskii, A. S. (1971). In *The Ageing and Stabilisation of Polymers.* English translation: Elsevier, Amsterdam, Chapter 3.
Nair, S. (1970). *J. Rubber Res. Inst. Malaya*, **23**(1), 76.
Pike, M. & Watson, W. F. (1952). *J. Polym. Sci.*, **9**, 229.
Porter, M. (1968). The chemistry of sulphur vulcanization of natural rubber. In *The Chemistry of Sulphides*, ed. A. V. Tobolsky. Interscience, New York.
Sekhar, B. C. (1962). *Proc. 4th Rubber Technol Conf.* (Institution of the Rubber Industry), p. 460.
Smith, J. F. (1974). British Patent 1 366 934.
Staudinger, H. & Boundy, H. F. (1931). *Ann. Chem.*, **488**, 153.
Staudinger, H. & Boundy, H. F. (1932). *Rubber Chem. Technol.*, **5**, 278.

Chapter 5

Styrene–Butadiene Rubbers (SBR)

5.1 INTRODUCTION

The world's rubber markets are dominated by two rubbers which between them share about 70–75% of the market; one of these is natural rubber, the subject of the previous chapter, the other SBR, the subject of this one. Styrene–butadiene rubbers were first prepared in Germany in 1929 but at the time they had no particular virtues over natural rubber and production did not commence until shortly before World War II when the material was introduced as Buna S in Germany. Details of the production process were supplied to the Standard Oil Company under an information exchange agreement and when the Japanese entered the war and overran Malaya, French Indo-China and the Dutch East Indies a crash programme was set up in the United States and Canada to produce the rubber, designated in the United States as Government Rubber—Styrene (GR–S). By the end of the war production under this programme was of the order of 1 000 000 tpa with some 100 factories in operation. At that time, however, the material was still clearly inferior to natural rubber and when the latter became freely available the demand for GR–S fell sharply.

However, the desire of many countries not to be strategically dependent on natural rubber, together with steady technical improvements, such as the appearance of the so-called 'cold rubbers' in the product, led to a re-emergence of the material which in the 1950s was redesignated SBR, now a universally accepted abbreviation. By the 1960s the material had captured a large part of the motor tyre market, particularly in North America. During the 1960s SBR became, in tonnage terms, the most important rubber, although in the early 1980s natural rubber recovered probably to become once again the biggest tonnage material. (Whilst global statistics for production and consumption of natural rubber are available, accurate data on SBR are only available for the non-CPEC countries.) This fall back from pre-eminence is due to a number of factors including the decline in tyre output in North America, the trends to smaller cars and hence to smaller tyres, the move to radial tyres and the growth of polybutadiene rubber as a blending material.

The bulk of SBR produced is prepared by emulsion polymerization, the product sometimes being referred to as ESBR to distinguish it from the solution-polymerized rubbers which first appeared in the 1960s and are sometimes designated SSBR. Whilst for long regarded as the SBRs of the future, solution SBR development has been retarded by the generally difficult economics surrounding general-purpose synthetic rubber production with production capacity greatly outstripping demand in recent years, thus discouraging the replacement of production plant.

The last 20 years have also seen the emergence of the thermoplastic butadiene–styrene rubbers, butadiene–styrene block copolymers referred to as SBS rubbers which have found their own specialized market and which are dealt with in Chapter 16.

5.2 PREPARATION

The two monomers for SBR production are today invariably produced from petroleum sources although in the past other sources have been used.

Butadiene may be obtained by direct cracking of petroleum fractions, by dehydrogenation of butanes and butenes or oxidative dehydrogenation of butanes or butenes. Whatever process is used, purification processes are then necessary and these are usually the dominant element in the cost of butadiene production. Such purification is usually achieved by an extractive distillation procedure using such chemicals as cuprous ammonium acetate, furfural, acetonitrile, *N*-methylpyrrolidone and dimethylformamide.

Styrene is produced by a two-stage process via ethylbenzene. In a typical process ethylene and benzene are reacted at about 95°C in the presence of a Friedel–Crafts catalyst such as aluminium chloride. To improve the catalyst efficiency some ethyl chloride may be added to the reacting mixture, the former producing some hydrochloric acid at the reaction temperatures. Ethylbenzene may also be produced via catalytic reforming processes. The reforming process is one which converts aliphatic hydrocarbons into a mixture of aromatic hydrocarbons. This may subsequently be fractionated to give benzene, toluene and a xylene fraction from which the ethylbenzene may be obtained.

Styrene is obtained from ethylbenzene by a variety of methods, of which that of dehydrogenation is the most well known. This usually involves passing the ethylbenzene over heated oxide catalysts in the presence of steam (which reduces the partial pressure and hence encourages the endothermic reaction).

$$CH_2{=}CH_2 + C_6H_6 \longrightarrow C_6H_5{-}CH_2{-}CH_3$$

$$C_6H_5{-}CH_2{-}CH_3 \xrightarrow{[-H_2]} C_6H_5{-}CH{=}CH_2$$

The bulk of SBR is produced by free-radically initiated emulsion polymerization. Before 1950 such polymerization was usually carried out at about 50°C using a water-soluble initiator such as potassium persulphate, with the average molecular weight being largely controlled by chain transfer agents such as *t*-dodecyl mercaptan. Conversion of monomer to polymer was taken to about 72%, at which point hydroquinone was added as a polymerization stopper. Such products, now known for reasons which will become apparent in the next paragraph as 'hot rubbers', are today of very limited use.

The second generation of SBRs appeared about 1950. These were polymerized at about 5°C using a redox initiating system and because of the low polymerization temperature became known as 'cold rubbers'. Two typical polymerization recipes are given in Table 5.1.

Coagulation is normally brought about by the addition of common salt followed by dilute sulphuric acid although an alum solution may be used to give products of better electrical insulation characteristics.

Whilst emulsion polymerization of butadiene and styrene leads to essentially random polymers this is not necessarily the case with solution polymers which are prepared using alkyllithium catalysts. If polymerization is carried out in non-polar solvents the butadiene polymerizes first to the virtual exclusion of the styrene, which only polymerizes once the butadiene has been consumed, thus leading to a block copolymer. To produce more random polymers the butadiene may be added incrementally so that it is always in much lower concentration than the styrene. Another approach is to add only small amounts of monomer mixture to the reactor at a time to produce a copolymer consisting of many short blocks of each monomer. Random polymers may also be obtained by adding ethers or

TABLE 5.1
Typical polymerization recipes for 'hot' and 'cold' emulsion-polymerized styrene–butadiene rubbers

Component	*Hot SBR (SBR 1000)*	*Cold SBR (SBR 1500)*	*Comments*
Monomer			
Butadiene	75	72	
Styrene	25	28	
Water	180	180	
Emulsifier system			
Fatty acid soap	4·5	4·5	May include rosin acid to improve tack
KCl		0·3	As electrolyte helps to stabilize emulsion. The $K_2S_2O_8$ has this additional role in 'hot' SBR formulation
Sodium naphthalene sulphonate		0·3	Auxiliary surfactant
Initiator system			
$K_2S_2O_8$	0·3		
p-Menthane hydroperoxide		0·06	Redox oxidant
$FeSO_4 \cdot 7H_2O$		0·01	Reducing agent
Sodium formaldehyde sulphoxylate		0·05	Reducing agent
Sodium salt of EDTA		0·05	EDTA = ethylene diamine tetra-acetic acid
Regulator			
t-Dodecyl mercaptan	0·28	0·2	
Polymerization stopper			
Hydroquinone	0·05		
Sodium dimethyl dithiocarbamate		0·05	
Polymerization temperature, °C	50	5	
Conversion, %	72	60	

amines to the reaction mixture, although in this case a high percentage of the butadiene adds on in the 1,2-position (see Chapter 6). The use of potassium butoxide as a randomizing agent instead of ethers or amines is said to give copolymers with a lower 1,2-content.

Further variations of the solution polymerization process may be used to control the molecular weight distribution and whether or not the copolymer is linear, T-shaped, X-shaped or star-shaped (Cooper & Nash, 1972).

5.3 STRUCTURE AND VARIATIONS OF EMULSION SBRs

Whilst styrene and butadiene may be copolymerized in any ratio, most commercial SBR rubbers produced outside the CPEC bloc employ a target bound-styrene content of 23·5%. Because of molecular weight differences between the two monomers this is equivalent to about one styrene unit per six butadiene units or one pendant benzene ring per 26 main chain carbon atoms. Thus the following formal formula for SBR polymer tends to overemphasize the importance of the benzene ring:

$$\text{-}(CH_2\text{—}CH\text{=}CH\text{—}CH_2)\text{-}(CH_2\text{—}\underset{\displaystyle C_6H_5}{\underset{|}{CH}})\text{-}$$

The glass transition temperatures of styrene–butadiene copolymers vary almost linearly with the monomer ratios in the polymer. Since polystyrene has a T_g of about +90°C whilst that of polybutadiene is about −100°C it follows that the glass transition of SBR will be of the order of −50°C and the material will be rubbery down to temperatures approaching this value.

A typical cold-polymerized emulsion SBR will have the following structural characteristics:

Monomer arrangement	Random
cis-1,4-Polybutadiene content, %[a]	9
trans-1,4-Polybutadiene content, %[a]	76
1,2-Butadiene (vinyl) content, %[a]	15
Gel content	Negligible–moderate
Molecular weight	$\bar{M}_n \sim 100\,000$, $\bar{M}_w$ 320 000–400 000
Molecular weight distribution	Broad ($\bar{M}_w/\bar{M}_n = 3$–5)

[a] The meanings of these figures are explained in the next chapter.

The principal variations between grades of emulsion SBR polymers (excluding oil-extended and black masterbatch grades—see Section 5.5) are:

1. Styrene content;
2. molecular weight;
3. molecular weight distribution;
4. emulsifier system used;
5. coagulation system used.

A target bound-styrene level of 23·5% was established for standard butadiene–styrene rubbers many years ago when it was considered to provide the optimum balance of properties. Most current grades employ this level although some higher-styrene rubbers (target styrene *c.* 40%) are available which are used in tyre compounds where a high road grip of the tyre tread is of prime importance. Even higher styrene levels are used in the so-called high-styrene resins used as fillers for shoe soling and other stiff compounds (see Chapter 20).

The non-oil-extended grades usually have $\bar{M}_n$ values of about 80 000–110 000 and $\bar{M}_w$ values in the range 320 000–400 000. Lower values lead to polymers showing excessive bale distortion on storage, whilst higher values can lead to processing difficulties, particularly as SBR does not break down on mastication. There is little published information on the effect of the molecular weight of the base polymer on vulcanizate properties although one treadwear study (Dacker *et al.*, 1969), using polymers with a range of Mooney values, showed that in the range under consideration molecular weight had little effect.

Where oil-extension is to be carried out (see Section 5.3.1) higher-molecular-weight rubbers are used ($\bar{M}_n$ 100 000–145 000, $\bar{M}_w$ 400 000–520 000), the stock viscosity being reduced by the presence of oil. Without modification such rubbers would have a Mooney ML 1+4 (100°C) viscosity of 110–130 instead of the more usual values in the range 45–55.

The molecular weight distribution of emulsion SBRs is broad ($\bar{M}_w/\bar{M}_n$ *c.* 3–5). There is much evidence (Tokita & Pliskin, 1973; Mills *et al.*, 1975; Mills & Giurco, 1976) that processability is dependent on the breadth of the distribution. Figure 5.1 shows the influence of the ratio $\bar{M}_w/\bar{M}_n$ (a measure of polydispersity or breadth of distribution) on die swell. A broad distribution is also useful in preventing break-up on mixing whilst the quality of dispersion of carbon blacks is also critically dependent on the breadth of distribution.

The choice of emulsifier system during polymerization may also affect

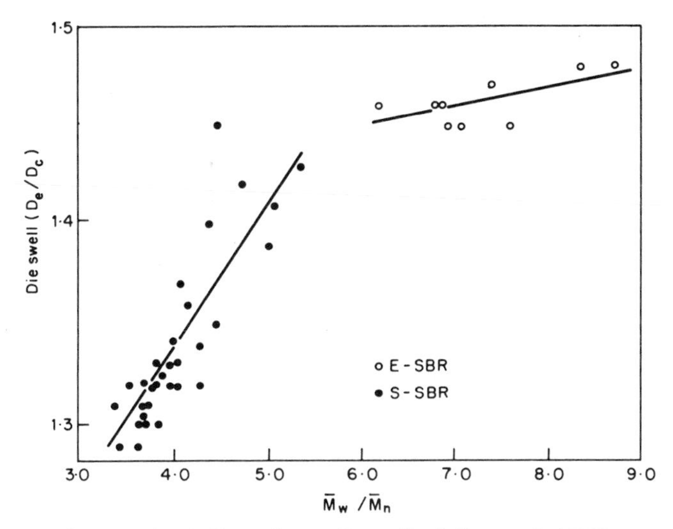

FIG. 5.1. Influence of polydispersity on die swell of oil-extended SBR compounds. D_e = extrudate diameter, D_c = capillary diameter. (Reproduced by permission from *Rubber Chem. Technol.*, **49** (1976) 291. Copyright American Chemical Society.)

the processing and product properties. If a fatty acid is used, the rubber will have little or no natural tack, will normally be fast-curing and will be non-staining. If a rosin acid is used instead, it will confer some tack but the rubber will be slow-curing and will show some staining. A compromise is obtained by using mixed fatty acid–rosin acid systems. All three combinations are used commercially.

As mentioned in the previous section, coagulation may be by a salt–acid system or by the use of alum. In the former case the rubber will be slightly hygroscopic and so where good electrical insulation properties are important alum may be used instead.

In order to prevent deterioration during polymer manufacture, shipment and storage, it is standard practice to incorporate a small amount of an antioxidant (but which is referred to when used in this context as a stabilizer). These may be non-staining or staining, the latter usually being somewhat more powerful. The level used is not sufficient for protection during the manufacture and service of the compounded product.

5.3.1 Oil-Extended Emulsion SBRs

A substantial proportion of the SBR marketed is supplied in the form of oil-extended rubber. In a non-extended rubber the SBR will have a broad molecular weight distribution with the lower molecular weight fraction acting as a processing aid and the higher molecular weight material giving strength to the final product. In the oil-extended rubbers the polymer is produced to a higher average molecular weight and then, before coagulation of the latex, is blended with a hydrocarbon oil, which takes the place of the lower molecular weight polymer fraction at lower cost. Naphthenic or aromatic oils (see Chapter 20) are normally used and for most grades a level of 37·5 pphr oil is used. (A few grades have a level of 50 pphr oil whilst some Russian grades have a target oil level of some 16–20 pphr.)

Besides improving processability, the presence of process oils may improve low-temperature properties, elasticity and flex life.

5.3.2 Cross-Linked and Other 'Hot' SBRs

There continues to be some use for 'hot' SBRs although in most respects the 'cold' rubbers are superior. The most important of these is Type 1009 which contains a small amount of divinylbenzene as cross-linking agent; the polymer may be represented by the structure

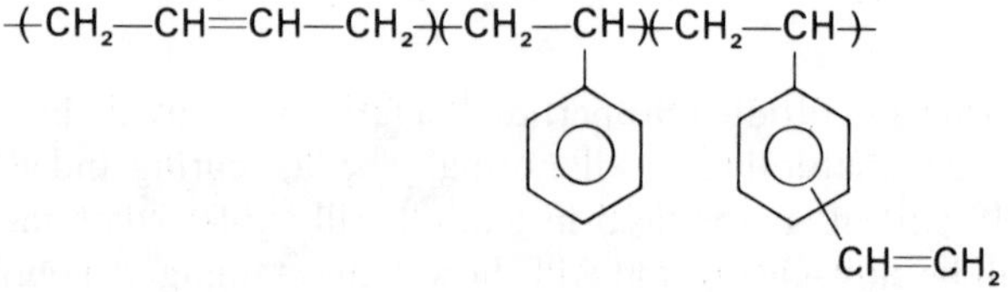

$$\text{-(}CH_2\text{—}CH\text{=}CH\text{—}CH_2\text{)(}CH_2\text{—}CH(C_6H_5)\text{)(}CH_2\text{—}CH(C_6H_4\text{—}CH\text{=}CH_2)\text{)-}$$

Added to a standard cold SBR at a level of 10–20% of the total rubber, it gives improved dimensional stability to extrusion and calendering compounds, reducing die swell and shrinkage. At one time Type 1006 was of some use for light-coloured moulded articles and also for the manufacture of high-impact polystyrene. For the latter application it has now been almost entirely replaced by polybutadiene rubber (see Chapter 6). Some other grades, such as Type 1011 which uses a rosin acid emulsifier and hence exhibits higher tack, find use in solution applications such as in the manufacture of pressure-sensitive tapes.

5.4 STRUCTURE AND VARIATIONS OF SOLUTION SBRs

In addition to the variations shown by emulsion SBRs, the solution polymers exhibit variations in molecular weight distribution, randomness and microstructure. These are discussed further in Section 5.6.1.

5.5 IISRP NUMBERING SYSTEM

The International Institute of Synthetic Rubber Producers has established the following numbering system for classification of emulsion-polymerized SBR:

	Series
Hot non-pigmented rubbers	1000
Cold non-pigmented rubbers	1500
Cold black masterbatch with 14 or less parts of oil per 100 parts SBR	1600
Cold oil masterbatch	1700
Cold oil black masterbatch with more than 14 parts of oil per 100 parts SBR	1800
Emulsion resin rubber masterbatches	1900

As an example SBR 1502 refers to a cold non-pigmented rubber incorporating a non-staining stabilizer, an emulsifier which is a blend of fatty acid and rosin acid, a target bound-styrene content of 23·5%, a nominal Mooney ML 1+4 (100°C) viscosity of 52 and which has been coagulated either by acid or salt acid.

In total there are about 90 rubber types with 'Regular Institute Numbers' whilst, in addition, various producers have been assigned ranges of code numbers for designating new semi-commercial dry rubbers or latices.

Of these individual types the most important is Type 1712, with over half of the SBR market and made by about two dozen producers. It is particularly widely used by the tyre industry but it is also used for mechanical goods where good end-use properties are required commensurate with an economy in cost. It is an oil-extended grade with a mixed acid emulsifier, a 23·5% target bound-styrene content, a nominal Mooney ML 1+4 (100°C) of 55 and using a highly aromatic oil at a level of 37·5 pphr.

Type 1502, described above, is probably the second most important whilst a third important SBR is Type 1500 which uses a rosin acid soap as emulsifier and is recommended where premium physical properties are desired.

(It may be noted that although IISRP in its classification system states that some of its grades will contain a staining as opposed to a non-staining stabilizer, at least one major producer, Shell Chemicals, uses only non-staining stabilizer in its SBR polymers.)

Solution SBRs are classified using the same numbering scheme as for solution-polymerized polybutadiene homopolymers, although within a particular series different individual numbers are used for SSBRs than for the homopolymer rubbers. The numbering scheme is:

Dry polymer	1200–1249
Oil-extended*	1250–1299
Black masterbatch	1300–1349
Oil-black masterbatch*	1350–1399
Latex	1400–1449
Miscellaneous	1450–1499

5.6 GENERAL PROPERTIES

The properties of SBR may be divided into two categories:

1. Properties in which they are similar to natural rubber;
2. properties in which they differ from natural rubber.

Like natural rubber, SBR is an unsaturated hydrocarbon polymer. Hence unvulcanized compounds will dissolve in most hydrocarbon solvents and other liquids of similar solubility parameters (see Chapter 3) whilst cured stocks will show extensive swelling. SBR will also be subject to olefin-type reactions such as oxidation, ozone attack, halogenation and hydrohalogenation although the activity and detailed mechanisms differ because in SBR there is no activating methyl group adjacent to the double bond. Both materials may be reinforced by carbon black whilst neither may be classed as heat-resistant rubbers.

The differences between the two materials may be considered under three headings:

1. Differences in the material supplied;
2. differences in processing behaviour;
3. differences in vulcanizate properties.

In comparison with the natural material, a grade of raw SBR is more uniform in several respects. It is more uniform in quality, since it can be produced under strictly controlled conditions and compounds are more consistent in both processing and product properties. It is also more uniform in the sense that it contains fewer undesired contaminants, whilst over the years it has been more uniform in price with smaller market

* These rubbers may contain any quantity of oil.

fluctuations (an exception being in 1974 following the sudden surge in the price of petroleum). It is, however, to be noted that the introduction of standard grades of natural rubber such as SMR (see Chapter 4) has led to greater uniformity of the natural product in most of these respects whilst the price of SBR also tends to limit large variations in natural rubber prices.

A major difference between SBR and natural rubber is that (with the exception of certain solution polymers) the former does not break down to any extent on mastication. It is thus necessary to supply the material at a viscosity that provides a compromise in allowing good dispersion of fine-particle-size fillers and ease of flow in calendering, extrusion and moulding. The absence of a need to masticate (now also possible with natural rubber where 'constant-viscosity' rubbers are used) not only leads to savings in energy consumption and time but means that it is easier to rework stock which has changed little in properties as a result of processing. Mill mixing with SBR is generally less easy than with natural rubber.

Vulcanization with SBR is somewhat slower than with NR and to maintain cure rates more powerful accelerator systems are usually required. Because of the lack of crystallinity the synthetic product is also deficient in tack and green strength and this is of consequence in tyre building.

The lack of crystallinity in SBR also has a significant influence on vulcanizate properties. Natural rubber tends to crystallize on extension and this leads to a good tensile strength even with gum stocks. Gum stocks of the amorphous SBR are very weak and it is necessary to add reinforcing fillers such as fine carbon blacks to obtain products of high strength. Black-reinforced SBR vulcanizates do, however, show very good abrasion resistance, generally being superior to the natural material above 14°C. On the other hand, they have lower resilience and resistance to tearing and cut growth. Whilst SBR tyres (particularly those made from oil-extended SBR) have a better wet grip rating than those of natural rubber, they do normally have a higher rolling resistance. However, at least one high-vinyl solution SBR may be used to produce tyres with both good wet grip and low rolling resistance. The absence of crystallization at low temperatures also allows the rubbers to exhibit good low-temperature properties. One grade of solution SBR, with a somewhat low styrene content of 18%, has a T_g of −75°C.

On ageing, SBR is quite different from NR in that it tends to cross-link rather than exhibit chain scission on oxidation.

The differences between SBR and natural rubber are summarized in Table 5.2.

TABLE 5.2
Summary of principal differences between SBR and natural rubber

Difference	*SBR*	*NR*
In raw material supplied	Greater product uniformity Fewer undesired contaminants Reduced market fluctuations in price	
In processing behaviour	Does not break down on mastication Requires more powerful curing system to maintain cure rates	Much better natural tack Better green strength
In vulcanizate properties	Amorphous Gum stocks are weak; reinforcing fillers required to produce strong vulcanizates Black-reinforced SBR has generally better abrasion resistance Better wet grip Greater rolling resistance	Crystallizable on cooling or stretching Better resistance to tear and cut growth Higher resilience

5.6.1 Comparison of Solution and Emulsion SBRs

Over the years solution-polymerized SBRs have been claimed to possess several advantages over the emulsion polymers, including:

1. Lighter colour of many grades, most of which contain non-staining stabilizers;
2. lower non-rubber content;
3. better dimensional stability of extruded products;
4. faster cure rates;
5. vulcanizates have better resistance to tear, flex cracking and groove cracking;
6. vulcanizates have better abrasion resistance;
7. vulcanizates have better low-temperature properties.

The solution polymers also exhibit some disadvantages, although these vary according to the type of solution SBR. In general an unmodified solution SBR with a random monomer distribution generally exhibits:

1. Poor processability (now recognized as being associated in part with a narrow molecular weight distribution);
2. lower tensile strength;
3. lower modulus.

The solution process does, however, provide considerable scope for modification of the molecular architecture and hence of the physical properties. Amongst the possible variations are:

1. Narrow or broad molecular weight distributions;
2. linear, T-shaped or star-shaped molecules;
3. variation of the *cis/trans* vinyl ratios in the butadiene segments;
4. incorporation of bonds that break down on mechanical shearing such as during mastication;
5. variations in the degree of randomness of the monomer units in the polymer chain. This can range from a statistically random arrangement, through tapered structures where one end of the chain may, for example, be rich in styrene, to block copolymers which may be of interest as thermoplastic rubbers (see Chapter 16).

Conventional emulsion and random solution SBRs do not break down on mastication. Hence a typical polymer may be too soft to give good filler dispersion yet too hard for some applications, such as sponge compounds. One way round this problem is to join the 'living' ends of three or four polymer chains with tin coupling agents to give T-shaped or star-shaped molecules (Uraneck & Short, 1970). The tin-based coupling site is somewhat weak and may be broken down on shearing the rubber in the presence of certain peptizing agents such as stearic acid. This concept of peptizable or mechanically degradable materials was actively developed by the Phillips company, many of whose Solprene materials (e.g. Solprene 1204) were of this type. (Phillips no longer markets these rubbers but licenses the process to a number of companies such as Negromex in Mexico.) The breakdown on shear of one of these materials is compared with that for an emulsion SBR in Fig. 5.2.

Too narrow a molecular weight distribution not only affects processability but can also adversely affect strength. In one series of experiments Duck *et al.* (1977) prepared a series of blends of solution SBRs of differing molecular weight which had similar average molecular weights, as expressed by a Mooney ML 1+4 (100°C) in the range 50 ± 10. It was found that a number of these blends gave tensile strengths significantly higher than those obtained with unblended narrow-molecular-weight distribution polymers of similar Mooney viscosity. This improvement has been related to the presence of some very large molecules of molecular weight $>10^6$ which rupture during processing, with the ruptured ends reacting with the surface of the carbon black particles to give superior reinforcement.

It has been common experience that those factors that tended to improve

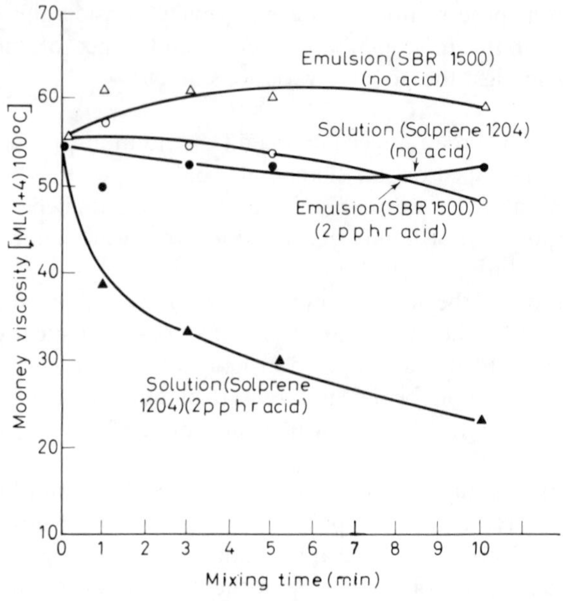

FIG. 5.2. Breakdown of solution SBR (Solprene 1204) and emulsion SBR (1500) in the presence of stearic acid (all stocks mixed at 120°C, 60 rev min^{-1}, in midget internal mixer).

abrasion resistance adversely affected processability and wet traction: this is indicated in Fig. 5.3.

It is, however, reported that a solution polymer with a butadiene/styrene ratio of 77/23 and a low pendant vinyl group content gave compounds with an abrasion resistance some 10% better than emulsion material. This polymer also had good processability, wet traction almost as good as the natural material but inferior tensile strength and tack.

It has also been common experience that rolling resistance of tyres, which it is normally desired should be low, tends to be linked with wet grip (see Fig. 5.4).

However, using the thesis that road grip is dependent on ultrasonic frequencies generated by the sliding tyre when breaking, whilst rolling resistance is dependent on lower frequencies generated during forward movement of the tyre, it has been found possible to produce tyres with good wet grip and low rolling resistance. One such rubber (Cariflex S-1215 from Shell) had a styrene content of 23·5% and a butadiene distribution of 20% *cis*/30% *trans*/50% vinyl. This particularly high vinyl content means that

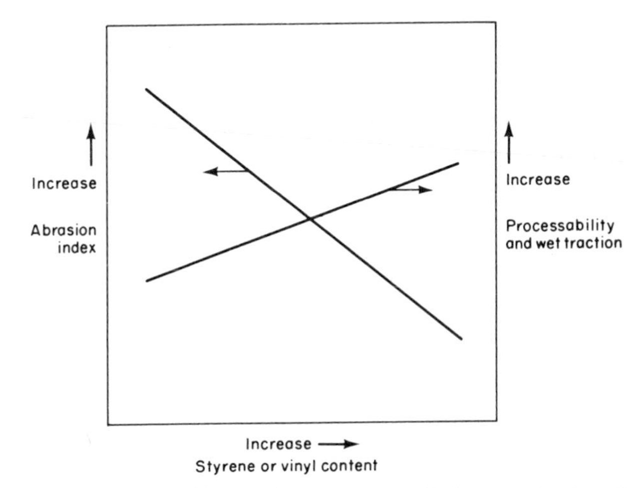

FIG. 5.3. General effect of increasing styrene or vinyl concentrations in solution SBRs on abrasion index, processability and wet traction. (From Haws *et al.* (1975), reproduced by permission of The Plastics and Rubber Institute.)

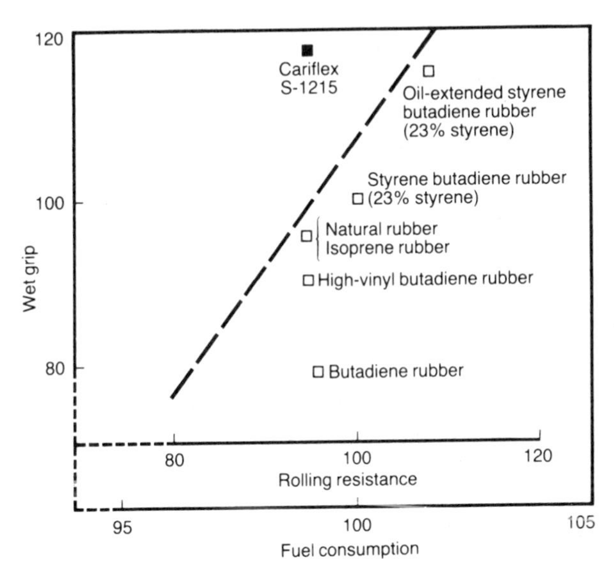

FIG. 5.4. Road performance of Cariflex S-1215. (From Krol, 1982.)

there are far more short branches (pendant vinyl groups and benzene rings) than in more conventional SBR materials and this leads to a somewhat higher-than-normal glass transition temperature of $-40°C$.

Although solution-polymerized SBR materials were introduced in the 1960s accompanied by claims that these were the general-purpose rubbers of the future, they have so far failed to capture a significant part of the market. This may be due in part to existing overcapacity of emulsion SBR plants discouraging investment in new plants, in part to the extra cost in manufacturing these materials and in part to the fact that many of the solution grades on the market did not show significant superiority to warrant the premium in price. There are, however, now indications that new-generation solution SBRs are becoming available that may make inroads into traditional markets although they face competition with some of the high-vinyl polybutadiene rubbers that are now also appearing on the market (see Chapter 6).

5.7 COMPOUNDING

Like natural rubber, SBR is a diene hydrocarbon rubber which is usually vulcanized using conventional accelerated-sulphur systems. With the different activity of the double bond, the slightly more aromatic structure and the absence of any ability to crystallize there are, however, a number of differences in details of formulation.

5.7.1 Vulcanization

SBR is somewhat slower-curing than NR and higher accelerator levels are necessary to achieve equivalent cure times, albeit with slightly lower sulphur levels. There is, however, a substantial difference in cure rates between grades, as illustrated in Fig. 5.5; this gives comparative accelerator loadings required for equivalent cure for 12 grades produced by one manufacturer, although these data do not take into account the variation in rubber hydrocarbon content between grades. It will be seen that a Type 1500 SBR requires a 50% higher accelerator loading than oil-extended Type 1712 rubber to achieve an optimum cure of 30 min at 144°C using a standard test recipe (Table 5.3).

Furthermore, when the compounds were adjusted to have similar cure times, their relative scorch times were quite different (Fig. 5.6).

As with other rubbers, variation in cross-link density can have differing

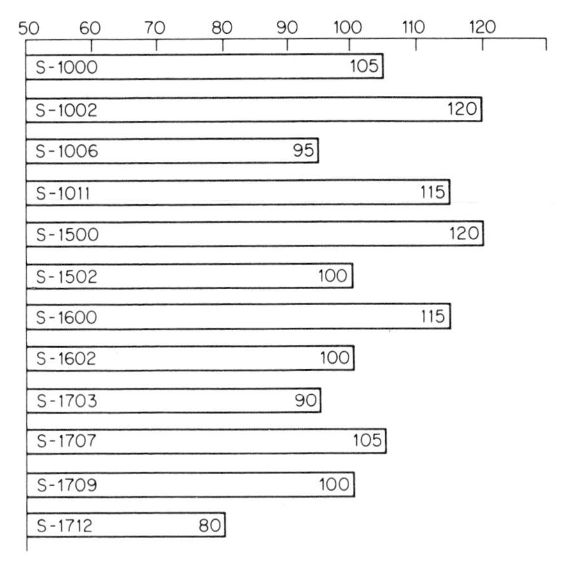

FIG. 5.5. Relative accelerator levels for various SBR types to give optimum cure in the same cure cycle (level for S-1502 represented by 100). Data are for Shell Cariflex grades which follow standard ASTM nomenclature:

1000 series: 'Hot' emulsion grades;
1500 series: 'Cold' emulsion grades;
1600 series: Black masterbatch of 'cold' grades;
1700 series: Oil-extended 'cold' grades;
1800 series: Oil-black masterbatch of 'cold' grades.

Oil-extended polymers are considered to be 100% rubber hydrocarbon.

effects on physical properties, as shown schematically in Fig. 18.5 (Chapter 18).

Physical properties may be significantly affected by changing the type and loading of the accelerator system, as shown in Figs 5.7–5.9. For example, where the tensile strength goes up with loading of a 100% CBS system (within the range studied), MBTS combined with TMTM, TMTD or DPG all decreased in tensile strength with increased accelerator loading.

The development of the so-called efficient vulcanization systems for use with natural rubber led to related studies on SBR (Rodger, 1979). The results may be summarized as follows.

1. Conventional SBR vulcanizates have a monosulphide cross-link content similar to that of an EV natural-rubber system (Table 5.4).

TABLE 5.3
Test recipe
(After Brydson, 1981)

	(*pphr*)	(*pphr*)
Polymer: Cariflex S-1000, S-1500, S-1700 series	100·0	—
Black masterbatch; Cariflex S-1600 and S-1602	—	150·0[a]
Zinc oxide	3·0	3·0
Stearic acid	2·0	2·0
PBNA	1·0	1·0
HAF black	50·0	—
Dutrex 20	5·0	5·0
Sulphur	2·0	2·0
Santocure 1·0, TMTD 0·1	Variable[b]	

[a] Cariflex S-1600 and S-1602 contain 50 pphr HAF black.
[b] The accelerator level was adjusted to produce an optimum cure at 30 min at 292°F, 144°C.

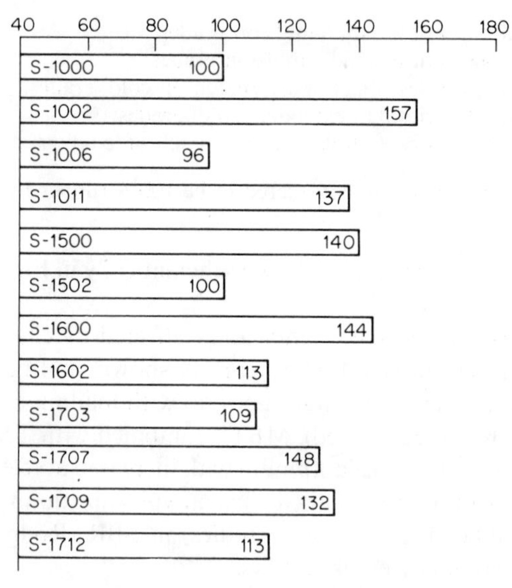

FIG. 5.6. Relative scorch times for SBR types (time for S-1502 represented by 100).

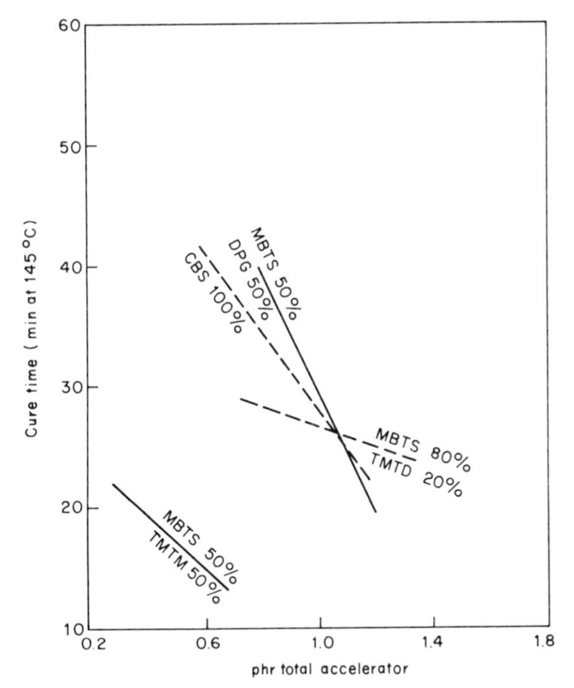

FIG. 5.7. Dependence of optimum cure time on accelerator type and concentration using a compound based on oil-extended 'cold' rubber (Type 1712) (data from Shell Chemical Co.). Formulation: Cariflex S-1712, 100; ZnO, 3; stearic acid, 2; phenyl-β-naphthylamine, 1; HAF black, 50; mineral oil (Dutrex 20), 5; sulphur, 1·75 pphr; accelerators—variable.

2. The replacement of a conventional curing system by an EV system in SBR increases the monosulphide content to about twice that of an NR EV system.
3. Whereas the cross-link densities of conventionally cured SBR vulcanizates increase on ageing at elevated temperatures (e.g. 110°C), the EV-cured material has a very stable cross-link density at the same temperature.
4. The reduction in fatigue life shown with natural-rubber EV systems is not duplicated with SBR (Table 5.5). SBR EV compounds after ageing show much better fatigue resistance than conventional compounds after ageing (Table 5.6).

It would appear that as a rule the use of EV systems in SBR leads to a lower aged modulus and hardness, better retention of elongation at break,

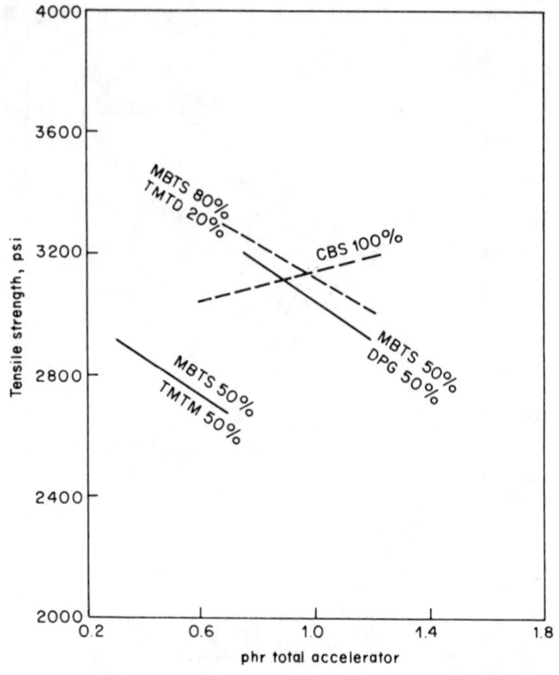

FIG. 5.8. Dependence of vulcanizate tensile strength on accelerator type and concentration. Formulation as for Fig. 5.7.

TABLE 5.4
Cross-link distributions for NR and SBR
(After Rodger, 1979)

Vulcanization system	*Cross-link type (%)*			
	NR		*SBR*	
	S_1	S_2+S_x	S_1	S_2+S_x
Conventional[a]	0	100	38	62
EV[b]	46	54	86	14

S_1 = monosulphide links; S_2 = disulphide links; S_x = polysulphide links.

[a] Conventional for NR: sulphur 2·5, MBS 0·6; conventional for SBR: sulphur 2·0, CBS 1·0 pphr.

[b] EV for NR: CBS 1·5, DTDM 1·5, TMTD 1·0; EV for SBR: CBS 1·5, DTDM 2·0, TMTD 0·5 pphr.

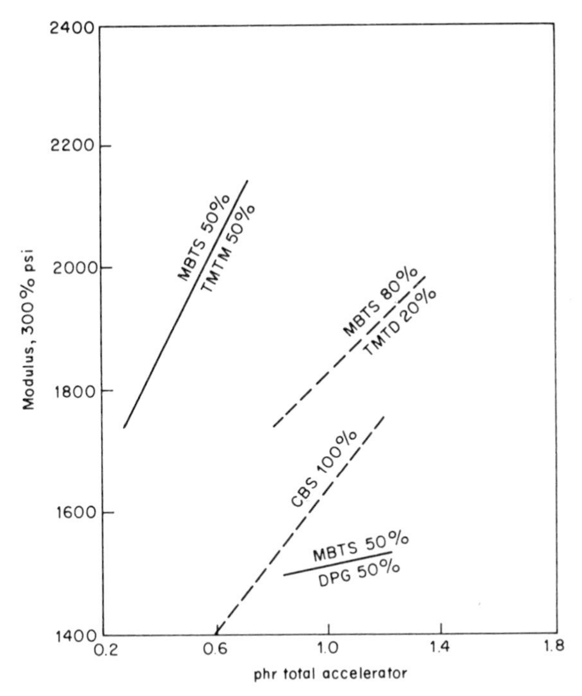

FIG. 5.9. Dependence of vulcanizate modulus on accelerator type and concentration. Formulation as for Fig. 5.7. (Reproduced by permission of Shell Chemical Co.)

TABLE 5.5
Fatigue performance of semi-EV in SBR
(After Rodger, 1979)

	Conventional	*Semi-EV*	
CBS, pphr	1·2	2·5	1·0
DTDM, pphr	—	—	1·0
Sulphur, pphr	2·0	1·2	1·2
Fatigue life—80% extension, kc to failure	323	275	416
Aged 3 days at 85°C, fatigue life—80% extension, kc to failure	61	81	117

TABLE 5.6
Properties of full EV systems in SBR
(After Rodger, 1979)

	Conventional	*Full EV*	
MBS, pphr	1·2	7·0	1·2
TMTD, pphr	—	—	1·2
DTDM, pphr	—	—	1·2
Sulphur, pphr	2·0	0·75	—
Aged 10 days at 90°C			
Increase in 300% modulus, %	115	65	60
Compression set			
22 h at 70°C, %	18	11	11
Fatigue (75% extension)			
Unaged, kc	645	750	800
Aged, kc	192	445	400

and a general reduction in compression set and heat build-up. Because of the high cost of accelerators, true EV systems are seldom justified but semi-EVs (i.e. systems using intermediate accelerator/sulphur ratios) provide a useful compromise between cost and performance.

5.7.2 Use of Fillers

Because SBR is not self-reinforcing like NR, it is essential that a reinforcing filler, usually carbon black, is used in the compound formulation. For the reinforcing blacks (HAF and ISAF) highest tensile and tear strengths are obtained at about 45 pphr black with the amount used being quite critical. In the case of the semi-reinforcing SRF blacks loading levels are less critical and at 75 pphr loading the tear strength using SRF may be higher than when the other blacks are being used (Figs 5.10 and 5.11).

Non-black reinforcing fillers can be used instead of carbon black whilst diluent fillers may also be employed simultaneously.

5.7.3 Plasticizers and Other Additives

Plasticizers may be added to unmodified SBR or they may be built in, in the case of oil-extended materials. Naphthenic and aromatic oils are usually employed to give the best mechanical properties, with paraffinic oils seldom being used.

Antioxidants and antiozonants are similar to those used in natural rubber.

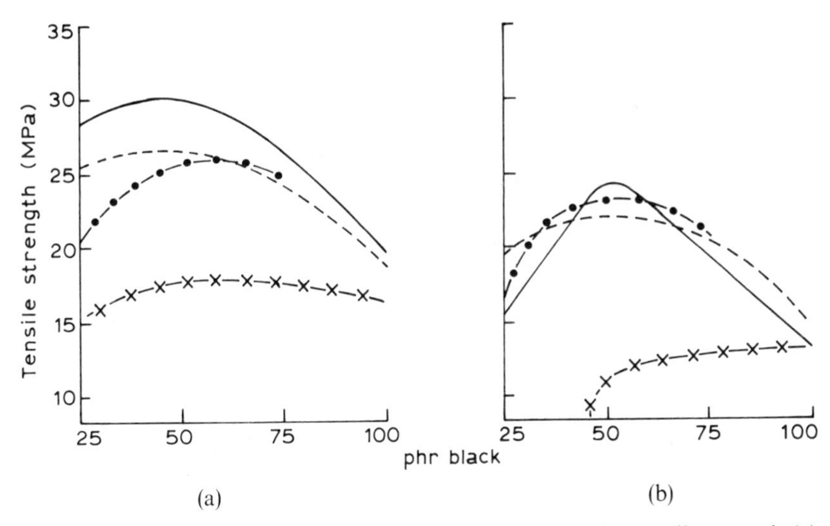

Fig. 5.10. Effect of carbon black in SBR. Carbon black level vs tensile strength. (a) S-1500; (b) S-1712. ——, ISAF; – – –, HAF; –·–, EPC; –×–, SRF.

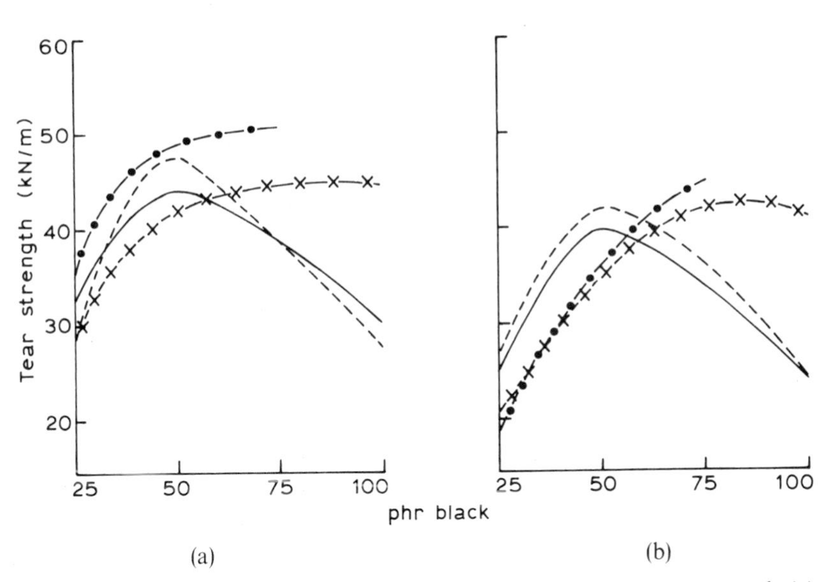

Fig. 5.11. Effect of carbon black in SBR. Carbon black level vs tear strength. (a) S-1500; (b) S-1712. ——, ISAF; – – –, HAF; –·–, EPC; –×–, SRF.

5.8 PROCESSING

Whilst SBR and NR are processed on the same equipment there are some differences in behaviour between the two materials.

Since the raw SBR polymer is supplied at the correct molecular weight for processing, and indeed cannot be broken down by mastication, the mastication step is not necessary. Mixing is preferably undertaken in an internal mixer. On a two-roll mill a rather ragged band is likely to form and mixing tends to be more difficult than with NR. It is usual to start with mill rolls at about 50°C although these may rise to about 110°C during processing.

Shaping operations such as extrusion are generally easier than for NR compounds, particularly as reworked stock has the same viscosity as original material. Shrinkage and die swell can be high and this may be reduced by compounding with some 'hot' rubber that is lightly cross-linked with divinylbenzene. Greatest improvements are obtained using Type 1009 rubbers but these have a severe adverse effect on mechanical properties. A compromise can be effected by replacing the Type 1009 rubber by a similar hot cross-linked rubber but with a higher styrene content, although this will be at some cost to resilience.

5.9 APPLICATIONS

About 70% (slightly more in North America) of solid SBR is used by the tyre industry. Not surprisingly the material has been more subject to the vicissitudes of that industry than most other rubbers. For example, there was a 25% drop in SBR production between 1979 and 1983.

Whilst there is an expectation of continuing growth in car and truck use the introduction of progressively more stringent legislation, particularly in the United States, aimed at reducing petrol consumption and further development of tyres with longer service life, does not encourage hopes of much growth in the tyre industry.

SBR does not have the heat resistance of saturated rubbers (e.g. EPDM, ACM) or even of polychloroprene and nitrile rubbers, the weathering resistance of EPDM, the oil resistance of the polar rubbers or the excellent 'rubberiness' of the natural material. However, because of its reasonable ageing resistance, very good abrasion resistance and good low-temperature properties it finds many uses in the general rubber goods industry. Examples include mechanical goods, shoe soling and carpet underlay. In the one-time important use in the manufacture of high-impact polystyrene plastics SBR has now been largely supplanted by polybutadiene rubber (see Chapter 6).

BIBLIOGRAPHY

Blackley, D. C. (1983). *Synthetic Rubbers—Their Chemistry and Technology.* Applied Science, London.
Glanville, L. M. & Bowman, I. J. (1977). Review paper—Progress of general purpose synthetic rubbers. *Progress of Rubber Technology*, **40**, 21, Plastics and Rubber Institute, London.
Saltman, W. M. (ed.) (1977). *The Stereo Rubbers.* John Wiley, New York.
Stafford Whitby, G. S. (ed.) (1954). *Synthetic Rubber.* John Wiley, New York. (A very useful source of information for work prior to 1954.)
Uraneck, C. A. (1968). In *Polymer Chemistry of Synthetic Elastomers*, Part 1, ed. J. P. Kennedy & E. G. M. Törnqvist. Interscience, New York.

REFERENCES

Brydson, J. A. (1981). In *Developments in Rubber Technology—2*, ed. A. Whelan & K. S. Lee. Applied Science, London, Chapter 2.
Cooper, R. N. & Nash, L. L. (1972). *Rubber Age*, **104** (May), 55.
Dacker, K. De, Dunnom, D. D. & McCall, C. A. (1969). *Rubber Age*, **96**, 53.
Duck, E. W., Bowman, I. J. W. & Wilson, C. A. (1977). *Preprints of International Rubber Conf., Brighton 1977*, Vol. II. Plastics and Rubber Institute, London, p. 33-I.
Haws, J. R., Nash, L. L. & Wilt, M. S. (1975). *Rubber Industry*, **9**, 107.
Krol, Ir. L. U. (1982). *Synthetic Rubbers Technical Manual*, **GPR8.1**, 1. Shell Chemical Co., Amsterdam.
Mills, W. & Giurco, F. (1976). *Rubber Chem. Technol.*, **49**, 291.
Mills, W., Yeo, C. D., Kay, P. J. & Smith, B. R. (1975). *Rubber Industry*, **9**, 25.
Rodger, E. R. (1979). In *Developments in Rubber Technology—1*, ed. A. Whelan & K. S. Lee. Applied Science, London, Chapter 3.
Tokita, N. & Pliskin, I. (1973). *Rubber Chem. Technol.*, **46**, 1166.
Uraneck, C. A. & Short, J. N. (1970). *J. Appl. Polym. Sci.*, **14**, 1421.

Chapter 6

Polybutadiene, Synthetic Polyisoprene and Other Unsaturated Hydrocarbon Homopolymer Rubbers

6.1 INTRODUCTION

In the last two decades of the 19th century rubbery materials were obtained by polymerizing isoprene (i.e. 2-methyl-1,3-butadiene) (**II**), 2,3-dimethyl-1,3-butadiene (**III**) and piperylene (i.e. 1,3-pentadiene) (**IV**). Each of these materials may be considered to be methyl-substituted derivatives of 1,3-butadiene (**I**). 1,3-Butadiene itself (referred to henceforth simply as butadiene) had also been polymerized into a rubber before World War I.

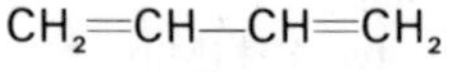

$$CH_2{=}CH{-}CH{=}CH_2 \quad \text{(I)}$$

$$CH_2{=}C(CH_3){-}CH{=}CH_2 \quad \text{(II)}$$

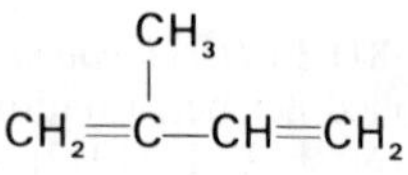

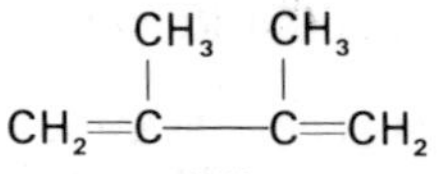

$$CH_2{=}C(CH_3){-}C(CH_3){=}CH_2 \quad \text{(III)}$$

$$CH_2{=}CH{-}CH{=}CH{-}CH_3 \quad \text{(IV)}$$

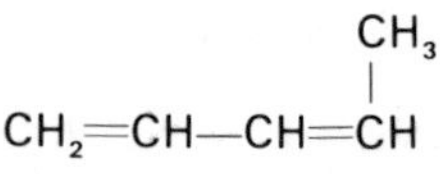

The early polymers had properties that compared very poorly with natural rubber and although there was some production of polyisoprene about 1910 and of methyl rubber from (**III**) towards the end of World War I (see Chapter 1) these materials did not remain on the market for long. The main reason was that manufacturing techniques then available did not allow close control of molecular structure. The polymers were highly branched, and even to some extent cross-linked, whilst the configuration of the monomer residues in the polymer chain was such that the polymers were too irregular to exhibit the level of crystallinity that is so beneficial to natural rubber.

Slightly better control was achieved when butadiene was copolymerized, for example with styrene, and, as was described in Chapter 5, such rubbers appeared in the 1930s, were forced on a reluctant industry during World War II and then accepted fully during the 1950s as the materials improved.

It was not until the exploitation of alkyllithium and Ziegler–Natta catalyst systems in the 1950s that homopolymers were produced with sufficient structural stereoregularity that they were a marketable commodity. Even today these materials have a role secondary to that of NR and SBR. In tonnage terms the most important is *polybutadiene*, used mainly in tyres and for making high-impact polystyrene. Synthetic *polyisoprene* does not have the structural regularity of the natural product but this deficiency can be put to use together with the high levels of purity possible. For strategic purposes the USSR has also been a major producer.

It was subsequently found that other interesting rubbers could be obtained by a ring-opening reaction involving catalysts related to the Ziegler–Natta materials. At one stage there were high hopes for a *trans*-polypentenamer as a tyre rubber but these were not realized. In due course, however, a *polyoctenamer* appeared on the market and has proved useful in blends. Somewhat related to these materials is *polynorbornene*, which has also found a specialized niche.

6.2 BUTADIENE RUBBER (BR)

In tonnage terms BR ranks third after NR and SBR with a nameplate capacity of around 1 750 000 tpa, of which about 375 000 tpa is located in the Centrally Planned Economy Countries (CPEC). In non-CPEC countries production is of the order of 1 000 000 tpa from about 30 plants situated in about 17 countries. Over three-quarters of the BR produced is used in tyres and tyre products with the bulk of the rest being used in the manufacture of high-impact polystyrene.

6.2.1 Manufacture

When butadiene is polymerized three types of isomerism can occur in the repeating unit:

1. 1,4-/1,2-isomerism;
2. *cis*/*trans* isomerism of 1,4-polymer;
3. stereoisomerism of 1,2-polymer.

If we consider only a repeat unit of the 1,2- and 3,4-structures it will be seen that these are identical and both may be considered as 1,2-polybutadiene units. However, this unit possesses an asymmetric carbon atom and this

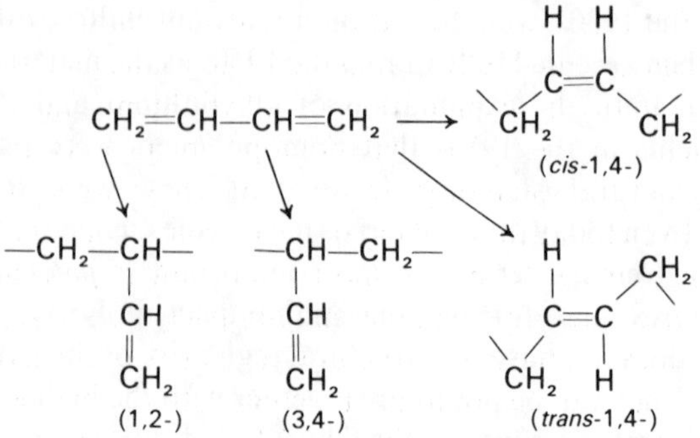

$$CH_2{=}CH{-}CH{=}CH_2 \longrightarrow$$

$$\text{(cis-1,4-): } {-}CH_2(H)C{=}C(H)CH_2{-}$$

$$\text{(1,2-): } {-}CH_2{-}CH(CH{=}CH_2){-} \qquad \text{(3,4-): } {-}CH(CH{=}CH_2){-}CH_2{-} \qquad \text{(trans-1,4-): } {-}CH_2(H)C{=}C(H)CH_2{-}$$

allows both isotactic and syndiotactic stereoregular and atactic stereo-irregular forms to occur.

There are thus four basically stereoregular forms of polybutadiene, namely:

1. *cis*-1,4-polybutadiene;
2. *trans*-1,4-polybutadiene;
3. isotactic 1,2-polybutadiene;
4. syndiotactic 1,2-polybutadiene.

Further variations of stereoregular forms could in theory also be obtained if the monomer units could be linked in a head-to-head/tail-to-tail as well as in the usual head-to-tail manner. Although there would be no difference with the 1,4-polymers different structures would be obtained by the 1,2-polymers.

In practice the synthetic materials are not 100% stereoregular and are structures containing varying proportions of the above forms in combination. These combinations may be more or less statistically random, occur in an alternating or other systematic repeating sequence or in blocks. Some cyclization may also occur.

Free-radical polymerization of butadiene gives polymers of low stereoregularity of little interest to the rubber manufacturer. Commercial polymers are prepared in solution using either alkyllithium or Ziegler–Natta catalysts.

The actual alkyllithium catalyst used appears to have little effect on the polymer properties. However, since the lower members are capable of distillation and those higher than methyl are hydrocarbon-soluble the greatest interest lies in the ethyl and butyl materials, particularly the latter. On the other hand, the choice of solvent is crucial since this affects both rate of reaction and polymer structure (see Table 6.1).

TABLE 6.1
Effect of solvent on polybutadiene microstructure

Solvent	*RLi*	cis-*1,4*- (%)	trans-*1,4*- (%)	*1,2*- (%)	*3,4*- (%)
Hexane	EtLi	43	50	7	
Benzene–Et_3N	EtLi	23	40	37	
THF	EtLi	—	9	91	

Polymer microstructure depends strongly on the ratio of monomer to organolithium concentrations. Conventional linear materials are prepared at temperatures in the range 24–40°C. Branching can occur at high temperatures whilst star-shaped or radial polymers may be produced by the use of trifunctional or tetrafunctional alkyllithiums. Radial polymers may also be produced by taking advantage of the fact that butyllithium-initiated polymerizations are 'termination-free' and that chain ends remain active. These may be coupled by polyfunctional agents such as divinylbenzene or methyltrichlorosilane to give radial or star-shaped polymers. Being more compact these polymers have lower melt and solution viscosities than linear polymers of equivalent molecular weight.

The Ziegler–Natta catalysts are usually considered to be systems prepared by combination of two materials:

1. An aluminium alkyl (or possibly an aluminium hydride, or an alkyl derivative of beryllium, zinc, magnesium, cadmium, etc.).
2. A derivative of a Group IV–VIII transition metal.

In the case of BR, preferred systems have been an aluminium alkyl halide such as aluminium diethyl chloride with a cobalt compound such as $CoCl_2$, or an aluminium alkyl halide and titanium tetraiodide.

Conventional butadiene rubbers are usually considered under three headings:

1. High-*cis* (*c.* 97%) 1,4-polybutadienes using the cobalt system.
2. Medium-*cis* (92%) polymers using the aluminium alkyl/titanium tetraiodide system.
3. Low-*cis* (*c.* 40%) polymers made using alkyllithium catalyst systems.

In 1987 Bayer introduced two polybutadiene rubbers (Bunas CB 22 and 23) which had been prepared using the transition metal neodymium as catalyst.

6.2.2 Structure and Properties of Polybutadienes

The melting point, glass transition and identity period of the four stereoregular polybutadienes are given in Table 6.2.

The very low glass transitions of the 1,4-polymers are useful rubbery characteristics but the high T_m of the *trans* polymer will inhibit rubberiness. Rubbery polymers must therefore be looked for either with a high-*cis* material or a non-crystallizing *cis*/*trans* copolymer.

Also of significance is the 1,2-content of the polybutadiene. This depends on the polymerization method (see Table 6.3) and has a direct effect on the glass transition temperature (Table 6.3 and Fig. 6.1).

The abrasion resistance of a rubber has been related to its ability to undergo rapid deformation. In turn this is a function of segment flexibility, which increases with an increase in the difference between the T_g and the test temperature. Since the polybutadiene rubbers have very low glass transition temperatures, the abrasion resistance of the vulcanizates is good in comparison with many other elastomers.

In this context it is interesting to note the effect of increasing the 1,2- or

TABLE 6.2
Melting point (T_m), glass transition (T_g) and identity period of stereoregular diene homopolymers

Polymer	T_m (°C)	*Identity period* (Å)	T_g (°C)[b]
Polybutadiene			
trans-1,4-	145	4·7/4·85[a]	−106
cis-1,4-	3	8·6	−107
Isotactic 1,2-	128	6·5	−15
Syndiotactic 1,2-	156	5·1	
Polyisoprene			
trans-1,4-	65	4·77/8·75/9·2[a]	−60
cis-1,4-	20–30	8·1	−73
3,4-	Amorphous	—	?
Polypentadiene			
trans-1,4-	95	4·82	?
Isotactic *cis*-1,4-	42	8·15	?
Syndiotactic *cis*-1,4-	52–53	8·5	?
1,2-	10–20	?	?

[a] The different identity periods for the *trans* polymers arise because different crystalline forms can exist.
[b] The data on the T_g values of the polybutadienes have been obtained by extrapolation (Bahary *et al.*, 1967).

TABLE 6.3
Effect of 1,2-content on the T_g of polybutadienes[a]

Polymer type	*All*-cis	*Commercial Ziegler type*	*Butyl-lithium*	*Emulsion*	*All-1,2-*
1,2-content, %	0	1–4	10	20	100
T_g, °C	−107	−103	−93	−78	−15

[a] Data from Alliger & Weissert (1968) except for the all-*cis* and all-1,2- data which were obtained by extrapolation by Bahary *et al.*, 1967. Alliger & Weissert quote similar figures as approximations.

vinyl-content of the polybutadiene rubber. As the 1,2-content increases so does the amount of abrasion in a laboratory test— a direct consequence of the dependence of abrasion resistance on T_g, which is in turn dependent on the 1,2-content (Blümel, 1964; Fig. 6.2).

Polybutadiene rubbers can show very high levels of resilience, this being well demonstrated by the children's toy, the superball. This high resilience may also be related to the high chain flexibility at normal ambient temperatures which is further related to the very low T_g. However, at high levels of deformation resilience becomes somewhat lower and in tyres and other applications involving high deformations a high level of heat build-up results; this is one of the reasons why BR is usually blended with NR or SBR.

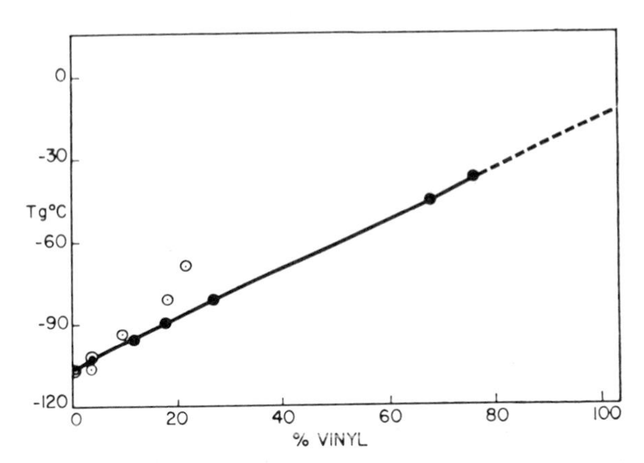

FIG. 6.1. Relationship between T_g and vinyl content of polybutadienes. (From data for two sets of work reported by Bahary *et al.* (1967), reproduced by permission from *Rubber Chem. Technol.*, **40** (1967) 1531. Copyright American Chemical Society.)

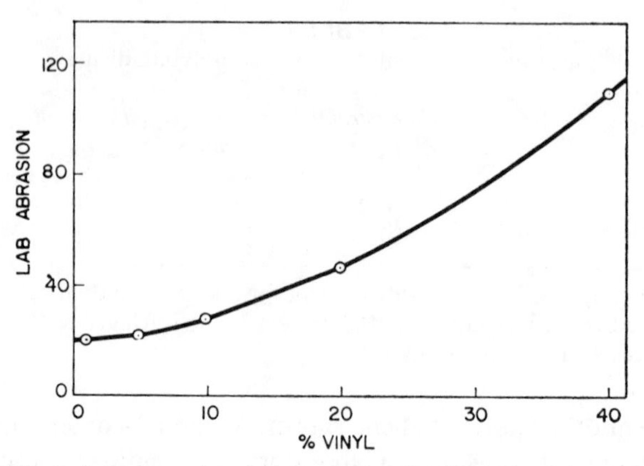

FIG. 6.2. Effect of vinyl content on laboratory abrasion of polybutadienes. (From Bahary *et al.* (1967) based on Blümel (1964), reproduced by permission from *Rubber Chem. Technol.*, **40** (1967) 1532. Copyright American Chemical Society.)

On heating, polybutadienes tend to isomerize, for example during vulcanization, and the view has been expressed that because of this the *cis/trans* ratio of the raw polymer is immaterial. This argument is based on the assumption that during vulcanization all polymers equilibrate towards a common microstructure. It has, however, been found (Blümel, 1964) that on heating a range of polybutadienes of different initial *cis* content the *trans* content increased in each case by an approximately equal amount (*c.* 6%) quite independently of the initial *cis* content and not enough to reach the reported equilibrium of 75% *trans* content. Blümel also pointed out that high-*cis* polybutadiene vulcanizates had higher tensile strength, tear strength, elongation at break and abrasion resistance than the low-*cis* compounds, which were faster-curing and showed less heat build-up in flexometer experiments. These differences were probably not due to variations in the *cis/trans* content but rather due to the higher 1,2- or vinyl-content, which tends to go hand in hand with the change in *trans* content.

Of the polybutadiene rubbers only the *cis*-polybutadienes of high steric purity are capable of crystallization and in the case of unstrained samples at normal ambient pressures this only occurs below 3°C. For this reason the influence of crystallization on tensile strength is so minimal that it may be ignored in the case of polybutadiene rubbers. It is thus rather surprising that some BR manufacturers of high-*cis* rubbers have claimed that their products have better resistance to cold flow (with raw polymer) than less regular materials.

The new neodymium-catalysed grades of Bayer are claimed to have improved green strength, tack, abrasion resistance and fatigue resistance, lower heat build-up under dynamic stress and better groove cracking resistance when used in tyres.

Commercial application of BR arises from the following properties, in which it shows superiority to both NR and SBR vulcanizates:

1. Abrasion resistance and groove cracking resistance of tyres;
2. low-temperature flexibility;
3. heat ageing resistance;
4. high resilience at low deformations;
5. ozone resistance, both static and dynamic;
6. ability to accept higher levels of filler and oil with less deterioration in properties than with either NR or SBR.

The main limitations are:

1. Poor tack;
2. poor road grip of tyre treads (a feature related to the high resilience);
3. poor tear and tensile strength as compared with both SBR and natural rubber. In some ways pure polybutadiene rubber vulcanizates may be regarded as brittle with the possibility of products being subject to chunking and gouging (breaking off of pieces of the vulcanizate during severe operational conditions).

A comparison of tread stocks using 50 pphr HAF black for BR, SBR and NR is given in Table 6.4.

TABLE 6.4
A comparison of tread stocks (50 pphr HAF black) based on BR, SBR 1500 and NR

Property	*BR*	*SBR 1500*	*Natural rubber*
Tensile strength, MPa	15·7	27·4	28·6
Modulus at 300%, MPa	8·3	18·1	18·6
Elongation at break, %	480	450	510
Tear strength (angle), $kN\,m^{-1}$	39	49	108
Resilience, Lupke, %			
At 23°C	60	40	50
At 70°C	70	55	60
Hardness, Shore	58	58	58
De Mattia cut initiation, kc	>500	>500	100
Cut growth, kc	10	20	100

In practice best results are obtained using blends of BR with either SBR or NR. Thus a typical 50/50 BR/SBR tread stock blend would have greater abrasion resistance than one of SBR but at some loss of wet grip. Tensile strength and modulus of the blends are slightly lower.

6.2.2.1 Medium-Vinyl Polybutadienes

Following the oil crisis of 1973 there was for a short period a severe shortage in styrene and a consequent huge increase in its price. This led several polymer suppliers to study the properties of medium-vinyl polybutadienes. Such materials may be considered as 'copolymers' of 1,4- and 1,2-structures with the pendant vinyl group instead of the pendant benzene ring in SBR.

$$-\!\!\left(CH_2-CH{=}CH-CH_2\right)\!\!\left(CH_2-\underset{\underset{CH=CH_2}{|}}{CH}\right)\!\!-$$

Experimental studies (Haws *et al.*, 1975) indicated, indeed, that medium-vinyl (45–55% vinyl) polybutadiene vulcanizates were similar to those of SBR. Best results were generally obtained with blends of 45% vinyl BR with SBR type 1702. Compared with all-SBR compounds, 50/50 blends, for example, showed improved abrasion resistance, higher resilience, lower heat build-up and greater resistance to tyre blow-out. There was little change in skid and traction on wet surfaces or in cut growth but some loss in tensile and tear strengths.

With renewed ready availability in styrene the properties of the medium-vinyl polybutadienes were not exploited. Recently, however, there has been renewed interest in these materials which could lead to significant developments in the next few years.

6.2.3 IISRP Numbering System

The International Institute of Synthetic Rubber Producers has introduced the following numbering system for stereoregular BR rubbers:

Dry polymer	1200–1249
Oil-extended[a]	1250–1299
Black masterbatch	1300–1349
Oil–black masterbatch[b]	1350–1399
Latex	1400–1449
Miscellaneous	1450–1499

[a] Dry polymer containing any quantity of oil.
[b] Black masterbatch containing any quantity of oil.

6.2.4 Processing

Butadiene rubbers, like SBR, do not break down on mastication like natural rubber. This is because both BR and SBR, being based on butadiene rather than isoprene, lack the activation of the double bond by the adjacent methyl group possible in the natural product. Commercial grades of BR are thus supplied at an appropriate molecular weight and hence Mooney viscosity.

If mill mixing, compounds based on 100% BR tight nips and as low a mill-roll temperature as possible are recommended. In practice BR is usually used blended with NR, SBR or IR, in which case the blending rubbers are usually added first to the mill, somewhat higher temperatures being used when blending with NR. Internal mixing and shaping procedures follow conventional practice.

Compound formulations are very similar to those used with SBR; for example, it is common to use slightly more accelerator and slightly less sulphur than with natural rubber.

6.2.5 Applications

Of limited use on its own, BR is a very useful blending material for the tyre industry, where its high resistance to abrasion and low heat build-up makes it very attractive. Major uses are as blends with SBR in passenger-car tyres and with NR in truck tyres. BR is also widely used as the toughening rubber in high-impact polystyrene. A smaller but well-known application, less in vogue today than a few years ago, is for highly resilient playballs. Its third place in tonnage terms after SBR and NR is a direct consequence of the importance of polybutadiene rubber to the tyre industry.

6.3 SYNTHETIC POLYISOPRENE RUBBERS

It is over 100 years ago that attempts were first made to produce the synthetic equivalent of natural rubber. Early attempts led to highly inferior products and it was not until the advent of stereospecific catalysts in the 1950s that the synthetic materials approached the performance of the natural product. Whilst in many ways the synthetic materials now available are superior, for example in purity and product consistency, stereoregularity remains less than for NR and this, together with some unfavourable economics, has resulted in limited development. The materials have been rather more successful where considerations of strategic supply have been paramount. Thus, whilst capacity for America, Western Europe and Asia is only about 240 000 tpa, that for the USSR is about 850 000 tpa. Other significant capacity exists in Romania and in South Africa, again for strategic reasons.

6.3.1 Preparation of Synthetic Polyisoprene Rubbers (IR)

In comparison with butadiene the presence of the methyl side group in the isoprene unit increases the range of possible structures in the basic repeating unit:

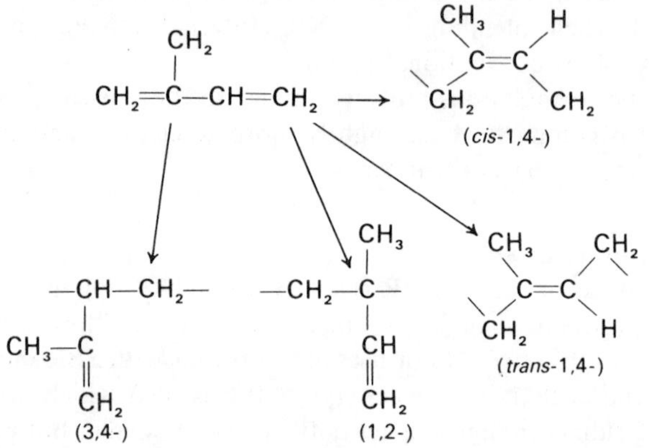

$CH_2{=}C(CH_3){-}CH{=}CH_2$ →

(*cis*-1,4-): $-CH_2(CH_3)C{=}CH(H)CH_2-$ with CH_3 and H on the same side

(*trans*-1,4-): $-CH_2(CH_3)C{=}C(H)CH_2-$ with CH_3 and H on opposite sides

(3,4-): $-CH(C(CH_3){=}CH_2)-CH_2-$

(1,2-): $-CH_2-C(CH_3)(CH{=}CH_2)-$

In this case the 1,2- and 3,4-structures are quite different and because each has an asymmetric carbon atom there are six stereoregular linear polyisoprenes, namely:

cis-1,4-polyisoprene,
trans-1,4-polyisoprene,
isotactic 1,2-polyisoprene,
syndiotactic 1,2-polyisoprene,
isotactic 3,4-polyisoprene,
syndiotactic 3,4-polyisoprene.

As with the polybutadienes, further variations are possible because of head-to-tail and head-to-head/tail-to-tail alternatives which could have a regular repeat unit. In reality in the synthetic materials the above forms co-exist in a variety of combinations and proportions. For synthetic rubbers, interest is confined to polymers with a *cis*-1,4-content in excess of 90%.

In order to achieve these levels of stereoregularity one of two catalyst systems is used:

1. Alkyllithium systems. Although these are capable of giving *cis* levels as high as 98% under ideal conditions, in practice commercial materials are usually about 92–93% *cis* and *c*. 7% 3,4. This system is only used by Shell to produce the Cariflex IR range.
2. Ziegler–Natta systems using a titanium halide such as the tetrachloride with a trialkylaluminium or a poly(*N*-alkylimino-alane). These give *cis* contents of some 96–98%, the residue again being largely 3,4-units.

6.3.2 Properties

The synthetic polyisoprene rubbers differ from natural rubber as a result of four factors:

1. Stereoisomeric purity;
2. molecular weight and molecular weight distribution;
3. presence of functional groups attached to the polymer chain;
4. impurities.

All of the synthetic materials are sterically less perfect than the natural rubber polymer, which is virtually 100% *cis*. Consequently they exhibit less crystallinity both on cooling and on stretching. Whilst this may be advantageous in lowering storage hardening, it does mean that unvulcanized compounds have lower green strengths and vulcanizates lower tensile strength, tear strength, modulus, cut growth resistance, abrasion resistance and fatigue life. On the other hand, compression set (e.g. after 22 h at 70°C) is significantly lower.

Molecular weights, and for that matter molecular weight distributions, for IR are usually lower than for NR, as indicated in Table 6.5 for some randomly selected grades (from Shuttleworth & Watson, 1981).

TABLE 6.5
Typical molecular weight data for natural and synthetic polyisoprene rubbers

Polyisoprene	$\bar{M}_w$	$\bar{M}_n$	$\bar{M}_w/\bar{M}_n$
Natural rubber (RSS1)	1 700 000	333 000	6·4
Natsyn 2200 (Goodyear)	763 000	333 000	2·3
Nipol IR2200 (Nippon Zeon)	775 000	233 000	3·3
Carom 2230 (Romanian)	330 000	98 000	3·3
Cariflex IR305 (Shell)	1 048 000	253 000	4·1

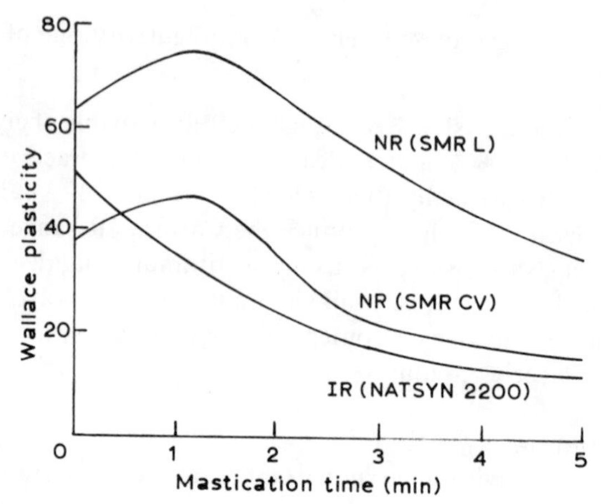

FIG. 6.3. Change in Wallace rapid plasticity with time during open mill polymer mastication (mill friction ratio 1:1:2, starting temperature 40°C). (From Shuttleworth & Watson (1981).)

A further difference between the natural and synthetic materials is the absence of 'aldehyde' groups in IR which have such an influence on such factors as mastication and storage hardening in the natural product.

All of the above factors (steric purity, molecular weight and lack of aldehyde groups) together affect mastication. Whilst IR does break down on mastication, it starts at a lower level than for conventional NR and does not show the initial increase in viscosity observed with natural rubber. Although constant-viscosity NR may start off with the lowest viscosity this advantage is offset by the initial increase (Fig. 6.3).

The non-rubber constituents present in synthetic polyisoprenes differ from those in natural rubber. In the synthetic material they consist largely of small catalyst residues together with added stabilizer so that both ash and acetone extract values are low and in the range 1–1·5%. As shown in Chapter 4, natural rubbers contain a variety of materials which may total up to 6%. Such additional materials, which include proteinous matter, may accelerate cure and have a positive effect on ageing. On the other hand, they do seem to be responsible for adverse effects on compression set and creep. Unless deproteinized, NR also has a lower electrical resistivity than the synthetic polymer.

Outside the Soviet bloc the range of synthetic polyisoprene rubbers available is restricted and may be generally classified as one of two types:

1. Low-*cis* (*c.* 92%) rubbers;
2. high-*cis* (*c.* 96%) rubbers.

The low-*cis* rubbers are significantly less crystalline than the natural product. They thus show, compared with NR, lower storage hardening due to crystallization, lower green strength, tensile strength, tear strength, fatigue resistance, abrasion resistance, cut growth and flexing resistance. The high-*cis* are intermediate between low-*cis* and natural rubber.

Besides showing improved compression set resistance the synthetic materials are also superior to NR in low-temperature flexibility, resistance to anaerobic ageing and creep resistance.

6.3.3 IISRP Numbering System

The International Institute of Synthetic Rubber Producers has the following numbering system for stereoregular isoprene rubbers:

Dry polymer	2200–2249
Oil-extended	2250–2299
Black masterbatch	2300–2349
Oil–black masterbatch	2350–2399
Latex	2400–2449
Miscellaneous	2450–2499

At the time of writing these numbers are only being used for dry polymers.

6.3.4 Processing

Because of the lower stereoregularity and hence crystallization level at normal ambient temperatures, IR shows little of the tendency to storage hardening shown by the natural product. On mill-mixing, initial banding is rapid and it is desirable to complete the mixing as soon as possible to avoid overmastication. This both makes dispersion difficult and, because of the high chain end concentration, leads to a low mechanical strength. However, because the material is less nervy, there are fewer voids in internal mixers, which may therefore operate to a higher effective capacity and a larger batch may be required to prevent the ram from bottoming. Premastication is usually unnecessary.

Care is needed to obtain a good dispersion. Amongst practices that may be used are:

1. Use of masterbatches in NR;
2. early addition of sulphur;
3. use of additives in liquid form wherever possible;
4. use of additives with low melting points and which may be dispersed in the liquid state.

In comparison with NR the synthetic material does not contain any fatty acids which have an activating effect on cure; thus rather higher additional levels may be required. The absence of natural accelerators also results in somewhat slower cures. It has been shown that the addition of DPG to an MBT-accelerated sulphur system has an even greater effect in IR than in NR, with the result that these synergistic accelerator systems give very similar results in NR and IR compounds.

Conventional black stocks made from IR instead of NR tend to show greater scorch safety, similar or slightly longer cure times and lower rheometer modulus. The addition of a small amount of a secondary accelerator such as 0·1–0·2 pphr of TMTD will usually bring the cure characteristics of the IR compound into line with that of NR.

6.3.5 Applications

As already mentioned the greatest use of IR is in the Eastern bloc of nations where it reduces dependence on natural rubber. Outside this area the use of synthetic polyisoprene rubber has become very limited.

At one time an important application area was in tyre compounds, usually in blends with NR. Advantage was taken of the improved chunking resistance and lower heat build-up in truck tyre shoulders and other tyre parts where lower heat build-up in processing is important. This advantage has not, however, been sufficient to merit the price differential of the synthetic over the natural material in recent years.

At the present time the main application of IR is believed to be a non-rubber one, namely in the preparation of chlorinated and isomerized rubbers for the surface coatings industry. IR is also of interest in the pharmaceutical industry because of the high purity of the product compared with natural rubber. More generally the material is of interest because of its purity, processability and uniformity.

Other rather specialized uses include golf ball threads, where the low modulus is of advantage, and for adhesive applications, where the light colour is a desirable property. Where low compression set and low creep are of importance in engineering applications IR may be considered, but at the time of writing it is not of great importance for largely economic reasons.

It has to be said that the commercial development of IR has been disappointing and many plants are now in a 'moth-balled' condition. This is due partially to unfavourable manufacturing economics and partly to the fact that the synthetic materials do not yet attain the steric purity and hence singular properties of the natural product.

6.4 POLYPENTENAMERS AND POLYOCTENAMERS

The polypentenamers and polyoctenamers provide two interesting examples of polyalkenamers prepared by a ring-opening process. The term *polyalkenamer* has been applied to hydrocarbon polymers of the type

$$\left[\mathrm{CH{=}CH{+}(CH_2)_x}\right]_n$$

This definition is usually extended to embrace polymers in which one or more of the hydrogen atoms in the basic polyalkenamer structure are substituted by alkyl, aryl, halogen or other groups. Thus polyisoprene and polychloroprene, as well as polybutadiene, may be considered as polyalkenamers.

Whilst these three materials are usually prepared by polymerization of a diene, polybutadiene may also be prepared by the ring-opening of cyclobutene. This is a comparatively rare example of a polymer that may be produced either by double-bond polymerization or by ring-opening (another is polyoxymethylene).

The development of the ring-opening process has enabled new polyalkenamers to be prepared. Those of interest have low T_gs, are regular and thus potentially crystallizable and furthermore are capable of sulphur vulcanization.

The ring-opening process involves a metathesis reaction in which two cyclo-olefinic double bonds are simultaneously cleaved, this being followed by transalkylidenation of the fragments. Thus for cyclopentene the process may be represented as

```
     H2             H2                H2                  H2
     C--CH     HC--C                  C--CH.....HC--C
    /    ||    ||   \                /     :      :   \
 H2C     ||    ||    CH2   --->   H2C      :      :    CH2
    \    ||    ||   /                \     :      :   /
     C--CH     HC--C                  C--CH.....HC--C
     H2             H2                H2                  H2
                                                |
                                                v
                                      H2                  H2
                                      C--CH=HC--C
                                     /            \
                                  H2C              CH2
                                     \            /
                                      C--CH=HC--C
                                      H2                  H2
```

In turn the resultant cyclodecadiene may take part in similar reactions leading to the production of progressively larger and larger ring structures (sometimes referred to as macrocyclic structures). The catalyst systems are reminiscent of Ziegler–Natta catalysts. In a typical system there are three components:

1. A halide or acetylacetonate of tungsten, molybdenum, tantalum or rhenium;
2. a trialkylaluminium or dialkylaluminium halide;
3. an activator such as epichlorhydrin or 2-chloroethanol.

Polymerization is carried out in the range −50°C to 0°C. The final products are not necessarily macrocyclic because of bond scission.

Because of the low chain stiffness and low interchain attraction the linear unsubstituted polyalkenamers have low T_g values. The occasional double bonds have the effect of increasing the flexibility of adjacent single bonds and overall this leads to a further reduction in the glass transition temperature. In the sequence from polydecenamer down to polypentenamer an increase in double bond concentration leads to a lowering of the T_g. The T_g of polybutenamer (i.e. polybutadiene) is however higher, presumably because the proportion of stiff links becomes significantly high to override the flexibilizing effects on the single bonds.

6.4.1 Polypentenamers

Initial interest in these polymers centred on polypentenamers with their very low glass transitions. As can be seen from Table 6.6, the melting point of the *cis* polymer is too low for the material to be self-reinforcing and main interest centred on the *trans* polymer.

The *trans* polymer showed high green strength, good building tack and generally good processability. Vulcanizates also showed good strength, abrasion resistance and, for a diene rubber, good resistance to reversion, ozone resistance and related crack growth as well as general ageing resistance. In the 1970s there were high expectations that the material would emerge as a major tyre rubber. These hopes were not realized and

TABLE 6.6
The major transitions of polypentenamers

	cis (*99% pure*)	trans (*85% pure*)
T_g, °C	−114	−97
T_m, °C	−41	+18
T_g/T_m, K/K	0·69	0·61

manufacture ceased after only a short period in production, apparently because of both some deficiencies in requirements for a tyre rubber such as low skid resistance and low road-holding characteristics of tyres made from the rubber, and of the uncompetitive costs of production.

6.4.2 Polyoctenamers

Whilst originally the polyoctenamers were considered to have little potential, it is these materials rather than the polypentenamers which have become established. First reported in 1966 (Natta *et al.*) they were introduced commercially by Hüls in 1980 under the trade name Vestenamer. Both the glass transition and the crystalline melting point (Fig. 6.4) are influenced by the *trans* content and at the time of writing grades available have values of 62% and 80%.

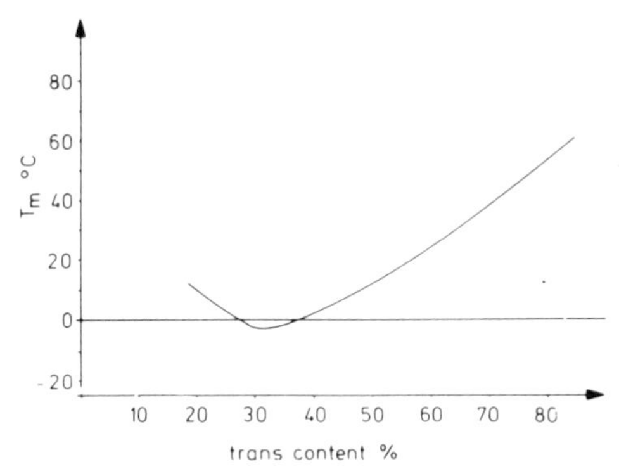

FIG. 6.4. Dependence of melting temperature on *trans* content. (From Dräxler (1983), reproduced by permission of *Kautschuk und Gummi Kunststoffe*.)

The original 80% *trans* grade has a T_g of −65°C and a T_m of 55°C. Such a high melting point results in substantial crystallinity at normal ambient temperatures and the material is a solid. If blended into diene rubbers, however, it provides blends of high green strength. The polyoctenamers available have low molecular weights and very low Mooney viscosities (ML 4 at 60°C of 12 and ML 4 at 100°C of 5 for Vestanemer 8012). This makes them difficult to process alone but it can be very useful in blends in order to improve the flow properties. In these respects the materials may be thought of as vulcanizable processing aids. In SBR the presence of the polyoctenamer increases modulus and hardness, and may increase resilience, abrasion resistance and tear resistance but with some loss of tensile strength. It follows that interest in the 80% *trans* materials arises from the following combination of properties:

1. Low melt viscosity which facilitates processing of blends, particularly those requiring a high vulcanizate hardness and which are required to be injection-moulded;
2. high green strength of blends where there is extensive handling of unvulcanized products, as in tyre building;
3. good physical properties, particularly modulus, tear resistance, abrasion resistance and resilience.

The 62% *trans* material has a low crystallinity and increases neither green strength nor vulcanizate hardness but does improve processability.

The main application areas being developed for polyoctenamers are:

1. As a blend component in tyre compounds;
2. hoses including reinforced high-pressure hose;
3. injection moulding of compounds of high vulcanizate hardness, e.g. rubber coatings on rollers;
4. microcellular rubbers of improved compression set resistance.

Polyoctenamers are usually used at levels of 10–30 pphr. Vulcanization rate and accelerator requirements are comparable with those of a slowly curing SBR.

6.5 POLYNORBORNENES

The polynorbornene rubbers were announced by the French company Société Chemique des Charbonnages (CdF Chemie) in 1975 following some 10 years of research on the Diels–Alder reactions of olefins and cyclopentadiene. The material is marketed as Norsorex with manufacturing capacity reported as 5000 tpa.

The monomer bicyclo[2.2.1]heptene-2 (usually known as norbornene) is produced by the Diels–Alder addition of ethylene to cyclopentadiene:

$$\begin{array}{ccccccccc} & CH & & & & & & CH & \\ CH & & & & CH_2 & & CH & | & CH_2 \\ | & & CH_2 & + & \| & \longrightarrow & \| & CH_2 & | \\ CH & & & & CH_2 & & CH & | & CH_2 \\ & CH & & & & & & CH & \end{array}$$

The monomer is polymerized by a ring-opening mechanism to give a very high-molecular-weight (>3 000 000) polymer with a repeating unit containing a ring structure and a double bond. Both *cis* and *trans* structures are obtainable according to the choice of catalyst used (Le Delliou, 1977):

```
          CH2
       __/   \        __CH=CH~~
~~CH          \    /
   |            CH
   |           /
  CH2         /
     \__ CH2
```

As might be expected for a structure with a five-membered ring within the main chain, the polymer does not show rubbery behaviour at normal ambient temperatures. It is only when the material is heated above the glass transition temperature of +35°C that the material begins to become rubbery. However, it may be plasticized with hydrocarbon oils and plasticized polynorbornenes are rubbery, with glass transition temperatures as low as −60°C. The plasticizer may also reduce the Mooney ML 1+4 (100°C) value from 150 to about 40.

In many ways polynorbornene is reminiscent of PVC. It is supplied as a white dry free-flowing powder and whilst it may be blended with additives on a mill or internal mixer it may also be compounded using dry blend techniques first developed for PVC. This not only reduces energy requirements but also the amount of polymer deterioration during processing. Very large amounts of plasticizer may be added and products can have a very low hardness (down to Shore A 10) yet retain reasonable tensile strength. The manufacturers have also placed particular emphasis on the dynamic properties which are appropriate for vibration damping and impact absorption.

In addition to being able to accept high levels of plasticizer, the polynorbornenes can also accept high levels of fillers such as carbon blacks, silicas, clays, whitings and even asbestos and cork. The suppliers have claimed that up to 1500 parts of filler may be added per hundred parts of polymer.

Whilst the polymer–plasticizer blends have a high green strength it is usual to vulcanize the rubber. Conventional accelerated-sulphur systems may be used, although the manufacturers usually recommend efficient vulcanization systems such as high CBS–low sulphur or a conventional accelerator with dithiomorpholine.

The mechanical properties of polynorbornene rubbers clearly depend on the type and amount of additive used. Some idea of the range of properties available is however obtained by considering the effect of changing the type and amount of filler in the following formulation.

	pphr
Polynorbornene	100
Zinc oxide	5
Stearic acid	1
Filler	200
Low-viscosity aromatic oil	180
Paraffinic oil	20
CBS	5
Sulphur	1·5

These effects are illustrated in Table 6.7.

The ageing properties of plasticized polynorbornene are highly influenced by the type of plasticizer oil used. Aromatic and naphthenic oils are rather volatile at elevated temperatures but the paraffinic oils are somewhat more resistant. Furthermore, it has been found that low-reinforcing fillers generally confer better ageing resistance than the more conventional reinforcing fillers. Ozone resistance may be improved by the use of a combination of antiozonant and a microcrystalline wax.

TABLE 6.7
Effect of fillers on the mechanical properties of Norsorex-based compounds (From Le Delliou (1977), reproduced by permission of the Plastics and Rubber Institute)

	HAF N330	*FEF N550*	*GPF N660*	*MT N990*	*Clay*	*Whiting*
Mooney viscosity, ML 1+4 (100°C)	95	85	80	45	40	39
Optimum curing time at 155°C, min	9	11	12	16	18	16
Tensile strength, MN m^{-2}	20·5	17·0	15·5	15·0	14·5	12·0
Elongation at break, %	330	350	380	510	520	580
Modulus, MN m^{-2}						
100%	2·5	2·5	2·2	0·7	0·8	0·5
300%	18·5	15·0	14·0	6·0	3·8	0·9
Hardness (Shore A)	55	53	50	32	33	25
Density	1·20	1·20	1·20	1·20	1·30	1·32
Tear resistance, MN m^{-1} × 10^{-3}	30	32	35	27	24	11
Rebound resilience at 20°C, %	18	26	32	55	55	59
Non-brittle temperature, °C	−38	−38	−38	−38	−38	−38
Compression set 22 h at 70°C, %	14	14	12	12	26	25

Interest in the material results largely from:

1. The possibility of obtaining solid products of very low hardness;
2. the ability to produce products of interesting damping characteristics;
3. the ability to offset the somewhat high polymer price by the use of very large quantities of oils and fillers and the use of dry blend mixing processes with their low energy requirements.

Applications include grommets, seals, engine mounts and bumpers. Other applications are for the production of filler masterbatches and as an aid for the absorption of oil and chemical spillages.

BIBLIOGRAPHY

Polybutadiene

Saltman, W. M. (1965). Butadiene polymers. In *Encyclopaedia of Polymer Science and Technology*, Vol. 2. Wiley, New York, pp. 678–754.

Tate, D. P. & Bethea, T. W. (1985). Butadiene polymers. In *Encyclopaedia of Polymer Science and Technology*, Vol. 2, 2nd edn. Wiley, New York, pp. 537–90.

Polyisoprene

Schoenberg, E., Marsh, H. A., Walters, S. J. & Saltman, W. M. (1979). *Rubber Chem. Technol.*, **52**, 526.

Shuttleworth, M. J. & Watson, W. W. (1981). In *Developments in Rubber Technology—2*, ed. A. Whelan & K. S. Lee. Applied Science, London, Chapter 8.

Polyalkenamers

Dall'Asta, G. (1974). *Rubber Chem. Technol.*, **47**, 511.

Polynorbornenes

Le Delliou, P. (1977). *International Rubber Conference*, Brighton, UK, May 1977. Preprints produced by the Plastics and Rubber Institute.

REFERENCES

Alliger, G. & Weissert, F. C. (1968). *Polymer Chemistry of Synthetic Elastomers*, Part 1, ed. J. P. Kennedy & E. G. M. Törnqvist. Interscience, New York, Chapter 3.

Bahary, W. S., Sapper, D. I. & Lane, J. H. (1967). *Rubber Chem. Technol.*, **40**, 1529.

Blümel, H. (1964). *Rubber Chem. Technol.*, **37**, 408.

Dräxler, A. (1983). *Kautschuk u. Gummi Kunst.*, 1037.
Haws, J. R., Nash, L. L. & Wilt, M. S. (1975). *Rubber Industry*, **9**, 107.
Le Delliou, P. (1977). *International Rubber Conference*, Brighton, May 1977. Plastics and Rubber Institute, London.
Natta, G., Dall'Asta, G., Bassi, I. W. & Carella, G. (1966). *Makromol. Chem.*, **91**, 87.
Shuttleworth, M. J. & Watson, W. W. (1981). In *Developments in Rubber Technology—2*, ed. A. Whelan & K. S. Lee. Applied Science, London, Chapter 8.

Chapter 7

Ethylene–Propylene Rubbers (EPM and EPDM)

7.1 INTRODUCTION

Between 80 and 85% of raw rubber produced (outside the Centrally Planned Economy Countries) consists of the diene rubbers discussed in the three previous chapters. The dominance of these materials is a result of a combination of such factors as suitability in tyre applications, comparatively low raw-material costs and well-established technologies. Outside the tyre industry, however, there are a number of rubbers which proportionally become more important. Prominent amongst these are the ethylene–propylene rubbers which with non-CPEC production of about 500 000 tpa only account for about 5% of total production but about 13% of the non-tyre market.

Such importance is a result of their possessing a number of properties in which they are superior, not only to the diene rubbers but also to the butyl rubbers (see Chapter 8), in respect of heat and chemical resistance and electrical insulation behaviour.

The ethylene–propylene rubbers originated from the work of Ziegler and of Natta on the development of stereospecific catalysts for the polymerization of alkenes. About 1954 high-density polyethylene was introduced using Ziegler catalysts whilst a modified catalyst system developed by Natta led to the introduction of polypropylene in 1957. Crystalline thermoplastics rather than rubbers, these two homopolymers could chemically be considered as high-molecular-weight alkanes (paraffins). As such they exhibited excellent resistance to a very wide range of chemicals, including a better resistance to oxidation than the diene rubbers, excellent electrical insulation characteristics over a wide range of temperature and frequency, and a resistance to polar solvents. Furthermore, they were produced from low-cost monomers, making the polymers amongst the cheapest of plastics materials. Unsurprisingly the two monomers were soon copolymerized and details of a rubber known as C23 Rubber were published in 1959. Produced by Montecatini, this product was the forerunner of the ethylene–propylene binary copolymers today designated as EPM rubbers.

Less regular in structure than either polyethylene or polypropylene, the copolymer was rubbery but because the copolymerization was somewhat non-random there was some residual crystallinity.

Because of the absence of double bonds, conventional accelerated sulphur vulcanization was out of the question and peroxides soon became the established curing agent. In the event sulphur was found by 1961 to have a useful role as a coagent for the peroxide although today it has largely been replaced by other chemicals.

The rubber industry was, however, reluctant to accept peroxide vulcanization on a large scale. Not only could stock contamination be a severe problem, but in addition compounds were scorchy whilst the peroxides then available had an unpleasant odour. It was therefore not surprising that by 1961 ternary copolymers (terpolymers) had been produced with a double bond in the polymer structure to facilitate a sulphur-based cure. What was particularly interesting was that the double bond did not form a weak link in the main chain but occurred either as part of a ring structure (**I**) or on a side chain (**II**):

This meant that if the double bond was attacked by oxygen or ozone the effect would be less catastrophic than with normal diene rubbers, particularly where the reaction led to chain scission. Today the terpolymers comprise the bulk of the ethylene–propylene rubbers that are marketed and are given the designation EPDM.

Overall there are about 13 manufacturers marketing around 200 grades, about 10% of which are EPM. The more important EPDM grades vary in such properties as ethylene–propylene ratio, type and amount of third monomer, molecular weight, microstructure and whether or not oil-extended.

In some respects the ethylene–propylene rubbers have failed to achieve early expectations. When first introduced there were grounds for believing that these rubbers, particularly the EPDMs, would achieve the status of general-purpose elastomers. The monomers were cheap and readily available; the rubber had an acceptably low T_g, was slightly crystalline and hence exhibited good tensile properties, and (by no means least) had very good ozone, oxygen and heat resistance. In the case of the terpolymers, more or less conventional curing systems could be employed and compounding presented no great difficulties. However, whereas with SBR global situations leading to shortages of natural rubber (World War II and the Korean War) stimulated technical and commercial development, the development of ethylene–propylene rubbers has been restrained by the ready availability of general-purpose materials produced under highly competitive conditions.

On economic grounds the cost of the termonomer has been a significant factor, as have certain aspects of production costs, so that EPDMs are significantly more expensive than natural rubber or SBR. Technical problems such as the lack of building tack have proved a deterrent to their use in tyre-building operations whilst problems of compatibility and covulcanization in the case of a far-from-impossible stock mix-up have also been significant.

As was pointed out in the first paragraph of this chapter, the rubbers have however become important in the non-tyre sector, particularly in what is known as the general rubber goods industry. In addition there has been generally steady growth production not suffering the serious decline experienced by the synthetic diene rubbers in the early 1980s. EPDM has in fact weathered the economic storms of recent years better than most other major rubbers.

7.2 MANUFACTURE

The ethylene–propylene rubbers are produced by a polymerization process using anionic co-ordinated catalyst systems developed by Natta and his research team at Milan. These systems are based on a combination of a derivative of titanium or vanadium as catalyst in conjunction with an aluminium alkyl as co-catalyst (e.g. $Al(C_2H_5)_2Cl$). Whilst titanium compounds are successfully used for the manufacture of polyethylene and polypropylene they give blocky copolymers with poor rubbery properties and were soon abandoned in favour of vanadium derivatives such as $VOCl_3$. More recently (Corbelli *et al.*, 1980) there has been some renewed interest in titanium-based systems.

Ethylene is much more reactive than propylene and unless special steps are taken the copolymer molecules formed at the beginning of the reaction will have a much higher content of ethylene units than copolymer molecules produced at the end of the reaction (Natta *et al.*, 1957; Lukach & Spurlin, 1964; German *et al.*, 1966). According to Natta *et al.* (1969) and

Ichikawa (1965), if the catalyst is the same then polymer composition is independent of polymerization time, catalyst ageing time, catalyst concentration and Al:V ratio. The nature of the organometal compound also plays a minor role as regards relative reactivities. Some reservations on the validity of these statements have been made (Baldwin & VerStrate, 1972).

Average molecular weight increases with reaction time, with decrease in reaction temperature, with decrease in catalyst concentration, with decrease in monomer concentration and with increase in ethylene/propylene ratio. It also depends on the nature of the alkyl aluminium compound and on the Al:V ratio.

Polymerization may be carried out in solution or in suspension. In the first case the monomers are added either to a hydrocarbon (e.g. hexane) or to a halogenated hydrocarbon which has been carefully freed from traces of water which would otherwise destroy the catalyst. Reaction begins on addition of the catalyst and the polymer dissolves in the solvent as soon as it is formed. After polymerization the solution is washed to remove residual catalyst and the polymer is separated from the solvent by boiling water. The polymer remains in suspension as a crumb in the water from which it is subsequently separated, dried and baled. Molecular weight is controlled by addition of regulators such as hydrogen or zinc alkyls.

In the suspension process one of the monomers, propylene, which is in liquid form, also acts as the reaction medium. It is not a solvent for the polymer, which on formation remains suspended in the propylene. Because of the absence of solvent and because of the high conversions possible there is more efficient use of reactor space and high specific productivities are attainable. In addition, there are no complications arising from the high viscosity of polymer solutions as in the solution process. Other advantages are the ease by which reaction heat may be removed by evaporation of the liquid propylene, the economical rate of use of catalyst, the ability to produce high-molecular-weight polymer and the facility to control structure and resultant properties (Corbelli, 1981).

7.3 STRUCTURE AND PROPERTIES

Like the diene rubbers, the ethylene–propylene rubbers are hydrocarbons. They are therefore resistant to polar solvents but dissolve (when unvulcanized) or swell (when vulcanized) in hydrocarbons. Being saturated, they are somewhat inert chemically and therefore have good resistance to oxygen, ozone, acids and alkalies. Besides being attacked by peroxide radicals they may also be halogenated.

The thermal properties are of interest both in theory and practice. Whilst an amorphous material begins to show rubbery behaviour above the T_g, in the case of crystalline polymers the high elasticity will depend on the degree of crystallization and may not become manifest until the T_m is approached. Both the T_g and the extent of crystallization will depend on the monomer ratio.

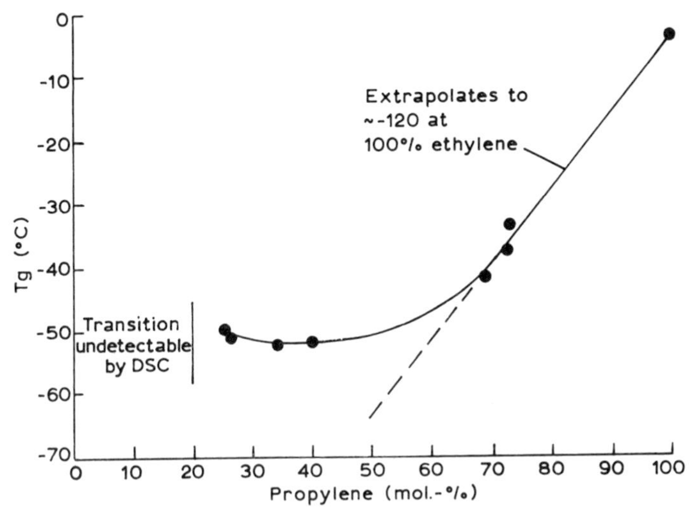

FIG. 7.1. Relationship between T_g and polymer composition. Data from differential scanning calorimeter (10°C min^{-1}) without scanning rate or thermocouple corrections. (Reproduced by permission from *Rubber Chem. Technol.*, **45** (1972) 709. Copyright American Chemical Society.)

The T_g is at its lowest with a propylene content of about 40% (Fig. 7.1). Since rubberiness at any temperature T tends to increase with an increase in the value of $(T - T_g)$ the rubbers with the lowest T_g tend to be the most rubbery at normal ambient temperatures.

As the proportion of ethylene units in the chain increases beyond parity ethylene sequences begin to occur in the chain which are capable of partaking in crystallization. The extent of crystallization will also depend on the manner in which the monomer units are distributed, i.e. whether they are random or blocky, but the general effects are clearly shown in Fig. 7.2.

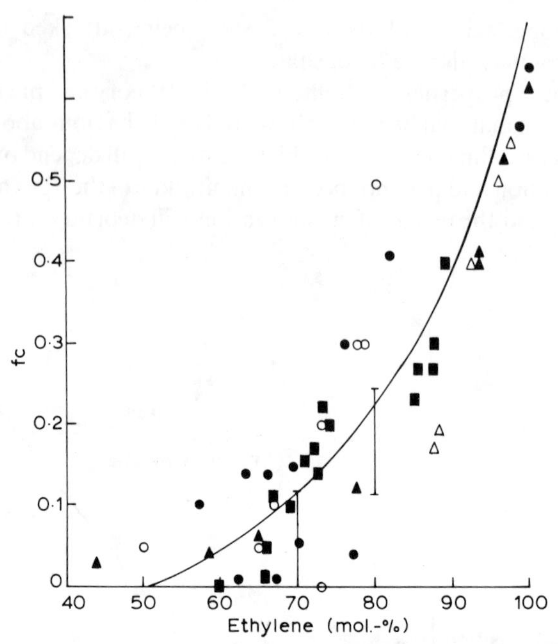

FIG. 7.2. Dependence of crystalline fraction (f_c) in ethylene–propylene rubbers on composition. The wide spread of results is ascribed to compositional heterogeneity and variations in sequence length distributions. Data collected by Baldwin & VerStrate from various sources. (Reproduced by permission from *Rubber Chem. Technol.*, **45** (1972) 709. Copyright American Chemical Society.)

Because of their measure of crystallization, the 'high-ethylene' rubbers are noted for their very good green strength.

In contrast, an increase in the proportion of propylene units beyond the point of parity does not lead to an increase in crystallinity because the catalyst systems in use do not normally lead to stereoregular polypropylene blocks in the chain.

As with the stereoregular diene homopolymers (e.g. natural rubber), crystallinity is induced by stretching, the crystallinity increasing with increasing degree of stretch. Whereas one school of thought (Bassi *et al.*, 1970) believes that crystallinity becomes independent of stress above a critical extension (Fig. 7.3), VerStrate & Wilchinsky (1971), using uncross-linked samples, arrived at results indicating that crystallinity only became dependent on stress above a critical extension (Fig. 7.4).

Uncross-linked ethylene–propylene gum stocks can have high tensile

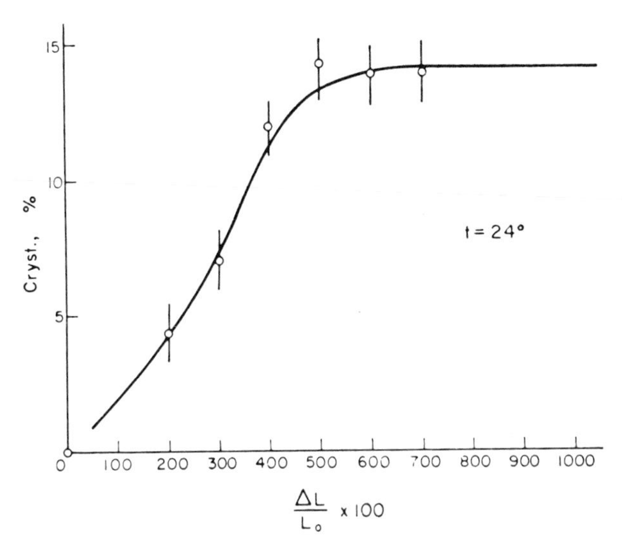

FIG. 7.3. Effect of extension on crystallinity according to Bassi *et al.* (1970). (Reproduced by permission of Pergamon Press.)

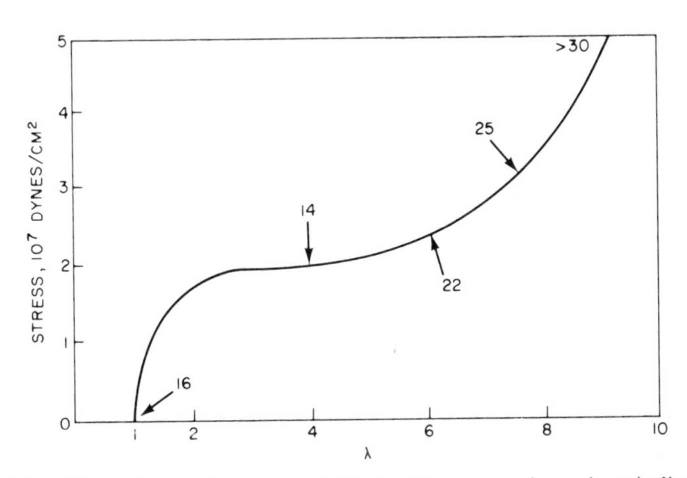

FIG. 7.4. Effect of extension on crystallinity. The arrowed numbers indicate the percentage crystallinity at various points on the stress–strain (λ) curve. (Reproduced by permission from *Rubber Chem. Technol.*, **45** (1972) 709. Copyright American Chemical Society.)

strengths (>2000 psi; 14 MPa), a reflection of the crystallinity in the polymer. The presence of crystallinity is also reflected in the observation that the maxima in the curves of tensile strength against cross-link density are shifted towards zero cross-linking as the amount of crystallinity increases. In certain instances (e.g. a 3:1 ethylene/propylene ratio rubber), the introduction of any cross-links at all reduces the tensile strength, the effect of loss of crystallinity not being fully offset by the insertion of cross-links.

Different curing systems give different strengths for equivalent degrees of cross-linking (Fig. 7.5). It would seem that this is due to two main factors:

1. The influence of curing systems on the scission/cross-link ratio, a high value generating a large proportion of ineffective chain ends thus altering the extensibility/modulus balance which will influence the strength;
2. the relative strength of the bonds in the cross-link.

In the case of terpolymers it has been found (Imoto *et al.*, 1968; Baldwin & VerStrate, 1972) that tensile strength and elongation at break increase

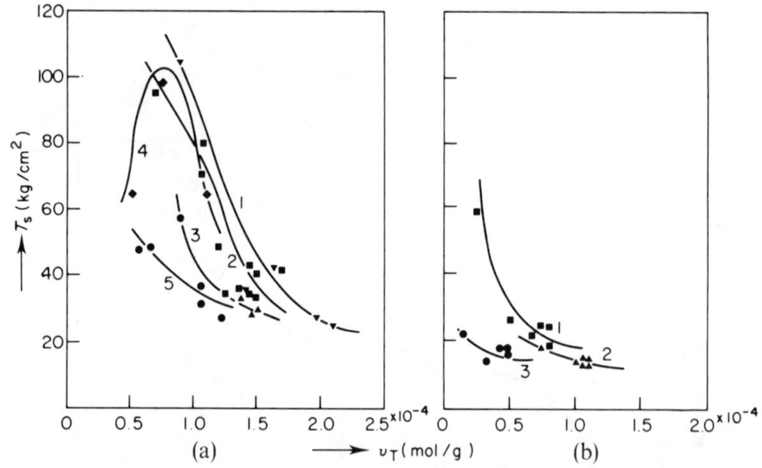

FIG. 7.5. Dependence of tensile strength (T_s) on cross-link density (ν_T) and cross-link structure for (a) EPDM (Royalene 301) and (b) EPM (Dutral). Tested using dumbbell type #3 at 250 mm min^{-1} elongation rate at 20°C. Cure systems: (a) EPDM (1) tetramethylthiuram disulphide, bis(benzothiazolyl) disulphide; (2) dicumyl peroxide, 1 phr sulphur; (3) dicumyl peroxide, 0·3 phr sulphur; (4) mercaptobenzothiazole; (5) dicumyl peroxide; (b) EPM (1) dicumyl peroxide, 1 phr sulphur; (2) dicumyl peroxide, 0·3 phr sulphur; (3) dicumyl peroxide. (From Imoto *et al.*, 1968.)

with an increase in polysulphidic linkage concentration, a result analogous to that observed with diene polymers.

7.4 VARIABLES BETWEEN GRADES

Commercial grades vary one from another, the principal variables being:

1. The ethylene to propylene ratio;
2. the type and amount of third monomer (in the case of EPDM);
3. molecular weight and molecular weight distribution;
4. microstructure;
5. nature of other additives such as oil and stabilizers incorporated by polymer suppliers.

The ethylene/propylene ratio may also be expressed in terms of percentage propylene content or percentage ethylene content, and the latter approach will be used here.

For optimum rubberiness and low-temperature flexibility the ethylene content should be about 50–60%. Higher values of ethylene content improve the green strength as a result of the greater ability to crystallize. Highest green strengths amongst commercial grades are obtained with an ethylene content of around 70%. Intermediate levels provide a compromise. It is of interest to note that whilst most suppliers make available elastomers with a range of ethylene/propylene ratios, different suppliers often have a different emphasis in their product range. Thus the bulk of DSM grades are rated as 'low-ethylene', the bulk of Hüls as 'medium-ethylene' while more than half of the Dutral products are of 'high-ethylene' content.

Over the years many chemicals have been considered as the *cure site monomers*, most of which have been unconjugated dienes. In order that any chemical reaction which causes bond scission did not have a catastrophic effect on the molecular weight, diene monomers are chosen so that the double bond is either part of an in-chain ring or a pendant grouping (see Fig. 7.1).

It is also clearly desirable that the double bond should not become involved in polymerization reactions leading to gelation and cross-linking during polymer manufacture. Additional requirements are that the diene should not seriously influence catalyst activity, should randomly polymerize with the two principal monomers, should be efficiently converted (or easily recycled, or both) during polymerization and should react efficiently during vulcanization.

Bearing in mind these requirements, many classes of unconjugated dienes have been investigated. These include alicyclic, monocyclic, fused-ring cyclic, bridged-ring bicyclic and tricyclic dienes. Of all the specific dienes investigated three have reached commercial status.

1. Dicyclopentadiene—inexpensive but containing methine hydrogens which may adversely affect cross-linking efficiency. (Generally known as DCP, but do not confuse with the curing agent dicumyl peroxide which is given the same abbreviation.)
2. Ethylidene norbornene (2-ethylidenebicyclo[2.2.1]-5-heptane; ENB)—which leads to rapid and efficient vulcanization.
3. 1,4-Hexadiene—which gives good recycling characteristics and ease of recycling (see Table 7.1).

Current practice is tending to increasing use of ENB although the other two dienes are still used commercially. The usual amount used is somewhere between 2 and 5% of the total monomer although up to 10% may be used in fast-curing grades.

The weight-average molecular weight, $\bar{M}_w$, is usually in the range

TABLE 7.1
Principal diene monomers used in EPDM manufacture

Monomer	*Predominant structure present in terpolymer*
1. Dicyclopentadiene 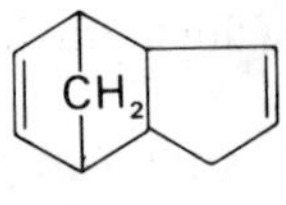 CH_2	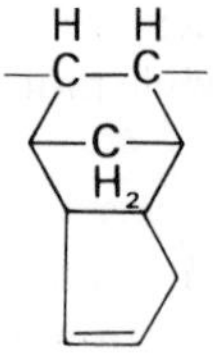 H H —C—C— C H_2
2. Ethylidene norbornene 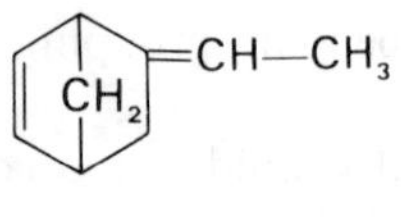 $=CH—CH_3$ CH_2	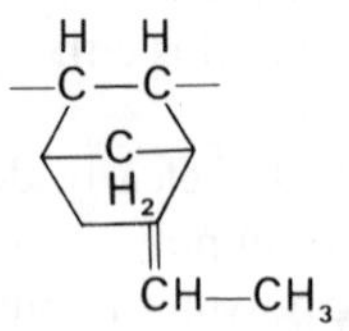 H H —C—C— C H_2 $CH—CH_3$
3. 1,4-Hexadiene $CH_2=CH—CH_2—CH=CH—CH_3$	$—CH_2—CH—$ $CH_2—CH=CH—CH_3$

1×10^5–2×10^5. In practice suppliers normally quote Mooney viscosity values as a measure of molecular weight. Unfortunately, strict comparison is difficult because different suppliers use different temperatures and time intervals for measurement. In addition, some grades are oil-extended so that in these cases comparison cannot be made with Mooney values of non-oil-extended grades to compare molecular weights.

The lower-molecular-weight grades tend to be used with lower amounts of filler to give high-quality products. These grades will typically have Mooney ML (1+3) 100°C figures in the range 20–60. Higher-molecular-weight compounds with Mooney values (ML (1+4) 125°C) up to 200 are frequently oil-extended, particularly at the highest molecular weights, and are used for production of lower-cost compounds.

Because of the method of polymerization, the molecular weight distribution is fairly narrow with values for $\bar{M}_w/\bar{M}_n$ in the range 3–5. The breadth of distribution is sometimes increased by blending grades of different average molecular weight in order to modify the rheological properties.

The properties of ethylene–propylene rubbers are influenced by the way the monomer units are distributed along the chain. In early grades quite long sequences of one monomer occurred and the polymer was referred to as blocky. In the case of long polyethylene sequences, crystallinity was facilitated to the improvement of green strength but to the detriment of rubberiness. There was also the possibility that chains formed early on in a reaction would have a different composition from those formed towards the end of a reaction. Suitable choice of catalyst has allowed a more random distribution as well as reduced variation between chains. It should, however, be noted that some grades are available in which the distribution is designated as sequential and others as block type.

7.5 GENERAL VULCANIZATE PROPERTIES

The ethylene–propylene rubbers find use because of their superiority to the diene rubbers in a number of respects. Amongst the most important are:

1. Excellent resistance to oxidation, ozone attack and, when suitably protected, sunlight;
2. very good heat ageing resistance;
3. excellent electrical insulation characteristics;
4. very good chemical resistance, particularly to alkalies and dilute acids.

The rubbers also exhibit generally good mechanical properties, particularly when reinforced with carbon blacks, but as with the diene rubbers they swell in hydrocarbon liquids. Their low-temperature properties are adequate for most applications.

The weathering resistance of ethylene–propylene rubbers is well established but it is necessary to protect against UV light. This may be done with finely dispersed carbon black which need not be at the high level of use normally employed for reinforcement. Non-black compounds are susceptible to surface damage but this may be reduced by the use of UV-absorbing pigments such as rutile grades of titanium dioxide or organic UV absorbers, although the latter are not very effective.

Ethylene–propylene rubbers perform well up to 125°C in continuous use, whilst suitably selected and compounded they may be used continuously up to 150°C. The highest heat resistance is achieved using EPMs, although peroxide-cured EPDM compounds have a similar performance which is superior to that of accelerated sulphur-cured EPDMs. It is generally considered that the 'high-ethylene' grades have the greatest thermal stability.

Whilst the electrical insulation characteristics of the raw polymer are outstanding, it has to be stressed that these will be adversely influenced by the presence of certain additives such as carbon blacks.

The chemical resistance of the rubber is a consequence of its paraffinic nature. Whilst the vulcanized rubbers are swollen by solvents of similar solubility parameter, such as hydrocarbons, they have good resistance to ketones and other polar solvents.

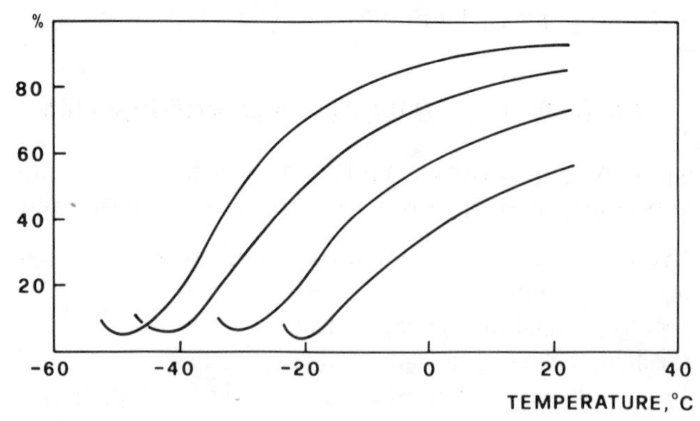

FIG. 7.6. Rebound of EPM with varying propylene content (propylene content increases from left to right). (From Corbelli, 1981.)

Whilst not particularly outstanding, the mechanical properties are more than adequate for most purposes. Thus black-reinforced compounds exhibit a tensile strength of over 3000 psi (20·7 MPa), good-to-excellent abrasion resistance and good tear resistance.

The rebound resilience is interesting in that although at about 5–10°C it is not very different from natural rubber, resilience is less temperature-dependent above this temperature and indeed over 25°C is almost independent of temperature. Values of rebound resilience vary with the ethylene/propylene ratio (Fig. 7.6). This diagram clearly also shows the relationship between resilience at a given temperature and the T_g (Natta *et al.*, 1962).

For many purposes the TR 50 test is often quoted instead of the T_g. In this test an elongated test piece is frozen at a very low temperature well below the T_g, then unclamped and the temperature allowed to rise. The temperature at which the rubber shows 50% recovery is noted as the TR 50 temperature. The effect of ethylene content on the TR 50 temperature is indicated in Table 7.2.

TABLE 7.2
Effect of ethylene content of an ethylene–propylene rubber on the TR 50 temperature

Approximate ethylene content (%)	*TR 50 temperature* (°C)
50	−40
60	−30
70	−18

Mention may also be made of the very low specific gravity of the raw polymer which at 0·86 is amongst the lowest for commercial polymers. At only 70% that of polychloroprene, the need to take into account volume cost is clear.

7.6 PROCESSING

As with many other synthetic rubbers, the ethylene–propylene elastomers are not subject to mastication. Thus flow properties are controlled by the molecular weight of the raw polymer and by the use or otherwise of mineral oils and fillers.

Mill mixing processability varies according to such factors as molecular weight and molecular weight distribution. Mill mixing is particularly difficult with high-viscosity materials. One problem is to ensure good dispersion of filler and in the case of fine fillers it is usually recommended that these be added before the oil. In the case of less fine filler, addition with the oil is possible. To prevent the band going on to the fast roll, the slower roll should usually be kept about 20°C colder.

Upside-down mixing techniques are frequently recommended with internal mixers in order to achieve good dispersion and rapid mixing. In this process oil and filler are first added together followed by the polymer and finally (providing that the mixing chamber is not too hot, i.e. above 120°C) the curatives. Alternatively the curatives may be added on a mill. If the bale is added complete with wrapper it will be essential to raise the temperature above 120°C to melt and disperse the polyethylene, and in this case the addition of the curing system should be a separate operation. Much shorter mixing times may be achieved by using the rubber in crumb, pellet or friable bale form with consequent savings in energy, faster throughput and hence better usage of plant and a greater facility for automation. Disadvantages include the higher cost of material and problems associated with transporting and storing materials which have a much higher volume per unit weight.

In extrusion EPM and EPDM behave like other rubbers in those factors which control characteristics such as die swell and form stability. High-ethylene stocks are usually the easiest to handle after shaping. It is desirable to feed with cool compounds and use a cold screw with low barrel temperatures, typically 66–68°C with dies in the range 90–120°C.

In calendering, typical roll temperatures would be a top roll 90–100°C, middle roll 60–80°C and bottom roll cold. In recent years there has been a substantial demand for uncured sheet and in order that such sheet should have good handling qualities a high green strength polymer is required. Low-density polyethylene may also be incorporated to improve both calendering characteristics and the strength of the finished sheet.

Extrudates and calendered sheet may be cured by any of the methods available generally to the rubber industry, including UHF, high energy and radiation.

7.7 COMPOUNDING

Compounding practice for the ethylene–propylene rubbers differs from that for the diene rubbers in a number of respects. These differences derive from the following peculiar characteristics of the ethylene–propylene materials.

1. EPM rubbers are saturated and are usually peroxide-cured. Such cures may be adversely affected by the presence of acidity and unsaturation in additives, whilst in addition many antioxidants will interfere with peroxide-curing reactions.
2. EPDMs have a low level of unsaturation and may be cured using accelerated sulphur systems. However, because the level of unsaturation is low, it is usual to employ powerful synergistic accelerator combinations. It is, once again, important that additives are free of unsaturation.
3. The saturated, or virtually saturated, hydrocarbon nature of the polymer leads to a low solubility parameter so that many additives are even more incompatible than with diene rubbers and may tend to migrate and bloom. Since, however, most additives have a small finite compatibility in the rubber it is often better, in terms of resistance to blooming, to use a mixture of similarly functioning additives rather than one alone and hence increase the overall additive level at which blooming occurs. Migration may also be reduced by the use of additives of higher molecular weight.

7.7.1 Curing Systems

7.7.1.1 Curing Systems for EPM

The absence of unsaturation in EPM requires that a non-sulphur curing system be employed. Although such techniques as chlorination and chlorosulphonation of the rubber to provide active cure sites, as well as high-energy radiation techniques, have been developed, the usual curing procedure involves the use of peroxides.

The earliest peroxides used suffered from such disadvantages as volatility (as with di-*t*-butyl peroxide) and unpleasant odour (as with dicumyl peroxide). Newer peroxides such as 1,1-di-*t*-butylperoxy-3,5,5-trimethylcyclohexane and 1,4-bis(*t*-butylperoxyisopropyl)benzene have proved more satisfactory (Fig. 7.7).

Peroxides can react with ethylene–propylene rubbers in a number of ways, not all of which lead to cross-linking; indeed, some lead to chain scission (see Chapter 18 for a further discussion). Early in the development of EPM it was found that incorporation of sulphur increased the cross-link/scission ratio, thus improving the efficiency of the peroxide. In this role sulphur has been known variously as a coagent, promoter and adjuvant.

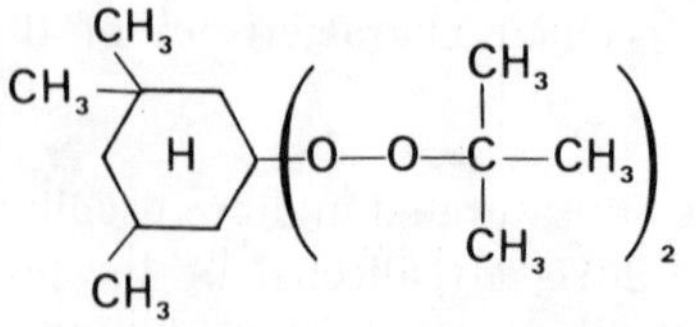

1,1-di-*t*-butylperoxy-3,5,5-trimethylcyclohexane

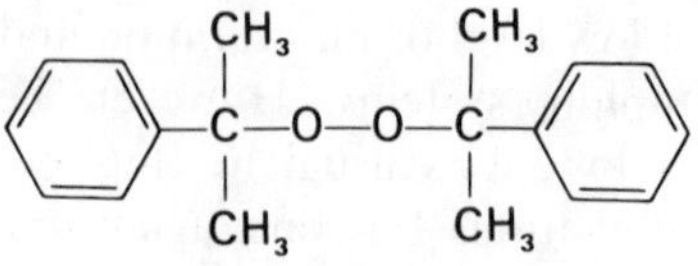

dicumyl peroxide

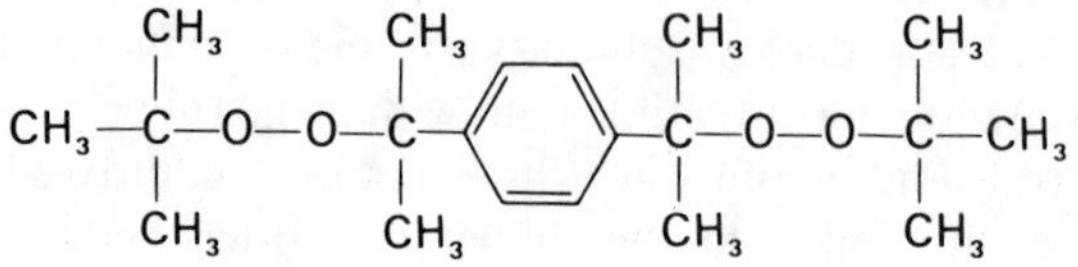

1,4-bis(*t*-butylperoxyisopropyl)benzene

FIG. 7.7. Some peroxides used as cross-linking agents.

Because of its tendency to impart an odour to peroxide-cured rubbers sulphur is no longer important as a coagent, particularly since several other classes of chemical have been found to be effective. These include di- and tri-acrylic and methacrylate esters, divinylbenzene, bismaleimides, cyanurates, polyallyl-substituted aromatic compounds and *p*-quinone

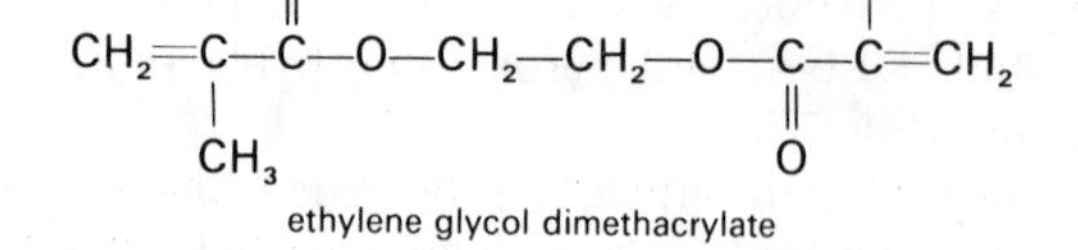

triallyl cyanurate

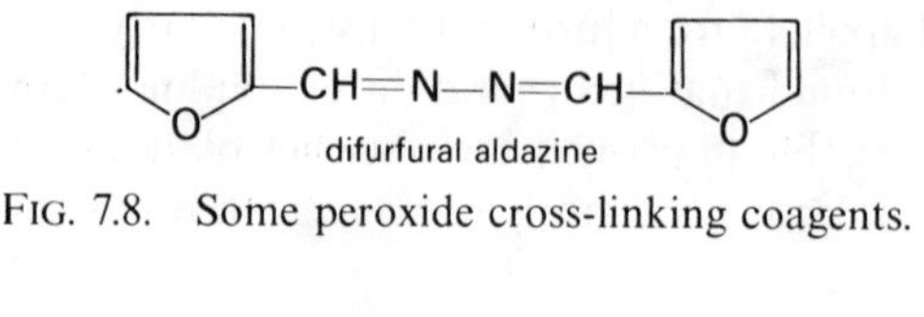

ethylene glycol dimethacrylate

difurfural aldazine

FIG. 7.8. Some peroxide cross-linking coagents.

dioxime. Three specific widely used coagents are given together with their chemical formulae in Fig. 7.8.

Zinc oxide has a necessary but imperfectly understood role in peroxide vulcanization. In part this is probably accounted for by the ability of zinc oxide to neutralize acidic impurities and thus prevent poisoning of the curing system. As with other rubbers, it is useful where optimum heat resistance is required.

Formulations vary according to the specific grade of EPM used and the choice of peroxide but a typical system, based on 100 parts of EPM, would be peroxide 7, coagent 2, zinc oxide 5.

7.7.1.2 Curing Systems for EPDM

EPDM rubbers were introduced primarily to avoid the necessity of peroxide curing with ethylene–propylene rubbers. It is therefore perhaps surprising that peroxide curing of EPDM is commercially practised. There are, however, a number of advantages including the higher temperature resistance, lower compression set, improved electrical insulation, non-staining behaviour and ability to reduce migration problems as compared with accelerated sulphur systems.

Compared with EPM, the cross-linking efficiency of peroxides is higher with EPDM so that less peroxide is required and for many grades the use of a coagent is either unnecessary or counterproductive. For high temperature applications, the main reason for employing peroxide, antioxidants of the type used with EPMs rather than those used with sulphur-cured EPDMs are recommended (see Section 7.7.2).

Accelerated sulphur curing systems require powerful synergistic accelerator systems to cope with the low level of unsaturation in the rubber. The actual system used will nevertheless depend considerably on the type and the amount of the third (diene) monomer used in the EPDM polymer. As has already been mentioned, polymers incorporating ethylidene norbornene are much faster-curing and may even be co-vulcanized with SBR.

One widely used curing system is:

	pphr
EPDM	100
Sulphur	1·5
Zinc oxide	5·0
MBT	0·5
TMTM	1·5
Stearic acid	1·0

A number of other systems have also been proposed involving three or four accelerators in combination.

Whilst EPDMs may be cured using maleimides, quinoid and phenolic resin-cured systems, they do not appear to have achieved commercial significance.

7.7.2 Other Additives

Apart from the high-ethylene grades with their high levels of crystallinity, EPM and EPDM gum stocks have quite low tensile strengths and the rubbers therefore normally require reinforcement with fine fillers. As with the diene rubbers, carbon blacks are widely used for this purpose and apart from some of the conductive blacks are relatively easy to incorporate and disperse.

For insulation purposes mineral fillers such as clays are used. Silicas and fine silicates may be employed to confer improved mechanical properties but may interfere with the cross-linking system and a small quantity of polyethylene glycol may be used to absorb preferentially on to the filler surface and render it inactive. The use of silica fillers may also make it harder to extrude satisfactorily so that it is common practice to use them in conjunction with china clay. Acidity in the silicas and silicates must be avoided where peroxide curing is being undertaken.

As with other rubbers, the use of coupling agents is widely undertaken to improve the performance of mineral fillers, with the choice largely depending on the curing system used. For example, vinyltri(1-methoxyethoxy)silane may be used with peroxides and mercaptopropyltrimethoxysilane with accelerated sulphur cross-linking systems.

The selection of mineral oils to act as plasticizers (softeners) has to be undertaken with care. In the case of peroxide cures naphthenic and aromatic oils may interact with the peroxide and, additionally, have poor compatibility, so that paraffinic oils are normally used. In the case of sulphur-cured EPDMs, both paraffinic and naphthenic oils may be used. Large quantities of oil, up to the weight of the rubber, may be incorporated providing the oil has a low polar impurity level and is of low volatility. Liquid polybutadiene has been found to be both a plasticizer and a suitable coagent for peroxide curing of EPM rubber.

Ethylene–propylene rubbers, particularly when peroxide-cured, have good heat ageing resistance and for many purposes antioxidants are unnecessary. Where further improvement is required for peroxide-cured systems, choice is restricted by the tendency of many antioxidants to react with the peroxides. One effective antioxidant is polymerized dihydroquinoline used in conjunction with zinc oxide and, sometimes, mercaptobenzimidazole. Nickel butyldithiocarbamate is one of a few antioxidants effective with EPDM rubbers that have been sulphur-cured although these would not normally be considered for optimum heat resistance.

As mentioned in Section 7.5, the best protection against sunlight is by carbon black or, in non-black compounds, titanium dioxide. Normal UV absorbers are ineffective.

7.8 APPLICATIONS

The status of ethylene–propylene rubber as the premier non-tyre rubber arises largely from its good ozone, oxidation, heat and chemical resistance and very good electrical insulation properties. The main factors limiting growth are poor hydrocarbon-oil resistance, greater cost than the general-purpose diene rubbers, and the poor building tack.

The largest field of use is in the automotive industry, where it is used in a wide diversity of applications which do not involve direct contact with fuels and lubricating oils. EPDMs have also been used, to some extent, as supplementary elastomers for tyre sidewalls to improve ozone resistance and in butyl-rubber-based inner tubes (see Chapter 8) to reduce cold flow and improve green strength.

Ethylene–propylene rubbers, particularly EPM, are widely used for cable insulation, particularly for low- and medium-tension work but more recently also in high-tension cables. The elastomers are also used for electrical cable terminals and for insulators.

In building, a wide range of products includes door and window seals and waterproofing sheets, both cured and uncured. In the field of appliances the rubbers are used for gaskets and other parts in washing machines, dishwashers and driers. Besides the advantage of good heat ageing resistance, the freedom from the tendency to stain enamelled and patented surfaces and their resistance to detergents and bleaches are important.

The blends with polypropylene and/or polyethylene to yield the so-called thermoplastic polyolefin rubbers are considered in Chapter 16.

BIBLIOGRAPHY

Baldwin, F. P. & VerStrate, G. (1972). Polyolefin elastomers based on ethylene and propylene. Rubber reviews for 1972. *Rubber Chem. Technol.*, **45**, 709 (1325 references).

Corbelli, L. (1981). In *Developments in Rubber Technology—2*, ed. A. Whelan & K. S. Lee. Applied Science, London, Chapter 4.

Natta, G., Valvassori, A. & Sartori, G. (1969). In *Polymer Chemistry of Synthetic Elastomers*, Part II, ed. J. P. Kennedy & E. G. M. Törnqvist. Interscience, New York, Chapter 7B.

REFERENCES

Baldwin, F. P. & VerStrate, G. (1972). *Rubber Chem. Technol.*, **45**, 709.

Bassi, I. W., Corradini, P., Fagherazzi, G. & Valvassori, A. (1970). *Eur. Polym. J.*, **6**, 709.

Corbelli, L. (1981). In *Developments in Rubber Technology—2*, ed. A. Whelan & K. S. Lee. Applied Science, London, Chapter 4.

Corbelli, L., Milani, F. & Fabbri, R. (1980). International Rubber Meeting, Nürnberg.

German, R., Hank, R. & Vaughan, G. (1966). *Kautschuk u. Gummi*, **19**, 67.

Ichikawa, M. (1965). *Int. Chem. Eng.*, **5**, 724.

Imoto, M. & 13 co-authors (1968). *Nippon Gomu Kyokaishi*, **41**, 1095.

Lukach, C. A. & Spurlin, H. M. (1964). In *Copolymerization*, ed. G. Ham. Interscience, New York, Chapter IVA.

Natta, G., Mazzanti, G., Valvassori, A. & Pajaro, G. (1957). *Chim. Ind. (Milan)*, **39**, 733.

Natta, G., Sartori, G., Valvassori, A., Mazzanti, G. & Crespi, G. (1962). *Hydrocarbon Processing*, **41**(9), 261.

Natta, G., Valvassori, A. & Sartori, G. (1969). In *Polymer Chemistry of Synthetic Elastomers*, Part II, ed. J. P. Kennedy & E. G. M. Törnqvist. Interscience, New York, Chapter 7B.

VerStrate, G. & Wilchinsky, Z. W. (1971). *J. Polym. Sci.*, *A-2*, **9**, 127.

Chapter 8

Isobutene–Isoprene Rubbers (Butyl Rubber or IIR)

8.1 INTRODUCTION

First introduced during the Second World War largely for use in inner tubes, butyl rubber and its derivatives have retained an important role as special-purpose rubbers with production similar to the ethylene–propylene rubbers and only exceeded by SBR, NR and BR. Its importance is largely consequent upon the low gas permeability for a rubber. Its other outstanding feature, low resilience, is usually more of a hindrance rather than an asset although it may be utilized in vibration-absorption applications.

Isobutene polymers were first reported in 1873 (Butlerov & Goryaniov) when low-molecular-weight materials were produced by heating the monomer with strong acids. Many years later, workers at IG Farben (Otto & Müller-Cunradi, 1931) found that the use of a Lewis acid such as BF_3 and a polymerization temperature of the order of $-75°C$ led to the production of high-molecular-weight polymer. The mechanism involved was the somewhat unusual one of cationic polymerization, use of which today is largely confined to the manufacture of isobutene polymers and copolymers. The polyisobutene, often abbreviated to PIB, was in due course commercially introduced as Oppanol (initially by IG Farben; post-1945 by BASF), Vistanex (originally Standard Oil, now Exxon Chemical Americas) and as SKP in the USSR. PIB has been used for such purposes as adhesive components, chewing-gum bases, motor oil additives and polyethylene additives.

Although it is rubbery, PIB had minimal use as a rubber since, being saturated, it could not be cured with sulphur, and attempts to cure with peroxides simply led to degradation. In consequence the materials suffered from unacceptably high levels of cold flow and creep. The deficiency was eventually overcome by Sparks & Thomas (1939) through the copolymerization of isobutene with a small quantity of a diene monomer to give a polymer which may be vulcanized by conventional sulphur-based systems. In early formulations butadiene was used and although it led to vulcanizates with slightly higher tensile strength and modulus and less susceptibility to overcure, isoprene is the standard diene monomer used today primarily because it was more convenient to use in the polymerization process (Welsh *et al.*, 1949).

The vulcanizable material was named Butyl Rubber in 1940 and commercial-scale production commenced in 1942. During World War II it was designated GR–I (Government Rubber—Isobutylene) but today the internationally recognized abbreviation is IIR. Outside the Soviet Union the market is controlled by two companies, Exxon (Esso) and associated companies and Polysar.

Because of their low unsaturation, butyl rubbers require powerful curing systems and this makes them incompatible for co-curing with conventional diene rubbers. In 1961 Exxon introduced the much faster-curing chlorobutyl rubbers (CIIR) whilst in 1971 Polysar introduced bromobutyl rubber (BIIR). Today both Exxon and Polysar make both types of halobutyl rubber.

Total nameplate capacity for IIR, CIIR and BIIR is about 600 000 tonnes with production at about two-thirds of this. The halobutyls have taken an ever-increasing share of this total in recent years and now account for roughly half of total production. In turn, the market share for BIIR has steadily increased and accounts for approximately one-half of the halobutyl total.

Cross-linked butyl rubber was introduced in 1967 but it was some years before its value in the manufacture of non-curing glazing tapes was recognized and usage became significant.

8.2 MANUFACTURE

Cationic polymerization is a highly complex process. As with free-radical polymerization, polymer chain growth involves a chain reaction although in this case the active species is a carbonium ion. Polymerization is facilitated by the presence of a Lewis acid such as BF_3 or $AlCl_3$ and in commercial practice the latter is usually used in conjunction with a small amount of a co-catalyst such as water. At temperatures of -85 to -95°C the reaction is extremely rapid and polymer is formed within one second of the monomer being fed to the reactor. For butyl rubber the monomers are fed into the reactor in solution in methyl chloride with a slurry of polymer in methyl chloride being produced. Temperature is controlled by the dissipation of heat by continuously evaporating ethylene in cooling jackets surrounding the reactor. The slurry then passes to a flash tank to boil off the methyl chloride, the catalyst residue is neutralized with sodium hydroxide and antiagglomerant and stabilizer are then normally added before screening, drying and compacting the raw rubber.

The lower the reaction temperature the higher the molecular weight down to about −100°C, below which further temperature reduction does not lead to a substantial further increase in molecular weight.

8.3 STRUCTURE–PROPERTY RELATIONSHIPS

In order to understand the properties of IIR it is helpful first to consider the structure and properties of polyisobutene.

PIB is a totally saturated linear hydrocarbon polymer. It therefore has low chemical reactivity, is resistant to polar liquids and has good electrical insulation properties. Like other polymers of asymmetrically disubstituted ethylenes, its T_g is much lower than for the corresponding mono-substituted polymer (polypropylene) with the T_g at −65°C (compare *c.* 0°C for polypropylene).

With its regular structure, it is to be expected that PIB should crystallize, and this it does on stretching but only with the greatest reluctance on cooling. Some crystallization is however observed if the polymer is held at −33°C for six months. This reluctance to crystallize on cooling may be related to the lack of chain flexibility brought about by the overcrowding effect of the pairs of methyl groups attached to and surrounding the polymer chain backbone. This sluggishness also leads to a lack of resilience over a wide temperature range and no doubt also contributes to the low gas permeability.

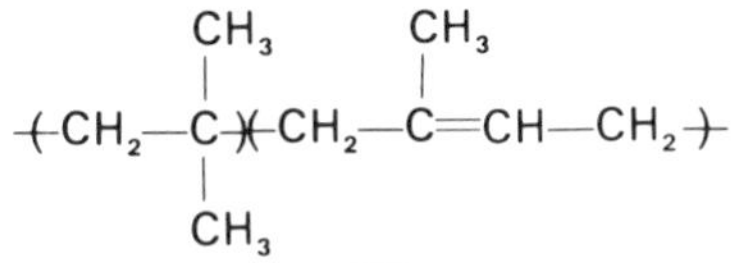

$$-\!\!(CH_2-\overset{\overset{\displaystyle CH_3}{|}}{\underset{\underset{\displaystyle CH_3}{|}}{C}})\!\!-\!\!(CH_2-\overset{\overset{\displaystyle CH_3}{|}}{C}=CH-CH_2)\!\!-$$

PIB

Butyl rubber, with a diene monomer content of 0·5–2·5 mol%, exhibits the characteristics expected of a marginally modified PIB. One noteworthy difference is that low-temperature crystallization appears virtually totally inhibited. A sample of butyl rubber containing as little as 0·5 mol% unsaturation failed to crystallize after a year at −33°C (Buckley, 1965) although it is on record (Kell *et al.*, 1958) that a butyl rubber with very low unsaturation did show signs of crystallizing after 18 months at −30°C.

The structure–property relationships of halobutyl and cross-linked butyl rubbers will be considered in Sections 8.9 and 8.10 respectively.

8.4 GRADES

Excluding the cross-linked and halobutyl polymers, there are about 30 grades of butyl rubber available, a somewhat smaller figure than for other large-tonnage rubbers but not surprising in view of the small number of suppliers.

The grades differ from one another in one or more of the following respects:

1. Level of unsaturation;
2. average molecular weight;
3. stabilizer present in raw polymer.

Commercial grades vary in unsaturation from 0·6 to 2·5 mol%. As a general guide, increasing the unsaturation allows more cross-linking but provides more sites for chemical attack, for example by oxygen and ozone. The effects of increasing unsaturation level are summarized in Table 8.1.

Most of the changes indicated in Table 8.1 are to be expected. The effect on heat resistance is apparently anomalous however. The explanation appears to lie in the fact that with the higher unsaturation grades a high initial network density is obtained on cure. In addition it has been suggested that the ratio oxidative cross-linking/oxidative scission is greater with the

TABLE 8.1
Effects of increasing unsaturation level in the range 0·6–2·5 mol% in butyl rubber

Properties improved[a]	*Properties adversely affected*
Rate of cure	Ozone resistance
State of cure	Weather resistance
Hardness	Flex crack resistance
Modulus	Elongation at break
Resilience	Damping constant
Heat resistance	
Compression set (lowered)	

[a] Numerical values increased unless otherwise stated.

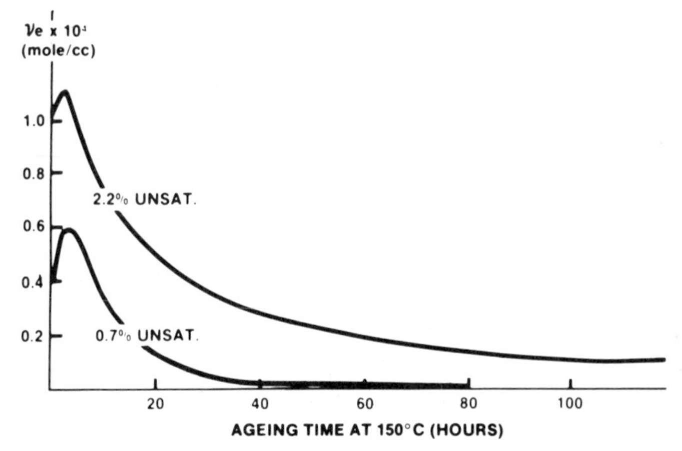

FIG. 8.1. Network density versus time—effect of unsaturation. Formulation of sulphur-cured gum vulcanizate (parts by weight): Rubber, 100; ZnO, 5; stearic acid, 1; MBTS, 1; tellurium diethyl dithiocarbamate (TDEDC), 2; sulphur, 1·5. Cured for 20 min at 145°C. (From Gunter, 1981.)

more unsaturated material. Thus, although the fall in network density after ageing is rapid, the starting point is a material with a higher network density and this difference is retained during ageing (Edwards, 1966) (Fig. 8.1).

Figure 8.2 indicates the general application range for polymers of different unsaturation levels.

Weight-average molecular weights are usually in the range 150 000–500 000 although much higher-molecular-weight materials have

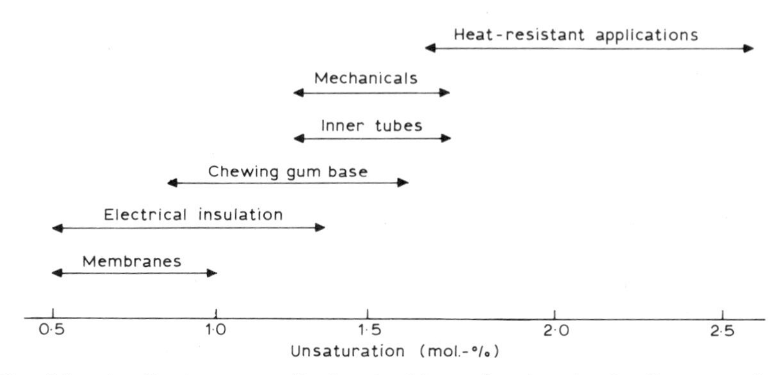

FIG. 8.2. Application ranges for butyl rubbers of various levels of unsaturation.

been prepared experimentally. As with other rubbers, the Mooney viscosity is used as a measure of molecular weight, but care must be taken in making comparisons because with different grades different testing temperatures may be used. Apart from some low-molecular-weight Russian grades, commercial polymers have Mooney ML 1 + 8 (100°C) viscosities in the range 42–82.

The lower-molecular-weight grades give better extrusion and calendering characteristics in moderately loaded compounds but high-viscosity grades are used to confer high green strength when highly filled and/or highly plasticized stocks are being processed.

As with other rubbers both staining and non-staining stabilized grades are available. In addition special grades which are free of stabilizer are available for chewing gum and food-contact applications.

8.5 GENERAL VULCANIZATE PROPERTIES

The key features of butyl rubber in approximate order of decreasing importance are:

1. Very low gas permeability for a rubber;
2. very low water absorption;
3. very good resistance to oxygen:
4. very good resistance to ozone;
5. very good resistance to weathering;
6. good heat resistance (maximum continuous service temperature of 120°C);
7. very good electrical insulation properties;
8. good gum stock tensile strength;
9. good tensile strength, tear strength and abrasion resistance of filled compounds;
10. good low-temperature flexibility;
11. high compression set;
12. poor resilience.

Butyl rubber is best known for its low gas permeability. It should however be noted that compared with most plastics its permeability is high and even the epichlorhydrin rubbers and some grades of nitrile rubber are also better.

Permeability involves solution of gas into a polymer on the high-pressure side, diffusion of the gas through the thickness of the film or sheet and evaporation at the low-pressure surface. Whilst gas solubility in IIR is similar to that of other hydrocarbon polymers, the diffusivity is low because this depends on the rapid motion of polymer segments to open up reasonably direct flow paths across the thickness. The low permeability is of obvious interest in inner tubes (and inner liners in the case of the halobutyls) but can be a disadvantage where gas or air is trapped during processing.

Figure 8.3 (Gunter, 1981) compares the permeability of several polymers to air. It is however to be noted that the permeability of nitrile rubber depends on its acrylonitrile content and some grades of NBR have lower permeabilities than butyl rubber.

The low water absorption of IIR is largely the result of a low level of electrolytic polymerization residues. Even when these are present, diffusion rates are low for the reasons given above. Where a very low water absorption is a critical requirement, tight cures raise the osmotic pressure

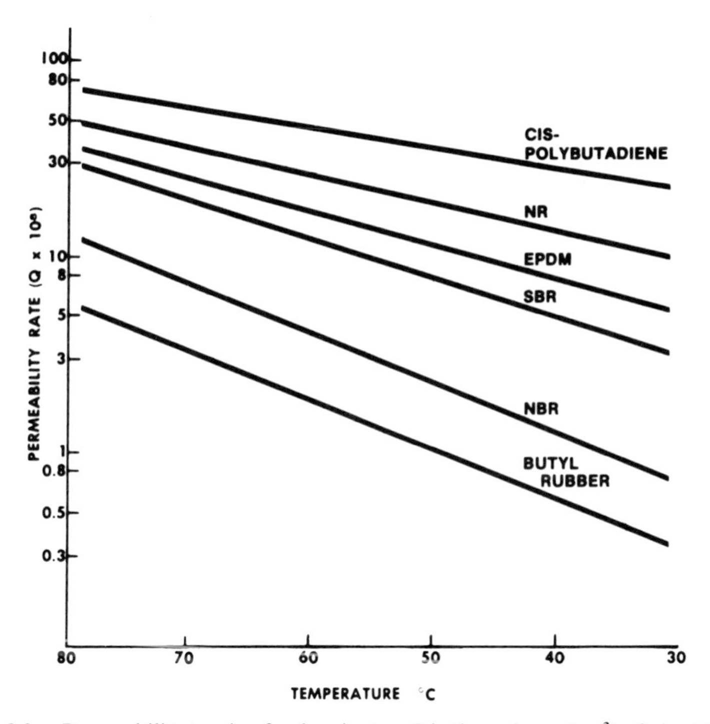

FIG. 8.3. Permeability to air of vulcanizates. Q is the volume (cm^3) of air at NTP which passes (per second) through a specimen 1 cm^2 in area and 1 cm thick when the difference in pressure across the specimen is 1 atm (0·1 MPa). (From Gunter, 1981.)

within the rubber arising from any trace amounts of water and reduce further absorption (Briggs *et al.*, 1963).

The electrical insulation properties also compare favourably with diene rubbers, largely because of the absence of electrolytes such as soaps.

The weathering, ozone resistance and heat resistance of IIR also compares favourably with the diene rubbers, but as with electrical insulation behaviour they are generally inferior to EPDM in these respects, and where these properties are of importance the latter class of rubber is usually preferred.

The ability of butyl rubber to crystallize on stretching helps to confer good mechanical properties even with lightly loaded stocks. Gum stocks can have tensile strengths of over 10·3 MPa (1500 psi) whilst black-loaded stocks have maximum tensile strengths of over 13·8 MPa (2000 psi). Such values of tensile strength together with good tear strength and abrasion resistance are ample for the applications in which butyl rubbers are normally used.

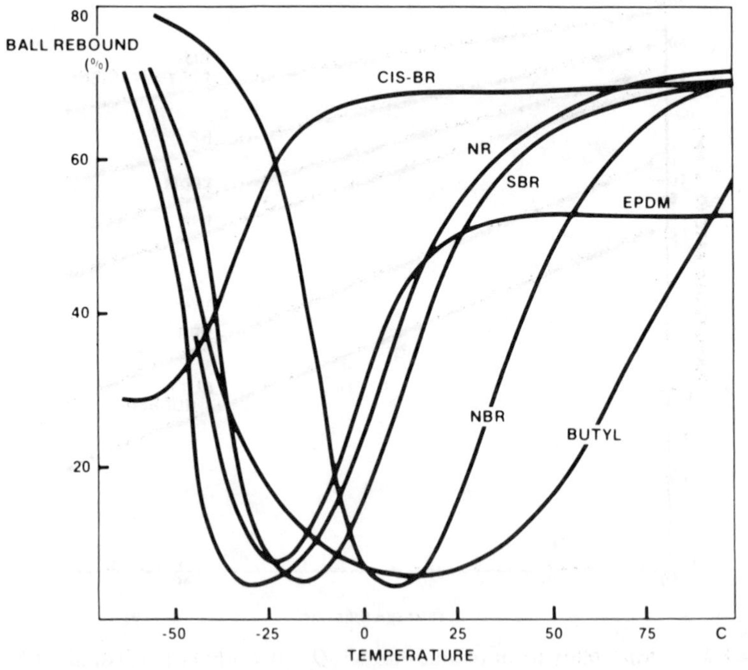

FIG. 8.4. Resilience (ball rebound)–temperature relationships for butyl rubber compared with other major rubbers. (From Gunter, 1981.)

T_g values in the range −65 to −73°C are usually quoted for butyl rubber. Whilst the rubbers are flexible above this temperature, the sluggishness of molecular rotation confers a low resilience on the material over a wider temperature range than for most rubbers. At typical ambient temperatures resilience is much lower than for the more common diene hydrocarbon polymers, although at temperatures above 70°C the resilience approaches that of a diene rubber (Fig. 8.4). The molecular sluggishness also confers strong vibration damping ability although it must be remembered that any absorption of energy is converted into heat which has to be dissipated.

8.6 PROCESSING

Three main points are to be borne in mind when processing butyl rubber.

1. The normal grades which contain stabilizer undergo very little breakdown on mastication. (Unstabilized grades used in food and drug applications do however break down in a manner similar to natural rubber.) Rubber of the correct initial viscosity should therefore be chosen, but if it is essential to reduce the molecular weight this may be effected by using a peroxide or mercaptan peptizer on a mill or internal mixer.
2. Contamination with highly unsaturated polymers such as diene hydrocarbon rubbers must be scrupulously avoided. Equipment must be free of any residual highly unsaturated materials from previous operations.
3. Because of the low gas permeability, special care must be taken to ensure that raw compounds and their vulcanizates do not contain trapped air.

When mixing by internal mixer it is recommended that batches should be 10% larger than those for SBR and fillers added as soon as possible to obtain good dispersion. Sulphur may also be added early. High mixing temperatures up to 105°C not only assist filler incorporation but help to drive off unwanted volatiles that may cause porosity. Upside-down mixing is recommended for compounds with large filler and/or oil loadings. In this process dry compounding ingredients plus half the oil are first added to the mixer followed by the polymer. The rest of the oil is added about one minute later and the curatives at the end of the mixing cycle or even in a second mixing stage.

When mill mixing, the raw polymer forms a lace-like web rather than a smooth band round the mill rolls, a smooth band only being formed after filler incorporation. In repetitive operations it is simplest to start with a portion of compound from a previous mix which will give a smooth band initially and then to add to this polymer and filler alternately. Where this is not possible, mixing should be initiated with as tight a nip as possible and a cold front roll (*c.* 32°C) and warm rear roll (*c.* 75°C), the rubber tending to adhere to the cooler roll. Any tendency to bagging at high filler loadings may be reduced by the use of lower roll temperatures.

The biggest problem with extrusion and calendering is the avoidance of trapped air. This can be minimized by the use of even-speed warming rolls to supply uniformly heated stock, and in the case of extruders by the use of equipment with a vacuum extraction facility. Where mouldings are required, the inherent nature of the injection moulding process by which air and volatiles tend to be pushed towards the hopper rather than the mould cavity makes the process generally more suitable than compression moulding. Where compression moulding is unavoidable, dusting agents such as talc must be avoided where two or more blanks are to be used in a given mould cavity, whilst bumping of the press can sometimes help to remove entrapped air.

8.7 COMPOUNDING

In addition to the selection of an appropriate polymer grade, considered in Section 8.4, attention has to be paid to the following aspects of formulation, which in each case show some differences from usage with diene hydrocarbon rubbers:

1. Vulcanizing systems;
2. fillers and promoters;
3. softeners and plasticizers;
4. antiageing additives.

Butyl rubber may be vulcanized using most of the types of curing system used with diene rubbers. Two points must however be made.

1. Because of the low unsaturation level compared with diene rubbers, very powerful curing systems are necessary. Butyl rubbers cannot be co-cured either mixed or in adjacent layers to diene rubbers because the latter are liable to react preferentially with the curing system, leaving the butyl rubber undercured.

2. Unmodified butyl rubber (as distinct from bromobutyl rubbers) cannot be cured with peroxides as these cause the polymer to degrade.

Whilst accelerated sulphur systems are used for most applications by virtue of their flexibility in formulation, good product properties and cost, butyl rubbers have been cured by non-sulphur systems more often than other sulphur-curable rubbers. Still important for heat-resisting applications are phenolic resin curing systems accelerated by halogen-containing additives.

In the case of sulphur vulcanization lower sulphur levels than those used with diene rubbers are usually employed because of the lower concentration of cure sites. More powerful accelerators are however used, and typically these are synergistic combinations of thiazoles in combination with thiurams or thiocarbamates. Sulphur donor systems based on dithiomorpholines and thiurams confer improved ageing and lower compression set.

Zinc oxide levels similar to those employed with diene rubbers are normally employed (5 pphr) but amounts up to 25 pphr are useful where it is desired to improve heat resistance by scavenging hydrogen sulphide produced during vulcanization. Stearic acid is not required as a cure adjunct although it may be used as a processing aid. Table 8.2 indicates typical curative system ranges with sulphur cures for butyl rubber.

Quinoid systems based on *p*-quinone dioxime, or the less scorchy dibenzoic acid salt, are fast curing, give improved ageing and thermal stability and have been used widely for continuous vulcanization of wire insulation. The decline in importance of this market for butyl rubber has reduced the importance of this process. Elsewhere the low initial strength of

TABLE 8.2
Typical curative range of sulphur and sulphur donor cures

Additive	*Accelerated sulphur system (pphr)*	*Sulphur-donor system (pphr)*
Sulphur	0·5–2·0	0–0·5
ZnO	5	5
Thiazole	0·5–1·0	0–0·5
Thiuram and/or dithiocarbamate	1·0–2·5	3–4·5
Dithiodimorpholine	—	1·5–2·0

TABLE 8.3
Typical quinoid curing formulations for butyl rubber

Curing formulation	*(pphr)*		
Butyl rubber	100	100	100
N-330 black	50	50	50
Zinc oxide	5	5	5
p-Quinone dioxime	1·5	2	—
Pb_3O_4	5	—	10
TMTD	4	4	—
Sulphur	—	2	—
Dibenzoate salt of PQD	—	—	6

quinoid-cured vulcanizates, their coloured nature, queries concerning health hazards, the development of the faster-curing halobutyl rubbers and the development of phenolic resin curing systems for optimum heat resistance have all led to the decline in quinoid systems.

Three typical curing systems involving quinoid cures are given in Table 8.3.

Exceptional heat resistance is obtained when butyl rubber is cured with dimethylol phenol resins and this system is widely used in the manufacture of tyre curing bags and bladders. Curing is however slow and also requires high curing temperatures (190°C). Some acceleration may be achieved by the use of halogen-bearing activators. Stannous chloride was at one time often used but is not suitable in moist environments. The use of halogen-containing polymers is now important and materials used include polychloroprene rubber, chlorosulphonated polyethylene, halobutyl rubber and brominated phenolic resin. Table 8.4 indicates typical recipes.

TABLE 8.4
Typical phenolic resin curing systems for butyl rubber

Curing system	*(pphr)*			
Active phenolic resin	10	7	10	—
Brominated phenolic resin	—	—	—	10
Polychloroprene	10	—	—	—
Bromobutyl rubber	—	10	—	—
Stannous chloride	—	—	2	—
Zinc oxide	5	5	—	5

Although gum stocks are stronger than those of synthetic diene rubbers, fillers are nevertheless used in IIR compounds. Fine particle size carbon blacks are, as with other polymers, employed to confer improved tensile strength, tear strength and abrasion resistance. Fillers are also necessary, in conjunction with softeners and plasticizers, to enable the rubber to be calendered and extruded successfully. For many purposes inert fillers such as whitings and clays may be used in conjunction with blacks to ease processing and reduce cost. Carbon black loadings are usually in the range 30–100 pphr and that of mineral fillers is 50–150 pphr.

Some years ago it was discovered that a process of *heat treatment* of butyl rubber–carbon black masterbatches in an internal mixer for a few minutes at about 150°C in the presence of a *promoter* led to vulcanizates with much improved tensile strength and modulus (Zapp & Gessler, 1953). This improvement appears to be due to promoter molecules forming bonds between polymer molecules and carbon black particles. On shearing the compound, the multiple attachments of carbon particle to polymer chains cause such stresses to be set up in the aggregates of carbon black particles that they disaggregate with resulting better dispersion of finer particles, a prerequisite for optimizing filler reinforcement (Rehner & Gessler, 1954). Nitroso compounds were found to be particularly effective as promoters but many have now been withdrawn because of concern over possible health hazards and so will not be discussed further here.

As with other rubbers, plasticizers (softeners) are used both to reduce viscosity on processing and to modify properties of the vulcanizate. They also may reduce cost. In the case of inner tubes, plasticizers are particularly important to ensure good resilience at low temperatures under severe winter conditions. Because of their saturated aliphatic hydrocarbon nature, butyl rubbers have low solubility parameters of the order of 16·1 $MPa^{1/2}$ (7·9 $(cal\,cm^{-3})^{1/2}$). In order to be compatible, aliphatic or naphthenic mineral oils are normally used with loadings in the range 20–30 pphr. For low-temperature applications aliphatic ester plasticizers such as dioctyl sebacate (DOS) and dioctyl adipate (DOA) are sometimes used. Miscellaneous other traditional softeners are also sometimes employed in IIR. Care must however be taken to avoid olefinic unsaturation in these additives since they may preferentially react with the curing system and retard cure of the rubber.

Because of its inherent stability, antioxidants and antiozonants are not normally necessary in IIR. Heat resistance is better controlled by proper selection of the curing system and paying attention to the ageing environment. For example, neither steam nor hot water will cause chemical degradation of networks at temperatures as high as 177°C in the absence of oxygen. The life of tyre curing bags and bladders has been prolonged by using de-oxygenated steam in contact with the bags and bladders in tyre moulding equipment.

In a similar manner ozone resistance is optimized by the use of low-unsaturation grades of polymer. If required, nickel dibutyl dithiocarbamate is a useful antiozonant, whilst waxes can also help. However, because of their similar solubility parameters, waxes are more compatible with IIR than with some other rubbers and larger quantities (usually in excess of 8 pphr) are needed to be effective. Ozone resistance can be further upgraded by blending with EPDM.

The weathering resistance of unpigmented IIR is not particularly good and the polymer degrades quite rapidly. Black-loaded compounds show very little deterioration whilst the more opaque white pigments such as titanium dioxide and zinc oxide can be useful in non-black formulations.

8.8 APPLICATIONS OF UNMODIFIED BUTYL RUBBERS

From its early beginnings the market for butyl rubber has been dominated by its use in inner tubes. During the first ten years of production virtually all IIR was used for this purpose, but with the advent of the tubeless tyre in the early 1950s other outlets were actively developed. These utilized such desirable characteristics as low water absorption, good electrical insulation properties, good weathering resistance of filled stocks, and good heat and chemical resistance. Applications resulting from these characteristics included wire insulation, white sidewalls for tyres and chemical process plant.

The subsequent development of the ethylene–propylene rubbers, with their superior ageing, heat and ozone resistance, led to a loss of a number of these applications to butyl rubber. Meanwhile, however, the market for inner tubes was maintained because the overall growth of tyre demand partly compensated for the fall in the proportion of tubed tyres and also because of the practice of using inner tubes in slightly damaged or repaired tubeless tyres.

More recently the market for inner tubes has declined but unmodified butyl rubber is now well established in a number of other areas. Amongst these is its long-standing use in tyre curing bags and bladders where the phenolic-resin-cured materials show excellent resistance to wet and dry heat.

Another application, and a rare example of the type of rubber used being made known to the general public, is its use as a waterproofing membrane for reservoirs, lakes and garden ponds. Whilst EPDM might be expected to have better weathering properties, IIR is preferred not only for such properties as low water absorption and permeability and flexibility over a wide temperature range but also because of the splice strength where sheets of necessity are bonded together. For this purpose low-unsaturation grades are used to optimize ozone and weathering resistance; in some instances up to 20 pphr of EPDM may be incorporated, although at the expense of splice strength and extensibility on ageing.

Other applications include sewer gaskets, butyl rubber being selected because of its low water absorption and permeability, caulking compounds, wax blends for coating paper and regenerated cellulose and for chewing gum bases.

Nameplate capacity for IIR and halobutyl rubbers in non-CPEC countries is about 600 000 tpa with production in recent years at about two-thirds capacity. Production is split roughly equally between unmodified IIR and the halobutyl rubbers, which have grown considerably in importance in the last two decades and which are discussed in the following section.

8.9 HALOBUTYL RUBBERS

The low cure rate of conventional butyl rubber when using conventional diene rubber formulations causes problems of co-vulcanization and co-bonding with diene rubbers, whilst contamination can cause serious problems. Halogenation of butyl rubber causes a molecular rearrangement so that the double bond becomes pendant to the main chain whilst the halogen attaches to the carbon which is allylic to the double bond. Not only is the double bond more active but alternatively cure may be effected via the halogen atom.

$$\sim CH_2-C(CH_3)_2-CH_2-C(CH_3)=CH-CH_2\sim + Br\cdot \longrightarrow \sim CH_2-C(CH_3)=CH-\dot{C}H\sim + HBr$$

$$\xrightarrow{Br_2} \sim CH_2-C(CH_3)=CH-CH(Br)\sim + Br\cdot$$

The first halobutyl rubbers introduced commercially were the chlorobutyl rubbers (CIIR) by Exxon in 1961. These were followed by the bromobutyl rubbers (BIIR) introduced by Polysar in 1971. Polysar commenced production of CIIR in 1979 and Exxon of BIIR in 1980. Commercial chlorobutyls contain about 1·1–1·3% chlorine, the bromobutyl rubbers 1·9–2·1% bromine, the difference being a direct reflection of the different atomic weights of the two halogens.

In general the halobutyl rubbers are *superior* to normal butyl rubbers in the following respects.

1. Greater cure versatility—they may be cured by chemicals such as peroxides and zinc oxide that do not cure IIR.
2. Much faster cure rate with lower curative levels.
3. Cure compatibility with other rubbers which has enabled the development of interesting blends and a reduction in problems arising from cross-contamination.
4. Good adhesion to themselves and to other elastomers co-cured in contact with each other. This is particularly important for tyre inner liner/carcass adhesion.
5. Better heat resistance.

The halobutyls retain most of the beneficial IIR characteristics and have *similar* properties to butyl rubber in the following respects.

1. Low gas permeability.
2. Excellent weathering resistance of filled compounds.
3. Excellent ozone and flex cracking resistance.
4. High hysteresis.

The halobutyls are *inferior* to unmodified butyl rubber in respect of:

1. Moisture resistance.
2. Cost.

Compared with chlorobutyl rubber, the bromobutyl rubber is more expensive; has a higher green strength; is faster curing; has better adhesion to unsaturated rubbers; has better flex cracking resistance both before and after ageing; may be cured with peroxides (chlorobutyl rubber cannot).

The different cure reactivities of the two halobutyl rubbers necessitate the use of different curing recipes and schedules which might otherwise result in undercuring (of CIIR) or scorching (of BIIR).

8.9.1 Vulcanization

Such is the reactivity of BIIR that both sulphur and zinc oxide will cure the polymer alone, although in practice this would not be recommended. Most commercial cures use either zinc oxide or sulphur in combination with other chemicals or with each other. Peroxides may also be used in pharmaceutical applications.

Two mechanisms have been suggested for the reaction of BIIR with zinc oxide:

Mechanism 1

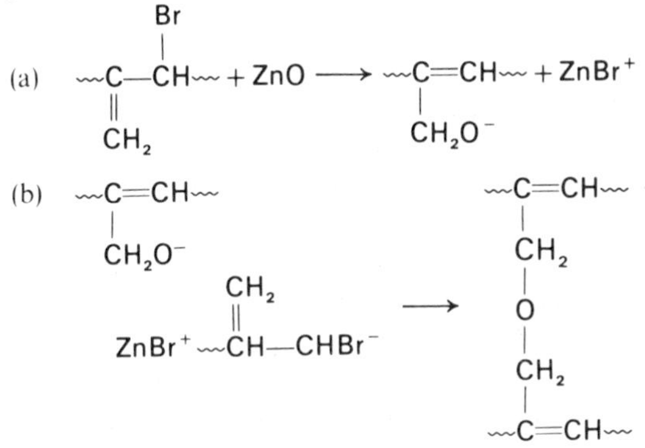

(a) $$\sim C(=CH_2)\text{—}CH(Br)\sim + ZnO \longrightarrow \sim C(CH_2O^-)\text{=}CH\sim + ZnBr^+$$

(b) $$\sim C(CH_2O^-)\text{=}CH\sim \quad ZnBr^+ \sim C(=CH_2)\text{—}CHBr^- \longrightarrow \sim CH\text{=}C(\sim)\text{—}CH_2\text{—}O\text{—}CH_2\text{—}C(\sim)\text{=}CH\sim$$

Mechanism 2

$$\sim C(=CH_2)\text{—}\overset{+}{C}H\sim + \sim C(=CH_2)\text{—}CH(Br)\sim \longrightarrow \sim \overset{+}{C}(CH_2\text{—}CH(\sim)\text{—}C(=CH_2)\sim)\text{—}CH(Br)\sim$$

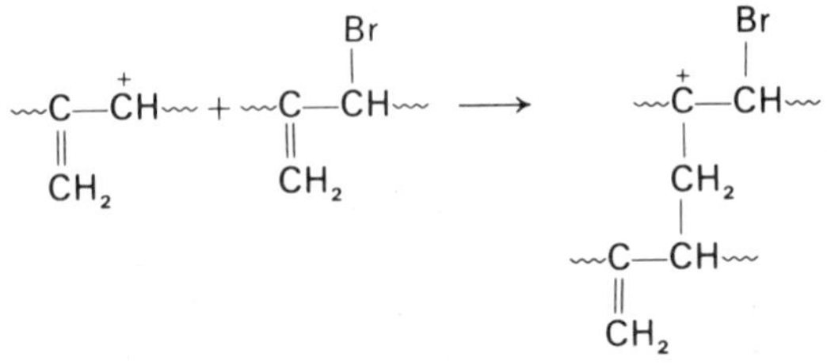

Using allyl bromide as a model system it was found on reaction with zinc oxide in a polyisobutene medium at usual cure temperatures that no allyl bromide, which would be expected from the first mechanism, was detected (Feniak *et al.*, 1974). The very good thermal stability is also consistent with carbon–carbon links which would be produced by the second mechanism.

Zinc oxide-based systems are used where heat resistance is particularly desirable, the zinc oxide usually being used at a level of 3 pphr. The systems include blends with a sulphur donor such as TMTD (0·2–0·5 pphr), with phenolic resins (0·6–7·0 pphr) or with a conventional accelerator (ZDC, 0·2–0·5 pphr). Where sulphur donor systems are used a small amount of

MBTS or a sulphenamide can act initially as a *retarder* to reduce scorch tendencies.

Zinc oxide/sulphur systems have lower heat resistance but within limits an increase in sulphur content will increase modulus. Some typical cure systems based on 100 parts polymer are:

A		B		C	
ZnO	3	ZnO	3	ZnO	3
MBD	0·75–1·5	MBTS	0·75–1·25	TBBS	0·75–1·5
Sulphur	0·5–1·0	Sulphur	0·2–0·5	Sulphur	0·5–1·0
				TMTD	0·1–0·3

System A is used for tyre sidewalls and inner liners (MBD, 4-morpholinyl-2-benzothiazole disulphide) whilst system B may be used for inner liners and general mechanical goods. System C may be used in blends with NR, the thiuram acting initially at this low level as a retarder. At higher levels the accelerator role dominates.

Whilst peroxides do cure BIIR, the degree of cross-linking is less than is obtained using systems based on zinc oxide, sulphur or phenolic resins. In order to obtain good stress–strain properties, low compression set and good heat resistance it is usual to use the peroxide in conjunction with a coagent such as *N,N′-m*-phenylene dimaleimide (both at *c.* 1·5 pphr).

As indicated previously, chlorobutyl rubbers cannot be cured by peroxides, whilst with other curing systems, usually require more powerful recipes.

8.9.2 Applications

The advent of a butyl-type rubber with very low air permeability and which may be cured alongside a diene rubber has led to a substantial market in tyre inner liners. Halogenated butyls possess the adhesion, flex and heat resistance together with the impermeability required of a high-quality compound. Not only is impermeability important in maintaining tyre pressures but by reducing the diffusion of air through the liner into the carcass there is a reduction in the intracarcass pressure. This reduction results in lower oxidative degradation of carcass components and thus a lowering of the resistance of the tyre to failure through flex fatigue or ply separation.

BIIR is normally selected where optimum performance is required, for example in inner liners for heavy-duty and premium-quality tyres. The less expensive CIIR is used where demands are less severe and it may be used in blends with NR.

Halobutyl rubbers are also useful in inner tubes because of their superior heat resistance, splice durability and tube growth. As with inner liners, BIIR is recommended for the most severe conditions.

Halobutyls may be blended with diene rubbers for use in tyre sidewalls, both black and white, where they confer improvements in ozone resistance combined with good adhesion and flex properties. They are also tending to replace butyl rubber in tank linings, hose covers and some pharmaceutical applications. They are not replacing the unmodified IIR in such areas as tyre curing bags, light tyre inner tubes, chewing gum bases and non-curing caulking compounds where the unmodified material has clear technical advantages and/or is more economic.

8.10 CROSS-LINKED BUTYL RUBBERS

Terpolymerization of isobutene, isoprene and a small amount of divinylbenzene leads to a polymer which is branched and also very lightly cross-linked. In addition there may be pendant unreacted vinyl groups:

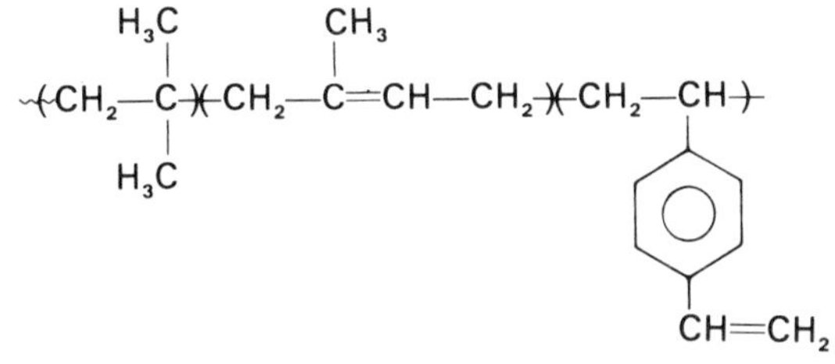

$$\sim\!\!(CH_2-C(CH_3)_2)\!\!-\!\!(CH_2-C(CH_3)=CH-CH_2)\!\!-\!\!(CH_2-CH(C_6H_4-CH=CH_2))\!\!\sim$$

Whilst the degree of cross-linking is sufficiently low for the rubber to be partially soluble in cyclohexane, the uncured rubbers have higher green strength, resilience, and resistance to sag and flow than the conventional linear materials. These properties have been particularly useful as the elastomeric base of non-curing sealing tapes used to seal, retain and cushion automotive windscreens and other fixed windscreens of vehicles. The material also finds some use for solvent-release caulks and pressure-sensitive adhesives.

If desired the pre-cross-linked butyl rubbers may be fully cross-linked using conventional curing systems. Furthermore, it has been found that, unlike with IIR, cross-linked IIR may be cured with peroxide. It appears that this peroxide cross-linking proceeds via the pendant vinyl groups and dominates the normal degradation reaction. Peroxide-cross-linked terpolymers have lower compression set and better ozone resistance than normal IIR of comparative unsaturation. Suitably stabilized, they also exhibit a heat resistance comparable with a phenolic-resin-cured conventional butyl rubber but with faster cure rates and higher compatibility with other types of rubber (Walker *et al.*, 1974). Some of these properties are shared with the halobutyl rubbers considered in Section 8.9.

BIBLIOGRAPHY

Buckley, D. J. (1959). *Rubber Chem. Technol.*, **32**, 1475.
Buckley, D. J. (1965). Butylene polymers. In *Encyclopaedia of Polymer Science and Technology*, Vol. 2. Wiley, New York.
Gunter, W. D. (1981). In *Developments in Rubber Technology—2*, ed. A. Whelan & K. S. Lee. Applied Science, London, Chapter 6.
Kennedy, J. P. (1969). In *Polymer Chemistry of Synthetic Elastomers*, ed. J. P. Kennedy & E. G. M. Törnqvist. Interscience, New York, Chapter 5A.

REFERENCES

Briggs, G. J., Edwards, D. C. & Storey, E. B. (1963). *Rubber Chem. Technol.*, **36**, 621.
Buckley, D. J. (1965). Butylene polymers. In *Encyclopaedia of Polymer Science and Technology*, Vol. 2. Wiley, New York.
Butlerov, A. & Goryaniov, V. (1873). *Ber.*, **6**, 561.
Edwards, D. C. (1966). *Rubber Chem. Technol.*, **39**, 581.
Feniak, G., Jones, R. H. & Walker, J. (1974). Lecture to Dayton Rubber Group, 15 Nov.
Gunter, W. D. (1981). In *Developments in Rubber Technology—2*, ed. A. Whelan & K. S. Lee. Applied Science, London, Chapter 6.
Kell, R. M., Bennett, B. & Stickney, P. B. (1958). *Rubber Chem. Technol.*, **31**, 499.
Otto, M. & Müller-Cunradi, M. (26 July 1931). German Patent 641 284, IG Farbenindustrie.
Rehner, J. & Gessler, A. M. (1954). *Rubber Age (NY)*, **74**, 561.
Sparks, W. J. & Thomas, R. M. (1939). US Patent 2 356 128.
Walker, J., Wilson, G. J. & Khumbani, K. J. (1974). *J. Inst. Rubber Ind.*, **8**, 64.
Welsh, L. M., Nelson, J. F. & Wilson, H. L. (1949). *Ind. Eng. Chem.*, **41**, 2834.
Zapp, R. L. & Gessler, A. M. (1953). *Rubber Age (NY)*, **74**, 243.

Chapter 9

Acrylonitrile–Butadiene Rubbers (NBR)

9.1 INTRODUCTION

A somewhat simplistic way of describing acrylonitrile–butadiene rubbers, commonly known as nitrile rubbers, is to say that they are the special-purpose rubbers with the conventional technology. Commercially available for over 50 years, they are known primarily for their resistance to liquid fuels such as petrol and other hydrocarbons.

They are the product of the work carried out over many years by Farbenfabriken Bayer, which was for part of the time a constituent of the industrial giant IG Farben. Early work on butadiene polymers first led to the production of Buna rubber by sodium-catalysed polymerization of butadiene. This was followed by the development of peroxide-initiated emulsion polymerization of butadiene about 1926 and the laboratory copolymerization of butadiene and styrene in 1929. Acrylonitrile–butadiene copolymers were first prepared in 1930 with pilot-plant production commencing in 1934 and full-scale production in 1937, the product being marketed as Buna N. Interestingly commercial production of Buna N preceded that of Buna S, the butadiene–styrene copolymer. Sometime about 1940 Buna N became known as Perbunan, the name retained by Bayer to this day.

During World War II production commenced in the USA, when the rubber was known as GR–A (Government Rubber—Acrylonitrile). By 1970 it was estimated that about 200 grades were being marketed by some 17 producers. By the mid-1980s this had grown to some 31 producers with about 500 grades (including 110 grades of nitrile latex and 18 grades of powdered rubber). Total world production, excluding that of the CPEC countries, has for most of the past 15 years hovered about the 200 000 tpa level, i.e. about 2·4% of the total rubber market with somewhat over one-third of this world total being produced in Western Europe, a rather higher proportion than for most synthetic rubbers.

Major producers are Bayer (Perbunan), Enichem (Europrene N),

Goodrich (Hycar), Goodyear (Chemigum), Nippon Zeon (Nipol) and Polysar (Krynac).

In recent years materials related to nitrile rubbers have been developed. These include carboxylated NBR, epoxy-modified NBR, hydrogenated NBR, copolymers containing isoprene and alternating copolymers.

9.2 PREPARATION

The normal grades of NBR are produced by emulsion polymerization of butadiene and acrylonitrile. The production of butadiene has already been considered in Chapter 5. A number of methods are used to produce acrylonitrile. One common method is the propylene–ammonia–air process which proceeds by reactions that may be summarized by the following equation:

$$CH_2{=}CH(CH_3) + NH_3 + 1\tfrac{1}{2}O_2 \longrightarrow CH_2{=}CH(CN) + 3H_2O$$

Various catalyst systems have been developed to increase the efficiency of the process, including bismuth phosphomolybdate and also a mixture of the oxides of cobalt, molybdenum, antimony and tin.

Another process involves the oxidation of ethylene to acetaldehyde with the aid of palladium salts and combination of the acetaldehyde produced with hydrogen cyanide to produce a cyanhydrin which on dehydration yields acrylonitrile:

$$CH_2{=}CH_2 \xrightarrow{Pd} CH_3CHO \xrightarrow{HCN} CH_3{-}CH(OH){-}CN \xrightarrow{-H_2O} CH_2{=}CH(CN)$$

A third process involves the reaction of propylene and nitric oxide:

$$4CH_2{=}CH(CH_3) + 6NO \longrightarrow 4CH_2{=}CH(CN) + 6H_2O + N_2$$

The one-time important route developed by Bayer which involved the addition of hydrogen cyanide to acetylene is no longer of importance for reasons of economy.

Acrylonitrile is a colourless toxic liquid boiling at 77·3°C. As with SBR, most NBR is cold-polymerized in the range 5–30°C with the reaction terminated at about 70–80% conversion. Unlike SBR, the two monomers have quite different reactivity ratios and in the usual range of ratios used the acrylonitrile tends to be consumed first. In order to maintain a reasonably uniform product it is therefore necessary to incorporate the acrylonitrile in stages during the reaction.

9.3 STRUCTURE AND PROPERTIES

Nitrile rubbers have the basic formula

$$-\!\!(CH_2-CH=CH-CH_2)\!\!-\!\!(CH_2-\underset{\displaystyle CN}{\underset{|}{CH}})\!\!-$$

There are two obvious features of this structure:

1. The double bond which facilitates sulphur vulcanization but which is also susceptible to oxidation and ozone attack;
2. the polar nitrile group which confers hydrocarbon oil resistance but tends to raise the T_g of the rubber with several other directly consequent effects.

There are a number of variables which lead to important differences between commercial grades. The following are important:

1. Acrylonitrile content—the dominant variable;
2. additional monomers or substitutes for acrylonitrile and butadiene;
3. average molecular weight;
4. molecular weight distribution;
5. branching;
6. microstructure;
7. stabilizers incorporated on manufacture.

9.3.1 Effect of Acrylonitrile Content

The T_g of polybutadiene is about $-100°C$, that of polyacrylonitrile about $+100°C$. Random copolymers have T_gs that are intermediate on an approximately linear interpolation so that, for example, a 50:50 copolymer will have a T_g of about 0°C. Thus increasing the acrylonitrile content will clearly raise the minimum service temperature down to which the rubber may be used. This deterioration in ability to remain rubbery at low temperatures is obviously a disadvantage. Higher acrylonitrile contents also lead to lower resilience, greater hardness and higher compression set, generally undesirable features although there is a slight increase in hardness, abrasion resistance and tensile strength at normal ambient temperatures (Fig. 9.1).

On the other hand, increasing the acrylonitrile content increases the petrol resistance and swelling in hydrocarbon oils. Since there is a very close relationship between T_g and swelling (Fig. 9.2), compromises have to be made. As a result most grades have nominal acrylonitrile contents between 32 and 35% but a wide range of materials are available with extreme levels for acrylonitrile content of 18 and 53% (both grades available, incidentally, from Nippon Zeon).

It may be noted in passing that since the two monomers have almost identical molecular weights, polymers with the common 33–34% acrylonitrile content will have about one monomer residue in three as acrylonitrile. Simple calculation will show that there will be about one pendant acrylonitrile group per 10 backbone carbon atoms, a much higher level than for pendant benzene rings in conventional SBR molecules.

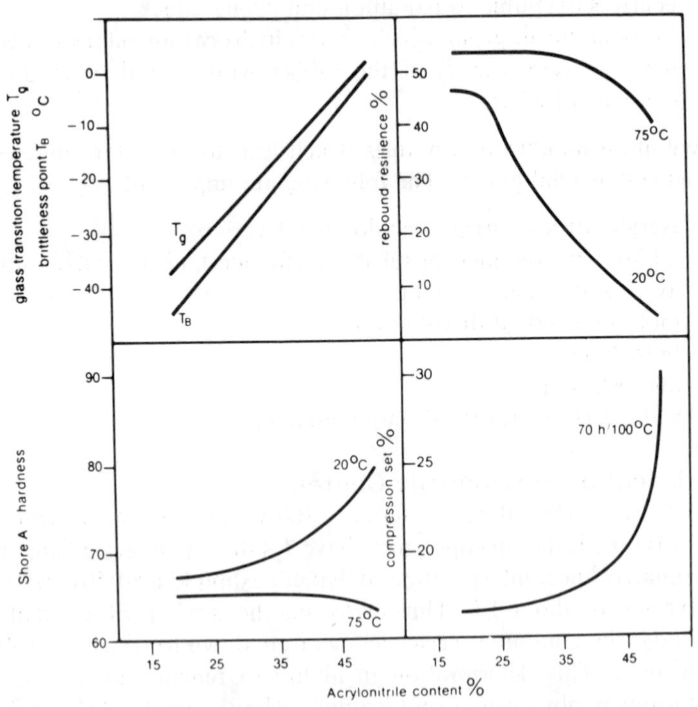

FIG. 9.1. Glass transition temperature (T_g), brittleness temperature (T_B), resilience, hardness and compression set versus the acrylonitrile (ACN) content. (From Bertram, 1981.)

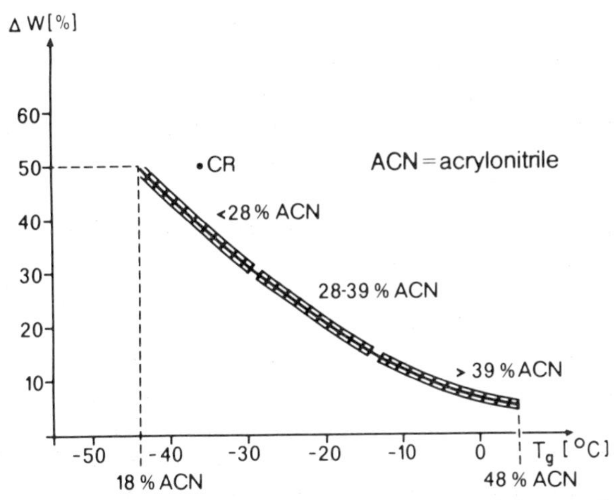

FIG. 9.2. Connection between acrylonitrile content, T_g and change of weight after immersion in ASTM Oil No. 2. (From Bertram, 1981.)

9.3.2 Use of Additional or Alternative Monomers

Substituted butadienes such as isoprene, piperylene and dimethylbutadiene have been used to prepare copolymers in place of butadiene. However, as would be expected of polymers containing methyl groups attached to the polymer backbone, these copolymers have higher T_gs than the corresponding butadiene–acrylonitrile copolymers. Isoprene–acrylonitrile copolymers have been marketed by Polysar and, although possessing some desirable properties such as good gum stock strengths, their high brittle point and poor rubberiness have led to their not achieving significant commercial importance. More recently Nippon Zeon has marketed ternary copolymers (terpolymers) based on butadiene, isoprene and acrylonitrile.

Substituted acrylonitriles such as methacrylonitrile have also been used to produce polymers with somewhat different flow and solubility characteristics. No commercial grades are known to be available at the time of writing.

Most important of the terpolymers are those containing methacrylic acid and which are known as *carboxylated nitrile rubbers* (XNBR). These may be cross-linked ionically using zinc compounds and exhibit higher abrasion resistance, tensile strength and tear resistance but poor compression set and scorch resistance.

As with NR and SBR, lightly cross-linked grades are also available, the third monomer being a cross-linking agent such as divinylbenzene or a difunctional acrylate. Terpolymers of butadiene, acrylonitrile and styrene are well known in the field of plastics (ABS polymers). Rubbery grades with a higher butadiene level have been prepared with improved flow properties compared with conventional nitrile rubbers.

9.3.3 Effect of Molecular Weight, Molecular Weight Distribution and Branching

Whilst number-average molecular weights are of the order of 100 000, i.e. similar to those for SBR, there are substantial variations between grades. The nominal Mooney viscosity ML 1+4 (100°C) is usually employed as a technological measure of molecular size although this value will be influenced by other features such as molecular weight distribution and branching. Whilst values for commercial grades range from 20 to 140, most grades are within the range 25–100, with a good choice of acrylonitrile content within each range.

Molecular weight can influence properties in three ways.

1. By influencing flow properties. Lowering the molecular weight lowers the viscosity and generally results in easier calendering and extrusion. Compounding is also generally easier with lower compounding heat build-up.
2. The higher-molecular-weight rubbers have better stability during open cure, can accept higher loadings of filler and plasticizer, and are less susceptible to air entrapment. These improvements are presumably the result of the higher green strength of the higher-molecular-weight polymer compounds.
3. Vulcanizate properties such as tensile strength and resilience are slightly higher with higher-molecular-weight polymers. This can be attributed to the fact that with low-molecular-weight polymers the proportion of non-load-bearing chain ends in the network is higher.

As might be expected, low-temperature properties and swelling are virtually unaffected by average molecular weight.

The molecular weight distribution of emulsion-polymerized nitrile rubbers is quite broad, due in some measure to the branching that occurs on polymerization. As with the SBRs, the cold-polymerized nitrile rubbers (being less branched) have narrower distributions than the earlier hot rubbers. It has been stated (Hofmann, 1968) that in commercial processes the mercaptan modifiers are introduced gradually during conversion in order to give a more uniform chain length. As with average molecular weight, the molecular weight distribution is strongly dependent on the degree of conversion since the increased tendency to branch, and even cross-link, at high conversions is bound to increase the breadth of the distribution.

The molecular weight distribution characteristics of NBR have been described in detail (Scholtan *et al.*, 1972). Attempts have also been made to relate long chain branching to viscoelastic behaviour (Nakajima & Harrell, 1980). The subject is however complex, in part due to the fact that although the molecular weight distribution depends on long chain branching it is not the only factor and it is difficult to separate out the effects of branching from the molecular weight distribution.

The microstructure (monomer sequence distribution and *cis*/*trans*/1,2-ratios) is influenced by polymerization conditions and can influence processing characteristics (Bertram, 1981).

In a typical NBR polymer, polymerized at 28°C and containing 36% acrylonitrile residues, the butadiene residue distribution was 77·6% *trans*-1,4-, 12·4% *cis*-1,4- and 10% 1,2-.

9.3.4 Stabilizers Incorporated on Manufacture

As is common with diene rubbers, some stabilizer (essentially antioxidant in nature) is added during manufacture to protect the raw rubber prior to full compounding, shaping and vulcanization. Most grades now contain non-staining or slightly staining stabilizers rather than the staining grades of earlier years.

9.4 GENERAL VULCANIZATE PROPERTIES

Although affected by variables in polymer structure, additive selection and vulcanization conditions, the key features of nitrile rubbers, in roughly decreasing order of importance, are as follows.

1. Excellent resistance to aliphatic hydrocarbon oils, fuels and greases.
2. Very good heat resistance in the absence of air.
3. Good hot air resistance (long term 90°C; 40 days 120°C; 3 days 150°C).
4. Very low gas permeability for a rubber.
5. Low permanent set.
6. Moderate low-temperature flexibility.
7. Moderate ozone resistance and poor resistance to sunlight ageing.
8. Poor electrical insulation properties.
9. Moderate tear and tensile strength.

For most of its applications NBR is employed because of its resistance to hydrocarbon fuels, oils and greases. Although there are rubbers with better oil resistance, others that may be used to higher temperature and yet others with better low-temperature properties, it is normally the first rubber to be checked out for suitability in this area because of its generally good all-round properties, easy processing and moderate price compared with the more specialized materials.

As has already been shown (Fig. 9.2), the hydrocarbon swelling resistance depends on the acrylonitrile content. It is important to appreciate that nitrile rubber does not have uniformly useful solvent resistance. Furthermore, although it is so widely used in applications in contact with petrol, it is not always suitable even here. For example, whilst resistance to aliphatic hydrocarbons is excellent, that to aromatic hydrocarbons may only be described as good. Therefore the trend towards high-aromatic petrols in some countries in order to avoid lead-containing anti-knock additives can lead to high levels of swelling which may in some instances be unacceptable.

Nitrile rubber vulcanizates have poor swelling resistance to oxygen-containing liquids such as ketones and some alcohols. Where methanol and/or ethanol have been used as partial petrol fuel substitutes, either to raise octane number or, when the price of petroleum is high, for reasons of economy, nitrile rubbers would not be expected to be the first choice of material.

There has also been some concern with the so-called 'sour gas' problem. This has occurred with overstored fuels and those consumed by engines with fuel injection and has been associated with hydroperoxides present that exert a hardening action on NBR. It has, however, been demonstrated (Dunn *et al.*, 1980) that use of compounding principles similar to those employed to optimize hot air ageing resistance will enable NBR to meet most specifications involving sour gas resistance requirements.

Finally, it is to be mentioned that nitrile rubbers are often unsatisfactory in contact with hydrocarbon oils containing sulphur-bearing additives which may be liable further to cross-link the rubber and it is this deficiency which largely led initially to the development of the acrylic rubbers.

In the absence of air, or more specifically oxygen, the diene rubbers have better heat resistance than is commonly appreciated and the nitrile rubbers are no exception.

No stress relaxation has been observed after prolonged exposure to temperatures of 150°C and it has also been reported (Grassie & Heaney, 1974) that no change occurred after 40 h exposure to 160°C. It has been found that ageing in oils purged with nitrogen improved the effective life at least threefold.

For many years it was considered that the ceiling use temperature of NBR in air was about 100°C. More stringent auto and other specifications introduced in relatively recent years have demanded resistance to higher temperatures. This resulted in intensive compounding studies by major suppliers which now allow the rubber to meet specifications calling for a long life at 125°C and up to a week at 135°C. Such compounding usually involves the use of silica fillers, magnesium oxide, ether–thioether plasticizer, selected antioxidant and a carefully selected sulphur/sulphur donor cure system (Dunn, 1984).

TABLE 9.1
Permeability to nitrogen and carbon dioxide
(After Stannett, 1968)

Polymer	*ACN (%)*	*($cc/cm/cm^2/sec/cm\ Hg \times 10^{11}$)*	
		P_{CO_2}	P_{N_2}
Polybutadiene	0	1 380	64·5
NBR (Hycar OR-25)	32	186	6·04
NBR (Hycar OR-15)	39	74·6	2·35
Butyl rubber	—	51·8	3·12

The very low gas permeability, for a rubber, of nitrile rubber has been widely overlooked. Permeability decreases as the acrylonitrile content increases (Table 9.1) and for many grades it is lower than butyl rubber for many gases. It is of interest to note that the solubility of most gases in NBR decreases with increasing acrylonitrile content but that with carbon dioxide an increase is shown.

The low permanent set of NBR is most important in sealing applications. Whilst in the past compression set after ageing has been the most important criterion, there has been increased interest in recent years in compression stress relaxation studies in measuring seal performance (Dunn, 1984). Both the compression set and the stress relaxation are considerably affected by compound formulation, although not always in the same order.

It was shown in Section 9.3.1 that although the high-acrylonitrile rubbers showed the best resistance to aliphatic hydrocarbon oil swelling, this was accompanied by a high T_g and cold flex temperature. Low-temperature performance may be improved by the use of plasticizers but these will also reduce hardness, tensile strength and modulus as well as affecting extractability.

The ozone resistance of NBR may only be described as moderate. Furthermore, antiozonants and waxes do not have much effect and where the use of NBR is essential in an application also requiring ozone resistance the use of PVC/NBR blends offers the best possibility of solution.

The poor electrical insulation properties discourage the use of the material for primary insulation although it may be used for sheathing.

9.5 PROCESSING

As with SBR, nitrile rubbers do not break down on mastication like natural rubber. Therefore, as with SBR, it is important to select a raw polymer of the most appropriate molecular weight, as assessed by the Mooney viscosity, for the process and application envisaged. Low Mooney grades absorb fillers and plasticizers more easily and exhibit less heat build-up on mixing, and it is claimed that faster extrusion rates may be achieved. It is also claimed that for soft compositions higher strengths may be obtained by the use of high-Mooney rubbers with larger quantities of plasticizer to reduce the hardness. When compounding, best dispersion occurs with moderate mixing temperatures (60–100°C) and high shear rates (Wijayarthna *et al.*, 1978).

Shaping and vulcanization processes are generally quite conventional. In the case of injection moulding, care must be taken to devise curing systems which do not show scorch tendencies at the high processing temperatures but which are still fast-curing. High-energy radiation and UV curing systems have been used for food contact applications, whilst room-temperature curing systems are available for rubber lining of pipes and large vessels.

The above comments apply to the cold-polymerized rubbers which now dominate the market, as is the case with SBR. In general the older hot rubbers, which were more branched, were more difficult to handle. However, as with SBR, a lightly cross-linked 'hot' rubber is available which can be added to cold rubber to improve extrudability.

9.6 COMPOUNDING

The nitrile rubber compounder needs to pay particular attention to four aspects: vulcanization, plasticization, reinforcement and stabilization.

Until the development of injection moulding for rubbers and the need for NBR to meet more rigorous temperature specifications, vulcanization systems were very conventional. Typical formulations used rather less sulphur than with natural rubber (about 1 pphr) with the magnesium-carbonate-coated grades often preferred to obtain adequate dispersion. Similarly, like SBR, rather more accelerator was used, e.g. about 1 pphr of MBTS. Conventional amounts of zinc oxide and stearic acid, about 5 and 1 pphr respectively, were the norm.

Substantial improvements in heat resistance were achieved by the development of the 'cadmate cure' systems. These are complex formulations typified by the following:

	pphr
Cadmium diethyldithiocarbamate	1·5–7
Cadmium oxide	2–5
Magnesium oxide	5
MBTS	0·5–2·5
Sulphur	0–1

Concern over the toxicity of cadmium compounds has now restricted the use of these systems.

Subsequent improvements have been the results of painstaking and systematic studies invòlving accelerator combinations. One such formula consisting of CBS 1·0, TMTD 1·5, OTOS 2·5 and S 0·4 pphr in a silane-treated silica-filled NBR showed good heat resistance to 135°C.

Rapid cure combined with scorch resistance has been achieved by the use of alkyl aryl thiuram disulphides in combination with a sulphenamide and a retarder. A typical injection moulding system (Bertram, 1981) is:

	pphr
Dimethyl diphenyl thiuram disulphide	2·0
N-morpholinyl-2-benzothiazyl sulphenamide	2·0
Retarder (Vulkalent E)	1·0
Pasted insoluble sulphur	0·8

In the same paper formulations have also been given for room-temperature curing systems. One system which cures in a few weeks at room temperature is S 2·4, MBTS 1·0, DPG 1·0 pphr.

Because of their polar nature, additives for reducing melt viscosity and vulcanizate hardness are much more akin to those used to plasticize PVC than those used to soften NR and SBR, so that the term *plasticizer* is usually used rather than the term softener. Whilst both materials act as spacers to separate the polymer molecules and impart greater molecular flexibility, interaction such as hydrogen bonding can occur between the plasticizers and NBR which can offset some of the flexibilizing effect.

Of commercial plasticizers, the phosphates exhibit the greatest interaction and have the least effect on depressing the brittle point of the rubber. They are, however, useful in respect of their flame-retarding properties and are typified by tritolyl phosphate and tri-isopropyl phenyl phosphate.

As with PVC, the phthalate plasticizers show the best all-round properties as well as being amongst the most economic of plasticizers. Widely used are the phthalate esters of mixed alcohols containing seven to nine carbon atoms and which are frequently known as dialphanyl phthalates. Showing less interaction than the phosphates, they have a greater depressing effect on the brittle point.

The greatest effect on depressing the brittle point is shown by aliphatic esters such as dioctyl sebacate and similar materials, and triethylene glycol dicaprylate. These materials are effective because of the low interaction between polymer and plasticizer. For low extractability, polymeric plasticizers such as polypropylene adipate, epoxidized oils and even liquid low-molecular-weight NBRs may be used. However, the liquid rubbers do generally have the most adverse effect on mechanical properties such as tensile strength.

Where heat resistance is important plasticizers are best avoided, but where this is not possible ether and thioether plasticizers have been recommended.

Aromatic mineral oils are sometimes used but because of their limited compatibility their use must be undertaken with care. Where plasticizers are used in NBR formulations, loadings are usually in the range 0–30 pphr according to the softness and low-temperature requirements.

Again, as with other butadiene copolymers, gum NBR stocks are mechanically weak, the irregular molecular structure preventing crystallization on stretching. For most applications semi-reinforcing fillers are adequate, with furnace blacks of the N550 and N770 types often being used, typically at levels of 40 pphr. There is, however, some evidence that carbon blacks of this type may adversely affect degradation, possibly by absorbing antioxidants, and this has led to interest in the use of silica fillers in conjunction with silane coupling agents. In particular, silanes have been shown to reduce compression set and improve sealing force retention in silica-filled vulcanizates.

As has already been indicated, heat and ageing resistance is very dependent on overall compound formulation but careful selection of antioxidants can give further improvement. The need for care in selection may be exemplified by the fact that it was found (Dunn, 1984) that the use of 1 pphr of a diaryl-*p*-phenylenediamine improved air ageing after exposure to ASTM Oil No. 3 at the expense of air ageing in the absence of oil. NBR grades containing network-bound antioxidants resistant to leaching and volatilization were introduced by Goodyear in 1981.

Antiozonants and waxes impart little improvement to NBR ozone resistance. Where this is an important requirement, blends with PVC and EPDM will give some improvement but at the expense of other commonly desirable properties.

9.6.1 Blends with Other Polymers

The blending of NBR with other polymers is a long-established practice. This has been done for various purposes such as to reduce cost, to modify rheological properties or to confer specific chemical or mechanical properties to the vulcanizate. Such blends may be with another rubber or with a plastics material such as PVC. Whereas in some instances the NBR is only a minor component introduced to toughen a plastics material, this section will be confined to a consideration of blends leading to rubbery vulcanizates.

As with all rubber blends due consideration should be paid to the following facts.

1. Compatibility on a molecular scale is rarely if ever achieved and multi-phase systems may be produced.
2. Additives, particularly those used for cross-linking, may have different levels of solubility in the components of the blend and may thus be unevenly partitioned between the polymers.
3. Curing rates, with a given system, vary between polymers.
4. Rubber–filler interactions vary between rubbers.

Blends of NBR with other polymers have been surveyed (Bertram, 1981) and only the more important will be considered here.

9.6.1.1 NBR and NR

In these cases the NR is incorporated to improve tack.

9.6.1.2 NBR and EPDM

This system is attractive in principle because good ozone resistance may be achieved with EPDM levels over 30%. Because of the differences in polarity, two-phase structures are produced and good ozone resistance is only obtained when the EPDM is finely dispersed in the NBR, so that blending conditions are critical. The different rates of cross-linking also cause problems, as does the fact that oil resistance decreases sharply with increase in EPDM levels.

9.6.1.3 NBR–NBR Blends

Where nitrile rubbers of widely differing acrylonitrile content are blended, heterogeneous phases may occur. Hopes that the presence of some low acrylonitrile rubber might lower brittle points without adversely affecting oil resistance have not been fulfilled.

9.6.1.4 NBR and PVC

NBR–PVC blends have been known since 1936. Because of similarities in their solubility parameters, reasonably homogeneous blends may be obtained providing that the two polymers are fluxed above the T_g of the polyvinyl chloride. For PVC contents below 50% the NBR is vulcanized.

The principal advantage of these blends is that, if the PVC content is at least 30%, they have excellent resistance to ozone. Because of this they have achieved technical importance in such areas as cellular thermal insulation, fire and irrigation hose, cable jackets and soft rollers. It is to be noted, however, that increasing the PVC level also raises the brittle point and compression set.

9.6.1.5 NBR and Phenolic Resins

As with PVC, the phenolic resins have similar solubility parameters to NBR, and homogeneous blends are possible. During processing the phenolics have a low viscosity and act as a processing aid. If, however, they are used with curing agents they will cross-link at the same time that the NBR is vulcanized to give hard products of low density. Under certain circumstances the phenolics may also act as vulcanizing agents for the NBR.

9.7 SPECIAL GRADES OF NITRILE RUBBER

9.7.1 Acrylonitrile–Isoprene Rubber (NIR)

A commercial NIR copolymer with 34% acrylonitrile content was available for a number of years but never acquired economic importance. As might be expected from structural considerations, the polymer had a higher T_g than the corresponding NBR materials and was consequently less rubbery and stiffer. Somewhat surprisingly the polymer was capable of strain crystallization, giving vulcanizates of good gum stock strength, whilst the uncured stocks exhibited higher tack and were easier to extrude. Terpolymers of butadiene, acrylonitrile and styrene are marketed by Nippon Zeon of Japan.

9.7.2 Hydrogenated Nitrile Rubber

One serious shortcoming of nitrile rubber is its limited heat resistance compared with some of the newer engineering rubbers such as the acrylics and fluoro-rubbers. This is generally ascribed to the double bonds present in the polymer. If these double bonds are removed by hydrogenation then it may be expected that improved heat resistance will be obtained. In the mid-1980s Bayer introduced such a hydrogenated NBR under the trade name of Therban, the initial grade having a nitrile content of 17%, corresponding to an acrylonitrile content of about 34% in conventional NBR.

As this is a saturated polymer it is necessary to use a non-sulphur curing system, and peroxides with triallyl cyanurate or triallyl isocyanurate are recommended. FEF thermal blacks are preferred for reinforcement whilst optimum ageing resistance is achieved by use of magnesium oxide and a diphenylamine derivative in conjunction with ZMBT.

In addition to improved heat resistance, hydrogenated NBR exhibits a number of other outstanding features. These include:

1. Excellent wear resistance;
2. very low brittle temperature (−52°C);
3. high hot-tear resistance;
4. high tensile strength;
5. low compression set;
6. very good weathering and ozone resistance;
7. good resistance to many oil additives and to H_2S and amines present in crude oil.

The first three properties are related to the low T_g of the rubber.

The material is competitive with fluoro-rubbers in oil drilling, nuclear power plant and automotive applications. Recent work has been reviewed by Milner (1987).

In 1987 Bayer announced the availability of grades that were only partially hydrogenated and could be sulphur cross-linked, and a further grade with an acrylonitrile content of 44%. They also announced that they were increasing capacity from 100 to 1000 tpa.

9.7.3 Carboxylated NBR (XNBR)

Terpolymerization of butadiene, acrylonitrile and either acrylic or methacrylic acid will yield a rubbery polymer containing carboxylic acid side groups. Known, somewhat loosely, as *carboxylated nitrile rubbers* and with the ASTM designation XNBR, these materials may be cured using normal sulphur vulcanizing systems. They may also, more importantly, be vulcanized ionically via the carboxylic acid side groups with, for example, zinc oxide.

The ionically cross-linked materials are of interest (Bryant, 1970) because of their high tensile strength, tear strength and abrasion resistance, and it is interesting to note that similar results may be obtained with carboxylated natural rubber that has been ionically cross-linked. The greater toughness of XNBR has led to uses with high-pressure seals.

Ionically cross-linked XNBR does, however, suffer a number of disadvantages. The cross-links are somewhat thermally labile and compression set resistance at elevated temperatures is only moderate. Scorching problems also occur, although these may be reduced by coating the zinc oxide particles with zinc sulphide or zinc phosphate (Hallenbeck, 1973) or by incorporating oligomerized fatty acids during polymerization to enhance scorch safety without affecting cure rate (Grimm, 1983, 1984).

Because of its limitations the use of XNBR in dry rubber technology has been small. Recently, however, XNBR has been increasingly recognized as an easy-processing abrasion-resistant material of use in a variety of industrial and military applications. Use in latex form has been more extensive. In such cases the problem of scorching does not arise and a variety of curing systems have been used. In addition to zinc oxide these include sodium aluminate and melamine–formaldehyde resins. The latex has been used in binders for non-woven fabrics, leather finishing and for oil-resistant coatings.

9.7.4 Cross-Linked NBR

These materials are analogous to cross-linked NR and SBR, and may be incorporated into NBR compounds as a rheological processing aid. Both hot- and cold-polymerized grades are available.

Cross-linked NBR is commonly made by adding about 1% of divinylbenzene during the polymerization, for example between 10 and 35% conversion. When blended with conventional NBR the cross-linked material provides spherical gel structures within the unvulcanized matrix of the conventional rubber; this improves calendering and extrusion behaviour as well as enhancing dimensional stability during open steam cure. The blends have lower compression set and better swelling resistance but many vulcanizate properties are adversely affected. For example, tensile strength, tear strength, elongation at break and resilience decrease whilst stiffness and hardness increase.

9.7.5 Alternating Copolymers

Solution polymerization of equimolar quantities of butadiene and acrylonitrile in hexane at 0°C and using a catalyst consisting of triethylaluminium, aluminium chloride and vanadyl chloride leads to a polymer exhibiting a high level of alternation in the monomer residues. Furthermore, the butadiene units are largely of the *trans*-1,4-configuration.

Because of this regularity of structure, both the raw polymer and the vulcanizates are capable of strain crystallization (this also implying some tactic homogeneity of the acrylonitrile units) so that the tensile strength of both cured and uncured stocks is higher than for conventional rubbers. Resistance to cut growth and creep tear are also said to be superior (Takamatsu *et al.*, 1973). Properties less dependent on structural regularity, such as oil resistance and low-temperature behaviour, are similar to those free-radically produced random copolymers.

Since alternation implies a fixed acrylonitrile content of about 48%, which is rather more than usually required, and because it is expensive to produce, alternating NBR has been largely of academic interest.

9.8 APPLICATIONS

Because of its moderate cost, processability and above all very good resistance to swelling by aliphatic hydrocarbons, NBR is the first material for consideration in respect of liquid fuel, hydrocarbon oil and grease applications. It is not always suitable in these applications because of limited resistance to aromatics and oxygen-containing liquids or where liquids contain sulphur-bearing additives or hydroperoxide impurities. The heat resistance may also be inadequate and for any one of these reasons it may be necessary to employ more expensive alternatives.

In spite of these limitations NBR is widely used in hydrocarbon-resistant applications, particularly for uses involving contact with petrol (gasoline). Consequently about 40% of the polymer produced is used by the motor vehicle industry. About 20% is used in machinery and a further 20% is used in food and textile equipment.

In terms of type of goods, about 40% is in the form of moulded technical goods, 25% hose and tube, 15% rubber rollers and 10% brake linings and clutches. The rest is distributed in such areas as cellular goods, coated fabrics and special footwear solings, together with a number of minor outlets.

The market for NBR may be described as mature and more dependent on the global economic situation rather than on new product development. In spite of the loss of some markets due to more rigorous technical specifications insisted on by users, in recent years nitrile rubber did not suffer as great a fall in consumption as did other rubbers such as SBR at the end of the 1970s.

BIBLIOGRAPHY

Bertram, H. H. (1981). In *Developments in Rubber Technology—2*, ed. A. Whelan & K. S. Lee. Applied Science, London, Chapter 3.

Dunn, J. R., Coulthard, D. C. & Pfisterer, H. A. (1978). *Rubber Chem. Technol.*, **51**, 389.

Hofmann, W. (1963). *Rubber Chem. Technol.*, **36**, 1.

Hofmann, W. (1968). In *Polymer Chemistry of Synthetic Elastomers*, Part 1, ed. J. P. Kennedy & E. G. M. Törnqvist. Interscience, New York, Chapter 4B.

Milner, P. W. (1987). In *Developments in Rubber Technology—4*, ed. A. Whelan & K. S. Lee. Elsevier Applied Science, London, Chapter 2.

REFERENCES

Bertram, H. H. (1981). In *Developments in Rubber Technology—2*, ed. A. Whelan & K. S. Lee. Applied Science, London, Chapter 3.

Bryant, C. L. (1970). *Journal of the IRI*, **4**, 202.

Dunn, J. R. (1984). Paper presented at PRI-sponsored conference, Rubberex, Birmingham, 12 March 1984.

Dunn, J. R., Pfisterer, H. A. & Ridland, J. J. (1980). *Gummi Asbest. Kunststoffe*, **33**, 296.

Grassie, N. & Heaney, A. (1974). *J. Polym. Sci., Polym. Lett. Ed.*, **12**, 89.

Grimm, D. C. (1983). US Patent 4 415 690.

Grimm, D. C. (1984). US Patents 4 435 535, 4 452 936.

Hallenbeck, V. L. (1973). *Rubber Chem. Technol.*, **46**, 78.

Hofmann, W. (1968). In *Polymer Chemistry of Synthetic Elastomers*, Part 1, ed. J. P. Kennedy & E. G. M. Törnqvist. Interscience, New York, Chapter 4B.

Milner, P. W. (1987). In *Developments in Rubber Technology—4*. ed. A. Whelan & K. S. Lee. Elsevier Applied Science, London, Chapter 2.

Nakajima, N. & Harrell, E. R. (1980). *Rubber Chem. Technol.*, **53**, 14.

Scholtan, W., Lange, H., Casper, R., Pohl, W., Wendisch, D. & Mayer-Mader, R. (1972). *Angew. Makrom. Chem.*, **27**, 1.
Stannett, V. (1968). In *Diffusion in Polymers*, ed. J. Crank & G. S. Park. Academic Press, London, Chapter 2.
Takamatsu, T., Onishi, A., Nishikada, T. & Furukawa, J. (1973). *Rubber Age*, **105**(6), 23.
Wijayarthna, B., Chang, W. V. & Salovey, R. (1978). *Rubber Chem. Technol.*, **51**, 1006.

Chapter 10

Polychloroprene Rubbers (CR)

10.1 INTRODUCTION

In Chapter 9 it was seen that the introduction of a polar nitrile group improved the resistance of a butadiene-based rubber to hydrocarbon liquids. A broadly similar effect may be obtained by, in effect, replacing the methyl group in the polyisoprene repeating unit to obtain a polymer of 2-chloro-1,3-butadiene, generally known as polychloroprene.

$$CH_2{=}\overset{\displaystyle Cl}{\overset{|}{C}}{-}CH{=}CH_2 \longrightarrow {-}CH_2{-}\overset{\displaystyle Cl}{\overset{|}{C}}{=}CH{-}CH_2{-}$$

The development of polychloroprene arose when a team of Du Pont scientists, following up earlier work by Father J. A. Nieuwland of the University of Notre Dame, was trying to purify divinylacetylene. They carried out an experiment which quite unexpectedly resulted in the production of what was subsequently identified as polychloroprene. This was in April 1930, and in November 1931 Du Pont announced the availability of the polymer. Although it was marketed initially as Duprene, within a short while the manufacturers changed the name of their product to Neoprene.

The original materials were highly branched and with a high gel content were difficult to process. In 1939 copolymers with sulphur were produced which could then be peptized, i.e. broken down into smaller molecules, by thiuram disulphides. These rubbers, marketed by Du Pont as the Neoprene G series, were much easier to process.

In 1949 the discovery that mercaptans were effective chain transfer agents made it possible to produce homopolymers of controlled molecular weight and thus reasonably easy to process. Without the sulphur atoms in the main chain, which are present in the chloroprene–sulphur copolymers, the homopolymers also had an improved heat resistance. They were marketed by Du Pont as the Neoprene W series. More recently grades combining many of the optimum features of the two types have been introduced as Neoprene GW rubbers.

Apart from some Russian production, which started in 1932, Du Pont remained the only major manufacturer until 1960 when Bayer commenced the manufacture of Baypren. Other manufacturers now include Distugil in France who produce Butachlor, Denka Chemical in the USA and Denki Kagaku Kogyo, Showa Neoprene and Toyo Soda in Japan.

The polychloroprene rubbers show a good combination of oil resistance, heat resistance, flame resistance and ozone resistance and have been important special-purpose rubbers for many years. However, in recent years they have met increased competition from other rubbers, for example EPDM, where better heat resistance than that obtainable from a diene rubber is required. There have also been some question marks over the toxicity of the monomer and over what is probably the best curing agent. For some years now world production has been of the order of 300 000 tpa.

10.2 PRODUCTION OF POLYCHLOROPRENE

For many years chloroprene was produced from acetylene via monovinylacetylene, a process in which explosions were not unknown to occur. Today manufacture is dominated by the butadiene route (developed by the Distillers Company in England) which is both safer and more economical. This process involves three stages: chlorination, isomerization and dehalogenation, as indicated below.

$$CH_2{=}CH{-}CH{=}CH_2 + Cl_2 \longrightarrow \underset{(60\%)}{CH_2Cl{-}CHCl{-}CH{=}CH_2} + \underset{40\%\ (cis\ \text{and}\ trans)}{CH_2Cl{-}CH{=}CH{-}CH_2Cl}$$

$$CH_2Cl{-}CH{=}CH{-}CH_2Cl \xrightarrow{\text{catalyst}} CH_2Cl{-}CHCl{-}CH{=}CH_2$$

$$CH_2Cl{-}CHCl{-}CH{=}CH_2 + NaOH \longrightarrow CH_2{=}CCl{-}CH{=}CH_2 + NaCl + H_2O$$

The monomer is a liquid boiling at 59·4°C.

Commercial polymerization is carried out in emulsion using free-radical initiators. Potassium and ammonium persulphate are commonly used as initiators and a typical polymerization is carried out at 40°C. The polymerization is stopped at the desired conversion by the addition of a free-radical inhibitor such as a thiuram. The resulting latex is then coagulated, for example by feeding it onto a freeze drum which is at about

−15°C. The resultant film is then dried, stripped off and shaped into a rope-like form. At one time the polymer was sold in short lengths of this rope-like form but nowadays this is cut up into small slices.

10.3 STRUCTURE OF CR AND STRUCTURAL VARIABLES

As with other rubbers, processing properties tend to be influenced mainly by differences in macrostructure such as branching and cross-linking, whilst microstructure is of importance in controlling the elastomeric properties.

In the absence of chain-transfer agents, gel polymer occurs at conversions from monomer to polymer at levels as low as 30%. Reductions in gel levels are therefore achieved either by copolymerizing the chloroprene with sulphur to produce in-chain sulphide links which may be broken down by thiurams to produce smaller molecules, or by using a chain-transfer agent such as a mercaptan compound.

Chloroprene polymers may in theory take up the same steric forms as those of isoprene. In practice the *trans*-1,4- structure is dominant with only small proportions of *cis*-1,4-, 1,2- and 3,4- monomer residues in the chain. The small amount of 1,2-material (*c.* 1·5%) does however have an important role to play in the vulcanization process. The various structures and typical proportions for a commercial polymer are shown in Fig. 10.1. It should however be noted that the amount of non-*trans* material increases regularly with increasing polymerization temperature, from a value of 5% at a polymerization temperature of −40°C to about 30% at 100°C.

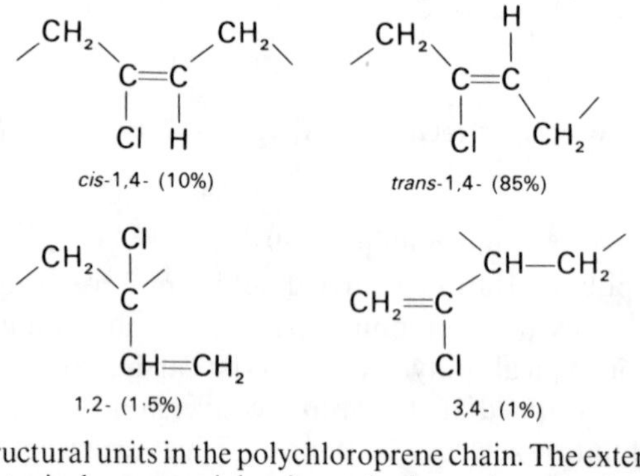

$-CH_2-C(Cl)=C(H)-CH_2-$

cis-1,4- (10%)

$-CH_2-C(Cl)=C(H)-CH_2-$

trans-1,4- (85%)

$-CH_2-C(Cl)(CH=CH_2)-$

1,2- (1·5%)

$-CH(C(Cl)=CH_2)-CH_2-$

3,4- (1%)

FIG. 10.1. Structural units in the polychloroprene chain. The extent of each type in a typical commercial polymer is shown in parentheses.

Although such a structure is somewhat less regular than that of the natural rubber molecule, such polychloroprene polymers are capable of crystallization, a typical polymer manufactured at 40°C being about 12% crystalline and with a T_m of about 45°C. Because of this ability to crystallize it is possible to obtain vulcanizates of good tensile strength without the use of carbon black. Since the level of non-*trans* material increases with increasing polymerization temperature, a rise in this temperature will reduce the polymer regularity and hence the ability to crystallize.

There are about 150 grades of CR on the market; the main variables between grades are the following.

1. The microstructure, particularly the level of non-*trans* material. One manufacturer at least produces a rubber with improved low-temperature flexibility by increasing this level and thus reducing the ability to crystallize. Conversely, some adhesive grades with so-called 'quick grab' characteristics are prepared by polymerization at lower temperature to increase crystallinity levels.
2. Molecular weight and molecular weight distribution. Whilst a typical peptizable sulphur copolymer, Neoprene GN, is reported to have a wide molecular weight distribution ranging from 20 000 to 950 000 with the greatest frequency around 100 000, a typical homopolymer (Neoprene W) has a much narrower distribution with a maximum frequency around 180 000–200 000.
3. Presence or otherwise of sulphur as a comonomer to give a peptizable polymer (this is more fully discussed in Section 10.4).
4. Presence or otherwise of a second monomer such as 2,3-dichloro-1,3-butadiene. This will reduce the regularity, reduce the tendency to crystallization and improve elastomeric properties at low temperature, and is an alternative to controlling crystallinity levels by varying polymerization temperature. Copolymers of this type include at least one of the most useful grades of polychloroprene elastomers.
5. The extent of gel. Whilst excessive gel makes processing very difficult, a small amount of gel in the rubber improves the processability. (Similar effects have already been observed with NR, SBR and NBR.)

The properties of the various types of CR resulting from these structural variables will be discussed in the next section.

10.4 STRUCTURE AND PROPERTIES

It is instructive to compare the structures of BR, CR and NR (Fig. 10.2). Ignoring differences in stereoregularity it is clearly seen that the difference between the three materials is the presence of the methyl group in NR and the chlorine atom in CR.

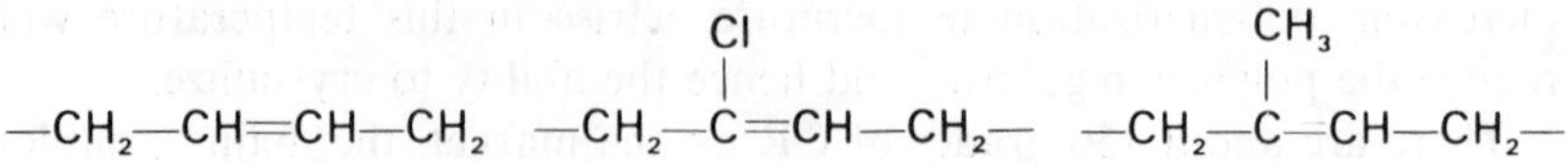

$$-CH_2-CH{=}CH-CH_2 \qquad -CH_2-\overset{\overset{\displaystyle Cl}{|}}{C}{=}CH-CH_2- \qquad -CH_2-\overset{\overset{\displaystyle CH_3}{|}}{C}{=}CH-CH_2-$$

(a) (b) (c)

FIG. 10.2. Repeating units of BR (a), CR (b) and NR (c).

Both the methyl group and the chlorine atom tend to stiffen the chain backbone and thus, when compared with polybutadiene, raise T_g and T_m in the case of CR (−43°C and +45°C respectively). In addition the presence of the chlorine atom has other effects, principally:

1. It increases the hydrocarbon oil resistance of the rubber;
2. it imparts a measure of flame resistance;
3. it reduces considerably the chemical reactivity at and around the double bond. One consequence of this is that the polymer has a better resistance to oxygen and ozone than natural rubber. At the same time it is no longer possible to vulcanize effectively with sulphur-based systems.

From what has been said above, it follows that the key properties of polychloroprene rubbers are as follows.

1. Better heat resistance than diene hydrocarbon rubbers (but inferior to IIR, EPDM, NBR and most speciality engineering rubbers);
2. very good resistance to oxygen, ozone and sunlight (but inferior to most of the saturated rubbers);
3. normally self-extinguishing, the flame resistance may be further improved by appropriate use of additives;
4. good resistance to hydrocarbon oils (although bettered by most non-hydrocarbon rubbers);
5. high resilience similar to natural rubber but less dependent on filler loading (Fig. 10.3);
6. high tensile strengths of both gum stocks (equalled only by NR) and black-loaded systems;
7. high abrasion resistance;
8. moderate low-temperature resistance.

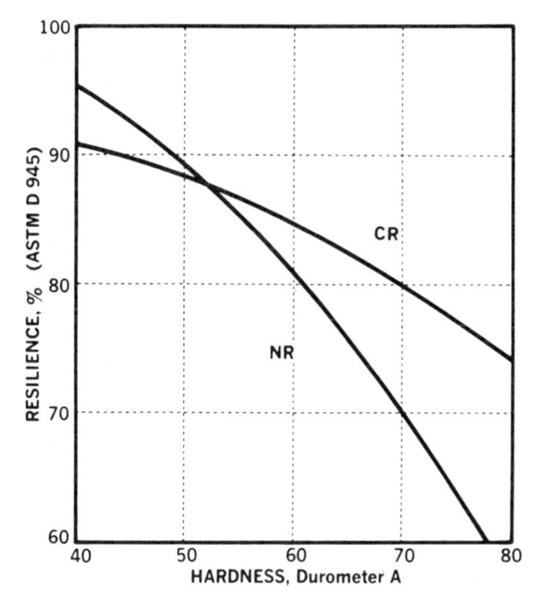

FIG. 10.3. Dependence of resilience of NR and CR on filler loading. (From Murray & Thompson, 1963.)

Commercial polychloroprenes may be divided into the following main classes:

1. Peptizable chloroprene–sulphur copolymers.
2. Homopolymers.
3. Copolymers of chloroprene and 2,3-dichlorobutadiene to give non-crystallizing polymers with good low-temperature properties.
4. Polymers of the above types but containing higher-than-normal proportions of gel to improve processing.
5. Adhesive grades that are fast-crystallizing with good 'quick grab' properties and which rapidly develop a high cohesive strength without curing.

The most important comparison is between the homopolymers and the peptizable copolymers, each of which have a number of distinctive properties, albeit somewhat blurred by the introduction of newer grades.

In terms of product stability and processing behaviour, the regularity of the homopolymer facilitates crystallization whilst the absence of a weak in-chain sulphide link gives both raw polymer stability and little mill-roll breakdown and roll sticking. For the same reason the homopolymers are

TABLE 10.1
Comparison of raw polymer properties and processing behaviour of CR homopolymers and sulphur-modified copolymers

	Homopolymers (mercaptan-modified)[a]	*S-copolymers (peptizable)*[b]
Raw polymer stability	Excellent	Fair–good
Crystallization rate	Fast	Slow–medium
Mill breakdown	Very little	Yes
Roll sticking	Very little	Yes
Nerve	Some	Low
Tack		More than homopolymer

[a] For example Neoprene W, Baypren 210 and Butachlor MC-30.
[b] For example Neoprene GN, Baypren 710 and Butachlor SC-20.

more 'nervy' and exhibit less building tack than the copolymers. These differences are summarized in Table 10.1.

Vulcanizates of the homopolymers are superior to the chloroprene–sulphur copolymers in the following respects:

1. Higher tensile strength;
2. lower compression set;
3. better heat resistance.

On the other hand, the peptizable copolymers are superior in respect of:

1. Tear resistance;
2. flex cracking resistance;
3. hot strength;
4. adhesion to NR and SBR substrates;
5. (higher) modulus and hardness.

In 1977 Du Pont introduced Neoprene GW which, although it is a sulphur-modified polymer, has a significantly lower sulphur content and gives vulcanizates of tensile strengths very similar to those obtained with homopolymers and compression set values closer to those of the homopolymers than the copolymers. Some typical properties of this polymer are compared with a typical homopolymer and copolymer in Table 10.2.

By copolymerizing chloroprene, or even chloroprene and sulphur, with 2,3-dichloro-1,3-butadiene, crystallization may be much reduced and low-temperature properties are much improved, the stiffening on cooling being

TABLE 10.2
Typical properties of vulcanizates from different polychloroprenes (After Bament & Pillow, 1981)

	Neoprene GW	*Neoprene GN-A*	*Neoprene W*	*Neoprene W*
Modifier	Sulphur	Sulphur	Mercaptan	Mercaptan
Ethylene thiourea	—	—	0·5	—
TMTD	—	—	0·75	—
Sulphur	—	—	—	1·0
TMTM	—	—	—	0·5
DOTG	—	—	—	0·5
Mooney viscosity 121°C (minimum value)	29	20	27	26
Cured 20 min at 160°C				
Modulus at 100% elongation, MPa	4·34	3·86	3·59	3·79
Tensile strength, MPa	19·72	18·34	19·66	21·52
Elongation, %	370	400	350	380
Hardness (Durometer A)	67	68	63	67
Tear strength (ASTM D624 'C'), $kN\,m^{-1}$	50	45	42	48
Compression set (D395 'B') 22 h at 100°C	36	71	25	48
De Mattia flex kc to 13 mm cut growth	120	50	5	17
Aged 7 days at 121°C				
Tensile strength, % retained	87	93	91	86
Elongation, % retained	50	37	50	40
Hardness, points change	+9	+13	+7	+13

controlled by the position of the glass transition temperature of the copolymer. This improvement in low-temperature flexibility may be accompanied by some loss in tensile and tear strengths at normal ambient temperature as well as being somewhat more nervy in processing.

The use of gel-containing CR rubber to facilitate processing has been practised since 1959, originally by adding a gel-containing polymer such as Neoprene WB, Baypren 214 or Distigul ME-20 to normal CR. In 1970 Du Pont introduced a range of medium-gel rubbers which are used directly, rather than as an additive to a standard grade (the Neoprene T series). Vulcanizates of these rubbers have similar mechanical properties to their homopolymer counterparts and in addition show less batch-to-batch variation than tends to be obtained when blending a high-gel rubber into a standard grade.

Many grades of polychloroprene rubber are used for adhesives, among them certain grades that have a high crystallization rate developed especially for solution adhesives; this results in quick grab characteristics and the development of a high cohesive strength without curing. As mentioned in Section 10.4 the high level of crystallinity is obtained by polymerization at lower-than-usual temperatures in order to increase the levels of *trans* material and hence the structural regularity.

10.5 PROCESSING

The processing characteristics of the homopolymers and the peptizable copolymers are markedly different, largely as a result of the presence of an unstable sulphide link in the main chain of the peptizable copolymer.

Whilst no polychloroprene rubber should be subjected to undue and unnecessary heating before vulcanization, particular care should be taken with the peptizable sulphur copolymers. If this is not done, compounds may be liable to scorching problems. In this context it is useful to consider the 'heat history' of the compound: this takes into account both temperature and time for which the material has been at a given temperature. It has been found that equal areas under a plot of an inverse logarithmic function of temperature against time are roughly equivalent in heat-history effect. For example, one day at 38°C is roughly equivalent to two days at 28°C or four days at 18°C. With such materials, heating should be kept to a minimum during processing and batch stocks should be sheeted out at not more than 1 cm thick and cooled as rapidly as possible.

On mastication the homopolymers show little breakdown; this is reminiscent of the butadiene polymers. On the other hand, the sulphur-containing peptizable copolymers do break down (with consequent breakdown in polymer viscosity), in this case reminiscent of natural rubber with a faster breakdown rate on a cold mill than a warm one (Fig. 10.4). Low-sulphur copolymers of the Neoprene GW type exhibit much less breakdown and approximate more closely to the non-sulphur types.

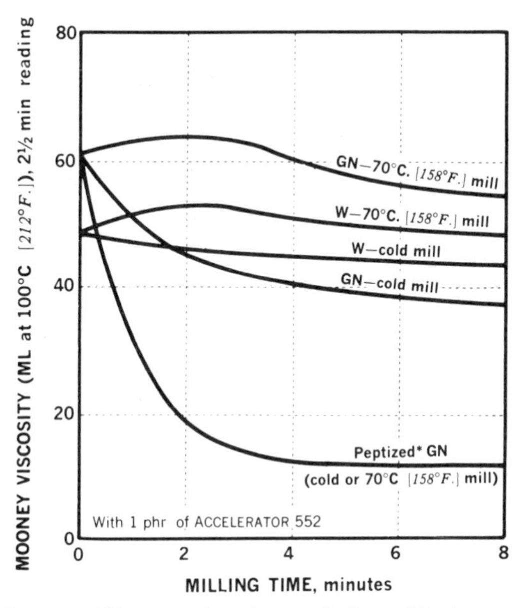

Fig. 10.4. Effect of milling on viscosity: variation with time and temperature: Data are for Du Pont Neoprene grades. (From Murray & Thompson, 1963.)

Breakdown of the sulphur-containing peptizable copolymers may be accelerated by the addition of a peptizer (in addition to the thiuram added during the polymer manufacture). Piperidinium pentamethylene dithiocarbamate at 0·25–1·0 pphr is the preferred material since although other materials such as DOTG, DPG and MBTS do function as peptizing agents they also reduce scorch safety. When peptizers are used the rate of breakdown is only marginally influenced by temperature.

When mixing, the best dispersion is obtained if additives are incorporated at a low temperature (<70°C) when the material is in an elastic phase. In practice it may be difficult to maintain such a low temperature but if it is possible to start adding fillers before this temperature is exceeded, the increase in stock viscosity caused by the presence of the filler will help the dispersion.

Moulding may be undertaken using compression, transfer or injection methods. Whilst typical temperatures are of the order of 170°C, curing may be carried out in the range 150–205°C.

When extruding, a cool barrel (40°C) and screw (60°C), warm head (70°C) and hot die (95°C) are usually recommended.

10.6 COMPOUNDING

The compounding of CR differs from the materials discussed in earlier chapters in that vulcanization has to be based on a non-sulphur system. The selection of other additives such as fillers, plasticizers and antidegradants is also somewhat specific.

10.6.1 Curing Systems

The chlorine atom in the polymer repeating unit rather deactivates the area around the main chain double bond so that this is not involved in curing processes. The cure site on the chain is in fact the highly reactive tertiary allylic chlorine present in the 1,2-units in the chain. As a consequence of this the curing system typically contains the following components:

1. Zinc oxide;
2. magnesium oxide;
3. accelerator;
4. retarder.

The effect of zinc oxide level on vulcanizate modulus is shown in Fig. 10.5(a). Most published formulations quote its use at a level of 5 pphr; increasing this value may improve heat resistance but will reduce scorch resistance. The zinc oxide is usually used in conjunction with 4 pphr of magnesium oxide, which not only helps to enhance modulus (Fig. 10.5b) but is also useful in improving scorch resistance. For best results a highly active grade of magnesia should be used, an increase in activity number raising modulus, tensile strength and scorch time. Such activity may be rapidly reduced by exposure to the atmosphere, so the magnesia should be stored before use in moisture-proof containers (Fig. 10.6). The sulphur-containing copolymers are more susceptible to the level and activity of the oxides used than are the homopolymers.

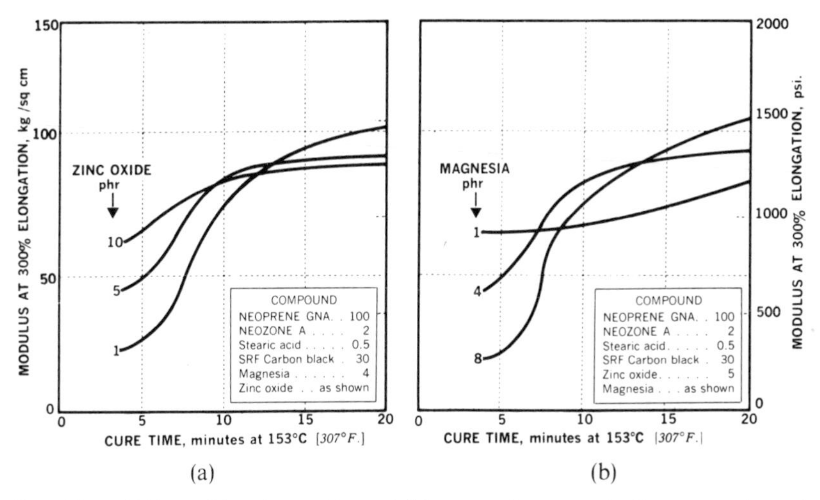

FIG. 10.5. Effect of (a) zinc oxide and (b) magnesia concentrations on rate and state of cure. (From Murray & Thompson, 1963.)

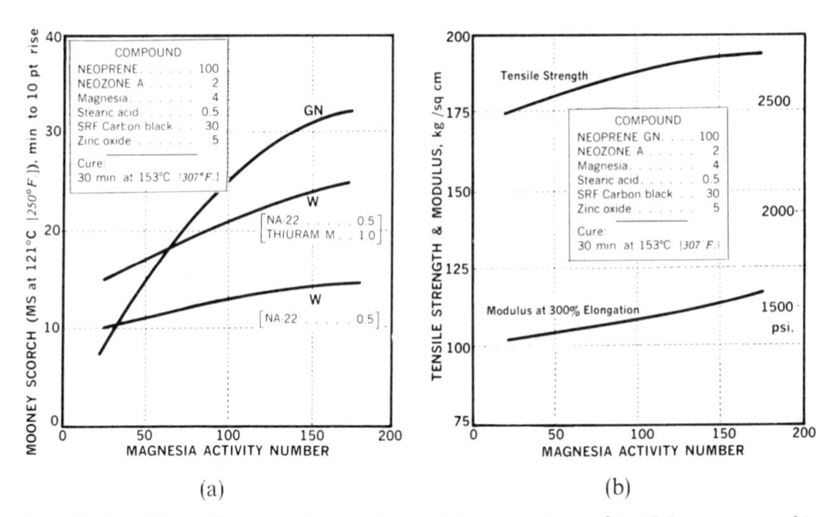

FIG. 10.6. Effect of magnesia activity on (a) processing safety (Mooney scorch), (b) tensile strength and modulus. Magnesia activity number as measured by iodine absorption (mg/g). (From Murray & Thompson, 1963.)

Where optimum water resistance is required the MgO/ZnO combination may be replaced with 20 pphr of red lead.

Whilst the sulphur-containing copolymers may be cured rapidly with the zinc and magnesium oxides alone, a high state of cure will require either longer cures or cures at a higher temperature or by the additional use of an 'accelerator'. In the case of the non-sulphur-containing polymers, such an accelerator becomes essential.

The most widely used accelerator is ethylene thiourea. The use of up to 1·0 pphr will lead to very substantial increases in modulus, tensile strength, hardness and resilience, together with a lower compression set and swelling in hydrocarbon oils. Unfortunately it has been found to have carcinogenic and other toxic effects on laboratory animals and this has led to widespread concern about its use. At the time of writing it continues to be widely used although normally handled only in masterbatch form, and prospective users should make themselves aware of the current situation. An alternative material is *N*-methyl-2-thiazolidinethione (Vulkacit CRV), which is very sensitive to magnesium oxide and to compound formulation. It also confers poor shelf-life to uncured compounds although it gives better scorch times than ethylene thiourea.

Other alternatives are ammonium hydrogen isophthalate (Vanax CPA), a good accelerator but rather expensive, and the DOTG salt of dicatechol borate. For uses where processing safety is important, an accelerator system comprising TMTM 1·0, DOTG 1·0, sulphur 1·0 pphr may also be used.

Since accelerators tend to shorten scorch times, it is quite common to consider the use of a retarder. TMTD, an accelerator for diene hydrocarbon rubbers, is usually preferred at a level of 0·5–1·0 pphr since not only does it act as a retarder at about 100°C but boosts cure at 150°C.

10.6.2 Other Additives

Whilst polychloroprene rubbers have good ageing resistance, performance may be improved by suitable choice of antioxidant. Octylated diphenylamine is widely used for this purpose (1–2 pphr) since it possesses the added virtue of not significantly reducing scorch resistance.

For applications which involve good flexing and/or ozone resistance an antiozonant should be used. For CR polymers diaryldiphenylamines are usually employed, but they do tend to be somewhat scorchy.

A wide range of materials has been used as plasticizers in CR. For low-cost applications coupled with high compatibility, naphthenic and aromatic petroleum oils are commonly used, the aromatic oils generally being more compatible but with increased staining and discolouration tendencies. Ester plasticizers such as butyl oleate can have a greater softening effect at normal ambient temperatures than the petroleum oils and have a greater depressing effect on the low-temperature stiffness point. However, this is partially offset by the fact that these esters, by increasing the polymer chain mobility at low temperatures, will induce additional crystallization, and their use must be considered with care. Polymeric esters are useful because of their low volatility whilst resins may improve tear and abrasion resistance, increase hardness and improve building tack. Low-molecular-weight CR will reduce viscosity during processing but co-vulcanizes with the normal CR and the vulcanizate hardness is hardly affected by their use. Somewhat similar effects may be obtained by the use of polymerizable acrylic esters such as tetramethylene dimethacrylate.

Unlike most synthetic rubbers, polychloroprene vulcanizates have a good gum stock tensile strength. Further improvement in many mechanical properties may be achieved with relatively large-particle-size blacks such as SRF with little, if any, further improvement if finer blacks are used. Where very good abrasion resistance is required, such as in high-quality conveyor belts, ISAF and HAF blacks may be employed, whilst for soft compounds MT blacks may be used.

10.7 APPLICATIONS

CR has now been commercially available for nearly 60 years and thus may be considered as a mature material. It has been widely accepted as an engineering material where its good strength, oil resistance, flame retardance, abrasion resistance, weathering behaviour and many other properties have proved of value. Typical applications include joint seals, bridge bearings, all-purpose hose, conveyor belting, cable sheathing, automotive moulded parts, truck tarpaulins and electrical connectors.

An approximate breakdown of areas of use is as follows.

Miscellaneous mouldings	50%
Latex and adhesives	20%
Wire and cable	15%
Belting	10%
Miscellaneous	5%

In recent years the technical specifications for many rubber components have become more demanding. As a consequence there are a number of one-time applications of CR rubbers that now use newer, and usually more expensive, rubbers. Nevertheless the polychloroprenes do have a good balance of properties and where it has been established that NR, SBR and EPDM rubbers are unsatisfactory for a given application they will usually be the next material to be considered.

BIBLIOGRAPHY

Bament, J. C. & Pillow, J. G. (1981). In *Developments in Rubber Technology—2*, ed. A. Whelan & K. S. Lee. Applied Science, London, Chapter 5.

Hargreaves, C. A. (1968). In *Polymer Chemistry of Synthetic Elastomers*, Part 1, ed. J. P. Kennedy & E. G. M. Törnqvist. Interscience, New York, Chapter 4C.

Hargreaves, C. A. & Thompson, D. C. (1965). 2-chlorobutadiene polymers. In *Encyclopaedia of Polymer Science and Technology*, Vol. 3. Wiley, New York.

Johnson, P. R. (1976). *Rubber Chem. Technol.*, **49**, 650.

Murray, R. M. & Thompson, D. C. (1963). *The Neoprenes*. E. I. Du Pont de Nemours & Co. Inc., Wilmington, Delaware.

REFERENCES

Bament, J. C. & Pillow, J. G. (1981). In *Developments in Rubber Technology—2*, ed. A. Whelan & K. S. Lee. Applied Science, London, Chapter 5.

Murray, R. M. & Thompson, D. C. (1963). *The Neoprenes*. E. I. Du Pont de Nemours & Co. Inc., Wilmington, Delaware.

Chapter 11

Acrylic and Other Rubbers with Ester Side Chains

11.1 INTRODUCTION

In Chapters 9 and 10 it was shown that the introduction of polar groups into the polymer structure gave enhanced resistance to swelling in hydrocarbon fluids as compared with hydrocarbon polymers. The materials discussed in these two chapters (NBR and CR) were, however, prepared using either butadiene or 2-chlorobutadiene so that double bonds are present in the polymer chain. These reactive sites, albeit somewhat less reactive in CR, left the rubber with some susceptibility to attack by oxygen, ozone and, in the case of NBR, to sulphur-bearing oils. Furthermore, the maximum service temperatures attainable were less than possible with less unsaturated hydrocarbon rubbers such as EPDM and IIR.

In consequence, over the years a number of rubbery polymers have been developed which contain polar groups to impart resistance to hydrocarbon oils but which do not contain reactive carbon–carbon double bonds. Such materials will form the subject matter of this and several subsequent chapters. This chapter will be concerned largely with the acrylic rubbers but will also deal with some ethylene–vinyl (or acrylic) ester copolymers which have many similar properties. Polyester rubbers in which the ester group is found in the chain backbone will be dealt with in Chapters 15 and 16.

11.2 ACRYLIC RUBBERS (ACM)

A large number of polymers and copolymers have been prepared using acrylic monomers and many have found commercial use in plastics, fibres, adhesives and surface coatings. These monomers may be considered as having been derived from acrylic acid (**I**). Many of these are acrylic esters of general formula **II** and are typified by ethyl acrylate (**III**). Closely related are methacrylic esters such as methyl methacrylate (**IV**) and the nitrogen-containing acrylonitrile (**V**). In the plastics industry the adjective acrylic is often used as a noun to denote polymethyl methacrylate (well-known under such names as Perspex, Plexiglas, Oroglas and Diakon), whilst the same adjective is used in the fibres industry as a noun to denote polymers and copolymers largely derived from acrylonitrile. The nitrile rubbers discussed in Chapter 9 were made using acrylonitrile in minor proportions. The acrylic rubbers to be discussed in this section have a polymerization feed dominated by acrylic monomers.

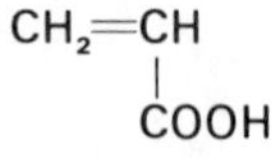

$CH_2{=}CH{-}COOH$ (I) $CH_2{=}CH{-}COOR$ (II) $CH_2{=}CH{-}COOC_2H_5$

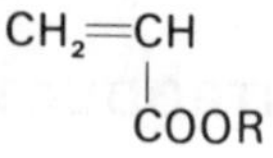

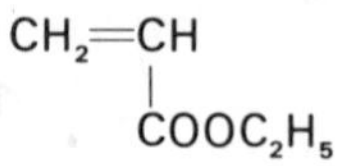

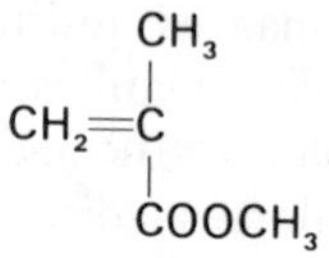

$CH_2{=}C(CH_3){-}COOCH_3$ (IV) $CH_2{=}CH{-}CN$ (V)

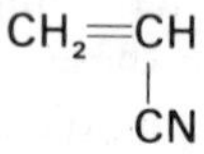

The acrylic rubbers were first introduced commercially in 1948 by B. F. Goodrich as a direct consequence of earlier work by the Eastern Regional Laboratory of the United States Department of Agriculture in the 1940s. This laboratory introduced two materials on the small scale: Lactoprene EV and Lactoprene BN. The EV was a copolymer of ethyl acrylate and 2-chloroethyl vinyl ether and the BN a copolymer of butyl acrylate and acrylonitrile. The Goodrich product Hycar PA-21, later renumbered Hycar 4021 and which for many years had an important market role, could be considered as a direct descendant of Lactoprene EV, whilst Hycar 2121X38 was a descendant of the BN grade.

In 1963 both the Thiokol Corporation and American Cyanamid commenced manufacture of acrylic rubbers and they have since been joined by a number of other manufacturers in the United States, Europe and Japan.

World production of dry ACM rubber is estimated to be about 10 000 tpa with about twice this amount used in latex form. In respect of the dry rubber about 80% of consumption, at least as far as the USA is concerned, is comprised of seals, O-rings and gaskets, largely for automotive use. Its use here arises not only because it is possible to operate at higher temperatures than with NBR but because it does not react with sulphur-bearing 'extreme-pressure' oil additives. With NBR such additives tend to cause embrittlement, presumably by a cross-linking mechanism arising from the presence of a double bond.

11.2.1 Chemistry

The designation ACM is given to 'copolymers of ethyl or other acrylates and a small amount of a monomer which facilitates vulcanization'. The polymerization may be by either free-radical or anionic mechanisms, the former being preferred for rubber manufacture, in part because it leads to less regular and hence more amorphous materials. Although polymerization in emulsion is preferred to bulk, solution and suspension methods, one problem is the readiness of the monomers to hydrolyse, particularly under basic conditions. For this reason soaps such as sodium oleate, widely used in the emulsion polymerization of other rubbers, are best avoided and the salts of long-chain sulphonic acids used instead.

If one considers the polymers of methyl, ethyl and *n*-butyl acrylate it will be clear that an increase in the length of the side group will have two important effects: the T_g will be depressed but the oil resistance will be reduced. Thus the T_g of polymethyl acrylate is $+8°C$, that of the ethyl polymer $-24°C$ and that of poly-*n*-butyl acrylate as low as $-54°C$. Whilst the low T_g (and the related low brittle temperature) of the butyl acrylate polymer is highly desirable in a rubber to be used for automotive and other uses liable to operate at low service temperatures, the polybutyl acrylate seriously lacks oil resistance.

The conflicting requirements of good elastic properties at low temperatures and good oil resistance led to considerable development work along the following lines of approach.

1. Use of a higher acrylate than ethyl acrylate in order to lower the T_g, together with a more polar monomer to improve oil resistance.
2. Distancing the polar group, such as a —CN group, from the main chain to maintain oil resistance but to have little influence on T_g.
3. Insertion into the side chain of —O— or —S— links which are highly flexible and which may lead both to a reduction in T_g and the degree of swelling in hydrocarbon oils.

Examples of the first approach are shown in Fig. 11.1 (Tucker & Jorgensen, 1968). It is interesting to note that *n*-butyl acrylate/acrylonitrile copolymers, used in early formulations, actually exhibit a higher swell for a given freeze point than *n*-butyl acrylate/ethyl acrylate copolymers.

The effect of distancing the polar group from the main chain is seen when cyanoethyl acrylate is used as one of the monomers. In view of developments which will be considered in the next paragraph it is interesting to note that good results were also obtained using 2-(2-cyanoethoxy)ethyl acrylate.

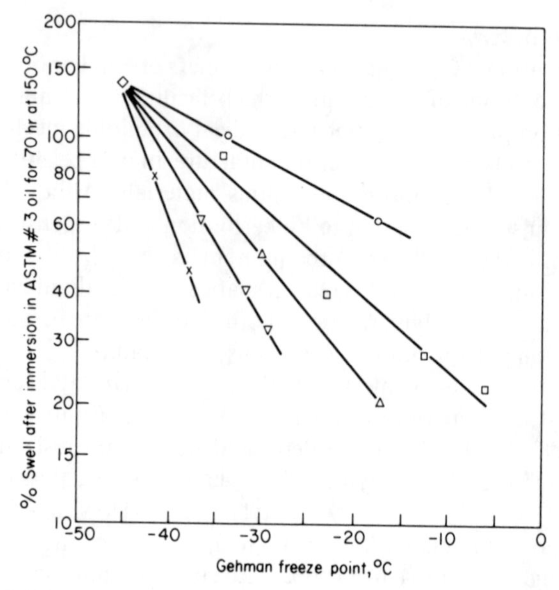

FIG. 11.1. Balance of oil resistance and Gehman freeze point for various copolymers of *n*-butyl acrylate. ◇, *n*-BA only; ○, *n*-BA/*N*,*N*-dimethylacrylamide; □, *n*-BA/acrylonitrile; △, *n*-BA/ethyl acrylate; ▽, *n*-BA/2-cyanoethyl acrylate; ×, *n*-BA/1,1,4-trihydroperfluorobutyl acrylate. (From Tucker & Jorgensen (1968), reproduced by permission of John Wiley & Sons, Inc.)

Insertion of —O— or —S— links into the side chain may both depress the T_g and improve the hydrocarbon oil resistance. This led to the commercial development of rubbers which are copolymers of either ethyl or butyl acrylate with an alkoxy alkylacrylate comprising some 20–50% of the total composition. Typical of such alkoxyalkyl compounds are methoxyethyl acrylate (**VI**) and ethoxyethyl acrylate (**VII**):

$$CH_2{=}CH{-}COO{\cdot}CH_2{\cdot}CH_2{\cdot}O{\cdot}CH_3 \quad \textbf{(VI)}$$

$$CH_2{=}CH{-}COO{\cdot}CH_2{\cdot}CH_2{\cdot}O{\cdot}C_2H_5 \quad \textbf{(VII)}$$

Commercial ACM rubbers today not only contain combinations of monomers to give various levels of oil resistance and brittle point, but in addition a monomer to provide an active cure site. Although homopolymers such as polyethyl acrylate can be cross-linked by peroxides, diamines and strong bases, the results are not satisfactory.

Many cure site monomers have been described in the literature but most of them fall into one of four groups:

1. dienes—to provide residual unsaturation;
2. halogen-containing monomers;
3. epoxides;
4. miscellaneous acrylic materials.

Of the dienes, butadiene appears to have been the first to have been studied. It made possible accelerated sulphur vulcanization but to the detriment of heat and oil resistance. Non-conjugated dienes, used with EPDM rubbers, have been quoted in the patent literature, such as tetrahydrobenzyl acrylate (**VIII**):

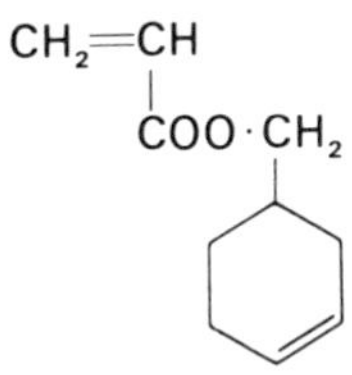

$$CH_2{=}CH{-}COO{\cdot}CH_2{-}C_6H_9 \quad \text{(cyclohex-3-enyl)}$$

(VIII)

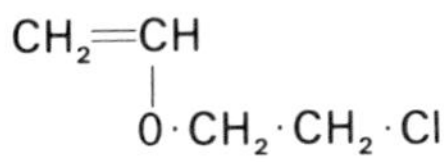

$$CH_2{=}CH{-}O{\cdot}CH_2{\cdot}CH_2{\cdot}Cl$$

(IX)

$$CH_2{=}CH{-}OOC{\cdot}CH_2{\cdot}Cl$$

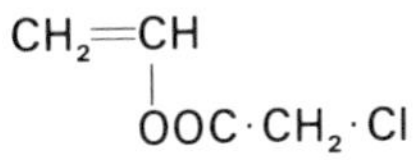

(X)

The commercial market is dominated by reactive halogen-containing cure site monomers. From 1948 to 1963 2-chloroethyl vinyl ether (**IX**) was used almost exclusively. ACM grades using this cure site monomer were generally amine-cured but gave rise to a number of problems such as limited bin stability, poor processing safety, sticking to mill rolls and mould fouling. In 1963 vinyl chloroacetate (**X**) was introduced. In this cure site monomer the chlorine is activated by the adjacent carbonyl group and has allowed the rubber to be vulcanized with cure systems that do not suffer some of the disadvantages of the earlier amine-cured vulcanizates. Subsequently several other activated halogen-containing monomers have been studied (**XI**, **XII**, **XIII**):

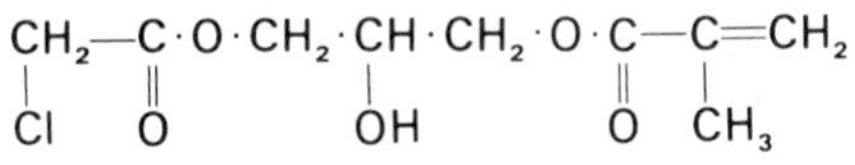

$$ClCH_2{-}C({=}O){\cdot}O{\cdot}CH_2{\cdot}CH(OH){\cdot}CH_2{\cdot}O{\cdot}C({=}O){-}C(CH_3){=}CH_2$$

(XI)

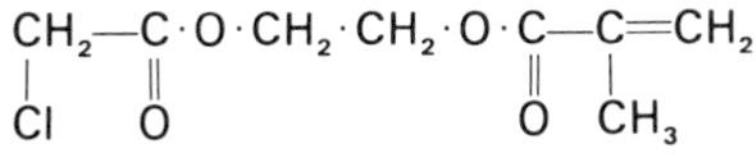

$$ClCH_2{-}C({=}O){\cdot}O{\cdot}CH_2{\cdot}CH_2{\cdot}O{\cdot}C({=}O){-}C(CH_3){=}CH_2$$

(XII)

$$ClCH_2{-}C({=}O){\cdot}O{\cdot}CH_2{\cdot}CH{=}CH_2$$

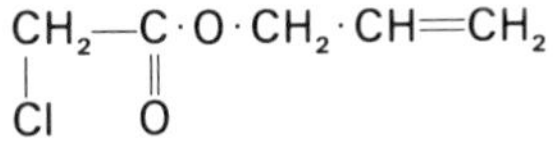

(XIII)

The introduction of epoxide groups into the polymer chain, for example by the use of allyl glycidyl ether, has also provided a cross-link site (**XIV**):

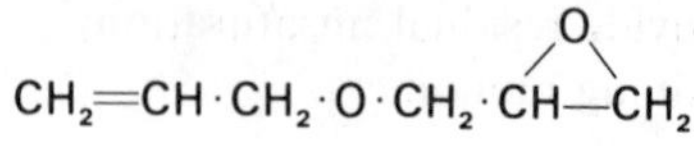

$$CH_2{=}CH \cdot CH_2 \cdot O \cdot CH_2 \cdot \overset{O}{\overbrace{CH{-}CH_2}}$$

(XIV)

Several acrylic monomers have been investigated as cure site monomers. These include acrylic acid, which may be cross-linked ionically via a divalent zinc ion. Self-curing acrylic rubbers have also been prepared. These are based on conventional recipes containing ethyl acrylate, and possibly an alkoxyethyl acrylate together with an acrylamide such as methylol acrylamide or an alkoxymethyl acrylamide. On heating, reaction occurs between amine and methylol groups. The rate of cure may be accelerated by the use of an acidic material such as phthalic anhydride and the cured rubbers are claimed to be stable up to 150°C.

Monomers bearing reactive cure sites usually represent about 5% of the polymer composition.

11.2.2 General Properties

ACM rubbers find use largely as a consequence of their resistance to heat, hydrocarbon oils and, in particular, oil additives, especially sulphurized types used for lubrication under extreme-pressure conditions.

They can withstand limited exposure to 204°C and one week at 175°C with relatively little damage. They are thus superior in this respect to most synthetic elastomers, the silicones and fluoro-rubbers being notable exceptions.

As a polar polymer the oil resistance of an ACM rubber is comparable with that of a medium–high-acrylonitrile NBR. As a saturated rubber it has good resistance to oxidation and ozone attack but as an ester it is somewhat prone to hydrolysis. ACM rubbers are not recommended for use with ketones, esters, phenols, some aromatic hydrocarbons and halogenated hydrocarbons.

Low-temperature flexibility depends on the grade of polymer used. Whilst some grades become brittle at the comparatively high temperature of −15°C, low-temperature grades are available that remain rubbery down to −40°C. This is, however, usually at the expense of hydrocarbon oil resistance.

Most of the newer grades have lower compression set than the early grades. A typical formulation cured for 8 min at 170°C and then post-cured for 24 h at 150°C had a compression set value of 24% after 70 h at 150°C and 37% after 70 h at 175°C (ASTM-D 395 Method B, Buttons).

Tensile strengths are quite good for an oil-resistant rubber, with post-cured compounds having a tensile strength as high as 16·6 MPa (2410 psi), but the corresponding values for elongation at break of such post-cured stocks are low, at about 150%. It is, however, possible with judicious choice of grade and cure schedule to achieve elongation at break figures in excess of 300%.

11.2.3 Vulcanization and General Compounding

Over the years a variety of curing systems have been used for ACM rubbers. Many are no longer used but still today there exists a diversity of systems. Before discussing these it is instructive to consider briefly some of the earlier systems.

The earliest acrylic rubbers did not have a special cure site and required cure by such materials as amines and metasilicates. The introduction of cure site monomers, such as 2-chloroethyl vinyl ether which was used almost exclusively from 1948 to 1963 led to the development of systems based on amines, ammonium salts and related materials. Amongst the amines used were triethylene tetramine, tetraethylene pentamine, Trimene Base (a proprietary reaction product of ethyl chloride, formaldehyde and ammonia marketed by Uniroyal), ethylene thiourea (in conjunction with a lead salt), hexamethylene diamine carbamate and *N,N′*-dicinnamylidine-1,6-hexane diamine.

With the introduction of vinyl chloroacetate and other cure site monomers more active than 2-chloroethyl vinyl ether, ammonium salts such as ammonium adipate and benzoate were introduced as curatives. These did not show the poor processing behaviour, lack of bin stability and other disadvantages of amine-cured stocks but did however tend to cause pitting of high-carbon steel moulds and in many cases it was necessary to use moulds made from stainless steel.

A substantial step forward occurred with the development of the soap/sulphur system (Holly *et al.*, 1965). This system provided good processing safety, economy, convenience and freedom from many of the problems of earlier systems. It consists of a mixture of some 3–6 pphr of an alkali metal salt of a fatty acid such as sodium oleate with about 0·25–0·35 pphr of sulphur, a soap/sulphur ratio of 10:1 being favoured. The soap is considered to be the curative and the sulphur the accelerator. The actual amounts of these two components and other adjuncts depend on the following considerations.

1. Cure rate is increased by increasing either soap or sulphur levels. Bases such as magnesium oxide will act as activators whilst acids such as stearic acid will retard cure. Faster cures are obtained with potassium salts rather than sodium salts.
2. Compression set is reduced by reducing sulphur levels, albeit at the expense of a reduction in cure rate.
3. Ageing resistance may be improved by an increase in soap level (or by replacing the sulphur by another accelerator).
4. If the soap is stearate there may be a reduction in tensile strength if more than 6 pphr is used.
5. The use of 1·5 pphr of phenylene-1,3-bismaleimide in addition to the soap and sulphur is claimed to increase cure rate, modulus and tensile strength without affecting scorch and compression set.

Whilst the soap/sulphur system provided an important stage of development, it has now been largely replaced, mainly because of the poor heat resistance of the vulcanizates. The specific choice tends to differ according to the grade of ACM used. One popular system replaces the sulphur with 2–6 pphr of 3-(3,4-dichlorophenyl)-1,1-dimethylurea (marketed as Poly-Disp T(KD)D-75). For many grades this is a good general-purpose system with good scorch safety, fast cure and low compression set.

Where scorch safety is particularly important the following system may be used.

		pphr
Curative	2,5-Dimercapto-1,3,4-thiadiazole	0·6–2
Accelerator	Tetrabutyl thiuram disulphide	2–3
Activator	Lead stearate	0·5–1·5
Retarder	Zinc stearate	0–2

Low compression set may be obtained using a system based on sodium stearate, tetramethyl thiuram disulphide and mercaptobenzothiazole disulphide. Another quite different system is recommended by one supplier (Enichem) and is based on a proprietary quaternary ammonium salt (Eveite B18) and *o*-tolylbiguanidine. Compared with the soap/sulphur system this confers better air ageing and water resistance as well as lower compression set (Lauretti *et al.*, 1984).

It needs to be stressed that since the nature of the cure site and the reactivity vary in commercial acrylic elastomers, they are not usually directly interchangeable from supplier to supplier in a given compounding formulation. Two general points may, however, be made.

1. The cure mechanism for most acrylic rubbers is alkaline in nature and hence cure is accelerated by bases such as magnesium oxide and retarded by acids such as stearic acid.
2. Despite significant advances in cure technology, all commercially available acrylic elastomers require a relatively long cure cycle or they must be post-cured to develop optimum compression set resistance.

In addition to the curing system (of curative, accelerator, activator and retarder), the following may be used in ACM rubbers.

1. Reinforcing fillers. FEF (N-550) black is most commonly used for this purpose although other blacks may be used. Silica fillers are less effective although their performance may sometimes be improved by the use of vinyl- and amino-type silane coupling agents.
2. Surface lubricants such as fibrous fillers and/or graphite may be used to reduce lip seal friction in rotary shaft seal applications.
3. Whilst ACM rubbers have inherently good oxidation resistance some further improvement may be achieved with 1–2 pphr of low-volatility diphenylamine antioxidants.
4. Plasticizers may be used to reduce the low-temperature brittle point but they must have a low volatility to withstand both hot-air post-curing as well as subsequent service at elevated temperatures. Wherever possible it is preferable to use a grade of ACM specifically formulated for low-temperature work where this property is important in the vulcanizate.
5. Lubricants such as 1–3 pphr of stearic acid in conjunction with proprietary process aids of undisclosed composition may be used to improve processing characteristics, but if they are used in excess problems of mould knitting, blooming and bonding may arise.

The following may be regarded as a typical formulation for an ACM compound.

		pphr
Rubber	ACM	100
Curative	Sodium stearate	4
Accelerator	Poly-Disp T(KD)D-75	2
Retarder	Stearic acid	1
Filler	FEF Black	65
Lubricant	TE-80 (proprietary material)	2

(No antioxidant is included in this formulation.)

11.2.4 Processing

Raw acrylic rubbers are relatively soft and, in order to achieve sufficient shear for good dispersion, low mill mixing temperatures should be used. Because of sticking problems, however, it is generally preferable to mix using internal mixers, once again using full cooling water.

Typical compression moulding cure schedules are 2 min at 190°C or 1 min at 200°C. At the same temperatures injection moulding cure schedules are usually about 25% less than this. Increasing cure time will increase hardness, modulus and tensile strength but reduce elongation at break and tear resistance. Post-curing is important to reduce compression set and is typically carried out in an air circulating oven for 4–8 h at 175°C. Increasing the post-cure time will also have similar effects to increasing cure time.

11.3 ETHYLENE–METHYL ACRYLATE RUBBERS (AEM)

AEM rubbers were introduced by Du Pont in 1975 under the trade name of Vamac. The rubbers are terpolymers of ethylene, methyl acrylate and a cure site monomer of undisclosed composition and may be represented as

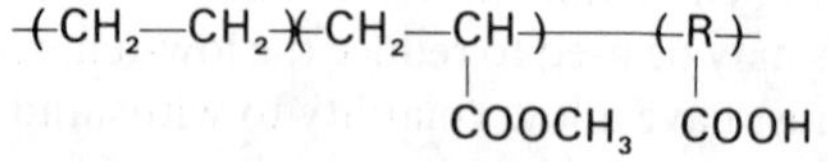

$$\left(CH_2-CH_2 \right)\left(CH_2-\underset{\displaystyle COOCH_3}{\underset{|}{CH}} \right)\left(\underset{\displaystyle COOH}{\underset{|}{R}} \right)$$

Du Pont remains the only supplier and production is estimated to be in the range 1000–2000 tpa. Whereas acrylic (ACM) rubbers tend to be used for specifications where nitrile rubbers fail to meet the temperature requirements, the ethylene–methyl acrylate rubbers are best considered for use where a chloroprene rubber fails to meet such a requirement.

11.3.1 Properties

As a saturated polar ester rubber of irregular structure, AEM exhibits good resistance to ageing and weathering, good resistance to aliphatic hydrocarbons, susceptibility to polar liquids, and poor resistance to strong acids and other hydrolysing agents; as amorphous polymers AEMs require reinforcing fillers to obtain a good tensile strength.

Their key properties in approximate order of importance may be summarized as follows.

1. Very good heat ageing resistance. The ASTM D2000/SAE J200 classification gives a 70-h upper service temperature rating of 175°C, which is more than 50°C higher than for NBR and CR and some 25°C higher than for polyacrylates, epichlorhydrin rubbers, EVA (see Section 11.4) and CSM, and is only bettered by silicone and fluoro-rubbers.

2. Very good weathering resistance. The influence of sunlight, water, oxygen and ozone has little adverse effect.
3. Good aliphatic hydrocarbon oil resistance, although highly aromatic fluids such as ASTM Oil No. 3 can cause extensive swelling (of the same order as for CR and CSM but much more than for ACM, NBR, T, FKM, FFKM or Q rubbers). Resistance to water and ethylene glycol is very good but some proprietary coolant additive packages can cause deterioration, mostly by embrittlement. AEM rubbers are not suitable for contact with non-mineral oil brake fluids, esters or ketones. Resistance to petrol (gasoline) is poor but diesel fuel and kerosene may be tolerated quite well.
4. Good low-temperature flexibility with medium-hardness compounds able to pass simple flex tests at −40°C.
5. Amine-cured rubbers (but not peroxide-cured) exhibit low compression set after post-curing.
6. Unmodified polymer will burn but compounds containing sufficient hydrated alumina have a good flame resistance, emit smoke of low density and do not give off corrosive gases.
7. A high capacity for vibration damping little affected by temperature over the range −30 to +160°C. In this respect it is superior to IIR at high temperatures and in the presence of oils.

11.3.2 Compounding

The first consideration to be made when compounding is the selection of polymer. At the time of writing the range of materials available is small and may be divided into two groups: masterbatches and gum polymers. The initial grades supplied were masterbatches containing fillers and antioxidants. Gum polymers are now available and since their use lowers the compound cost they are now the normal basis for further compounding. Two grades are available, one with a higher green strength which facilitates the use of compounds with high plasticizer levels.

The usual curing system consists of a combination of hexamethylene diamine carbamate (marketed as Diak No. 1) and a guanidine. For low compression set it is common to use 1·5 pphr of the carbamate with 4 pphr of DOTG, whereas for flex resistance 1·25 pphr of the carbamate is recommended with 4 pphr of DPG. Alternatives to hexamethylene diamine carbamate quoted in the literature are triethylene tetramine, a proprietary ketimine (Epicure H2) and another proprietary amine (Vulcofac CA 64). For very thick mouldings the guanidine is replaced by a proprietary secondary amine (Armeen 2C).

Peroxide cures may be used for cable applications where fast cures under continuous pressurized vulcanization are required but they are not recommended for moulded articles where poor hot strength and mould sticking may cause difficulties. Somewhat unusually, compression set resistance is poor when peroxide cures are used.

Carbon black, usually SRF, is the preferred filler, whilst mineral fillers must be selected with care, fumed silicas having the least effect on ageing. Aluminium hydroxide is a useful flame-retarding filler but its use has an adverse effect on heat ageing and also may over a period of time form ionic cross-links with the polymer.

Particular care has to be taken to avoid contamination of these rubbers with many metallic oxides because of the tendency of the latter to cause premature cross-linking by ionic reactions, although small amounts of iron and chrome oxides may be tolerated as pigments.

The choice of plasticizer if required must also be made with care. A normal ester plasticizer, although depressing cold-flex temperatures to about −50°C, is volatile at high temperatures and reduces upper service temperatures to about 125–135°C. Polyesters have much lower volatility but have little effect on low-temperature properties. Most commonly used are proprietary mixed ether/esters such as Thiokol TP759 and Nycoflex ADB-30, which may be used in the range −45 to +165°C.

Whilst antioxidants are beneficial, antiozonants are unnecessary.

To prevent sticking to processing machinery, release agents should be used, one widely accepted system being a combination of stearic acid, octadecylamine and an alkyl phosphate, the latter to be minimized in non-black and peroxide-cured systems.

The above principles are encapsulated in the following compound for use in automotive bellows for a constant-velocity joint close to hot exhaust components and therefore requiring a greater heat resistance than is obtainable with polychloroprene rubber:

	pphr
AEM rubber (high green strength)	100
Diphenylamine antioxidant	2
Stearic acid	2
Alkyl phosphate	1
Octadecylamine	0·5
APF Black N683	70
Ether/ester plasticizer	35
DOTG	4
Hexamethylene diamine carbamate	1·5

11.3.3 Processing

The low viscosity and green strength are the governing factors controlling processing operations. Mixing should be carried out at temperatures as low as possible with release agents added at the commencement of mill mixing and upside-down techniques being used in internal mixing.

In extrusion the main problem is the low collapse resistance. This may be reduced by use of high-green-strength grade reinforcing fillers (silica or FEF) and minimum levels of plasticizer. The major problems in moulding are mould sticking, minimized by the use of hard chrome-plated moulds cleaned between cycles, and air trapping due to the low viscosity. Post-curing, typically for 4 h at 175°C, improves tensile strength, modulus and resistance to compression set, and stabilizes physical properties before commencement of evaluations of ageing behaviour.

11.3.4 Applications

Applications of AEM rubbers split fairly evenly into cable applications and automotive uses. In cables it is particularly important for ignition wire sheathing, and for medium-voltage cables with its high temperature rating and zero halogen content it helps to meet low toxicity and smoke-generating levels required in recent more stringent regulations for cables used, for example, in ships and submarines.

Automotive parts include shaft seals, coolant and power steering hose, high-temperature spark plug boots and the automatic bellows for constant-velocity joints, for which a formulation was given in the previous section.

11.4 ETHYLENE–VINYL ACETATE RUBBERS (EAM or EVA)

(The reader should take care not to confuse the ASTM designations EAM and AEM. In this section the more commonly used abbreviation EVA will be used.)

Ethylene and vinyl acetate may be polymerized in any ratio, and indeed a very broad spectrum of ethylene–vinyl acetate polymers are commercially available. Where only a small amount of vinyl acetate is present (e.g. 3%) the vinyl acetate serves to reduce the regularity of the polymer structure and hence its ability to crystallize, and the polymer may be considered as a softer and more flexible extension of the polyethylene range. At levels of 10–15% the degree of crystallinity is more limited and the polymer is

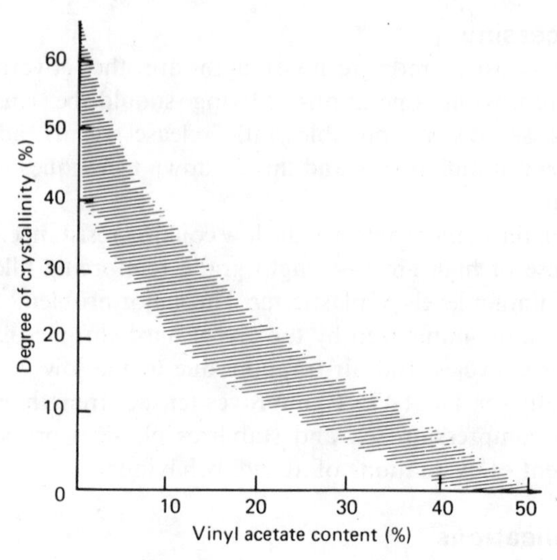

FIG. 11.2. Relationship between degree of crystallinity and vinyl acetate content. (From Gilby, 1982.)

similar in flexibility to a plasticized PVC compound. In the range 40–60% vinyl acetate, the materials are substantially amorphous and are elastomeric. The relationship between degree of crystallinity and vinyl acetate content is illustrated in Fig. 11.2.

In addition to the effect on regularity and hence the level of crystallization, an increase in the vinyl acetate content also increases the

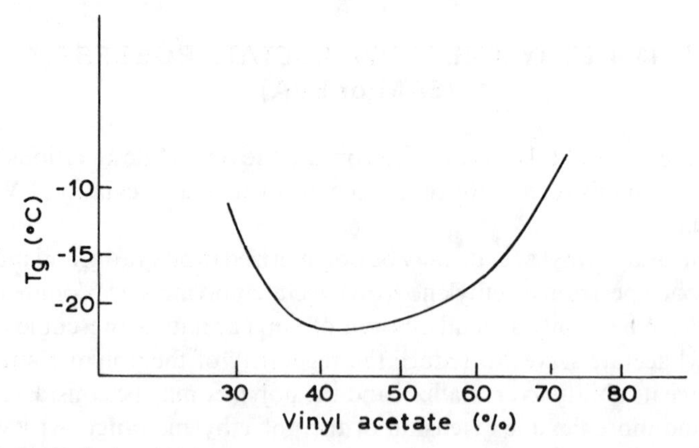

FIG. 11.3. Relationship between T_g and copolymer composition of EVA rubbers.

polarity of the copolymer. In the elastomeric range the main effect of this polarity is to improve resistance to hydrocarbon oils and to increase dielectric loss factor. In the range where the polymer is totally amorphous the stiffening temperature will be related to the T_g rather than to the crystalline melting point; as increases in vinyl acetate content lead to a higher T_g, copolymers with a vinyl acetate content above 60% are not practical rubbers (Fig. 11.3).

A third effect of increasing the vinyl acetate content is the increase in chain stiffness and this contributes to the increase in T_g alongside the increase in polarity.

EVA elastomers may be subdivided into two groups:

1. Substantially amorphous polymers normally requiring vulcanization (referred to below, somewhat arbitrarily, as *vulcanizable EVAs*);
2. partly crystalline polymers with various levels of rubberiness rather reminiscent of plasticized PVC and which are not normally vulcanized, referred to below, again somewhat arbitrarily, as *thermoplastic EVA rubbers.*

Useful reviews of these materials have been provided by Bartl (1972) and Gilby (1982), although the first is in German and the second largely deals with the thermoplastic rubbers.

11.4.1 Vulcanizable EVA Rubbers

Commercial grades of vulcanizable EVA rubbers are marketed by Bayer (Levapren) and USI (Vynathene). Most grades have vinyl acetate contents in the range 40–60%. Normally produced by solution polymerization using free-radical initiators, the mean number-average molecular weights have been estimated to be in the range 15 000–40 000 and this gives rise to low Mooney viscosities, usually with a nominal value for ML 1+4 (100°C) of 20.

Usage of the vulcanizable EVAs largely arises from the good dry heat, ozone and weather resistance. Heat resistance is usually rated as being slightly better than for EPDM, similar to ACM, slightly inferior to ethylene–acrylic (AEM) rubber and distinctly inferior to the fluoro- and silicone rubbers. Typical data suggest an indefinite life at 120°C, one year at 140–150°C and a few weeks at 180–200°C.

They are inferior to most polar rubbers, including polychloroprene and chlorosulphonated polyethylene rubbers, in terms of hydrocarbon oil resistance. Unvulcanized compounds dissolve in benzene, toluene and methyl ethyl ketone and swell in acetone, ethyl acetate and petroleum. Low-temperature properties may be quite good, with ASTM D-746 brittleness values below −50°C. Tensile strengths of up to 25 MPa and breaking elongations of 200–600% are reported. Hardness ranged from 60 to 85 Shore.

11.4.1.1 Compounding and Processing

The preferred vulcanizing agents are peroxides, the actual type selected depending on such factors as stability, volatility, price and odour. Typical examples are 2,5-bis(*t*-butylperoxy)-2,5-dimethylhexane, which is very stable but rather expensive, and di-*t*-butyl peroxide, which also has low odour and high stability but is rather volatile. Triallyl cyanurate is commonly used as a co-agent. Curing may also be brought about by high-energy irradiation.

Reinforcing fillers are necessary for good mechanical properties. In the case of mineral fillers it is important to check that acidity, for example in china clay, does not interfere with vulcanization.

Antioxidants are required only for the more rigorous applications; many of them interfere with cure in approximate proportion to the amount used. Among special antioxidants available are those of the polycarbodiimide type.

Both ester and mineral oil plasticizers may be used, but as with other additives it is important to check that these do not interfere with cure. Stearic acid may be used as a lubricant.

Low processing temperatures, e.g. 60–80°C for extrusion, may be operated and care should be taken with the very low melt viscosity at elevated temperatures when compared with other elastomers. Curing schedules strongly depend on the type of peroxide used. Post-curing may be undertaken to remove peroxide decomposition products.

11.4.1.2 Applications

These rubbers have found use where good heat resistance but not oil resistance is required and include moulded and extruded seals in heating installations and automobile applications. Such applications tend to fall just outside the operating range for EPDM rubbers but do not call for the more expensive fluoro or silicone materials.

11.4.2 Thermoplastic EVA Rubbers

The thermoplastic EVAs, some of which are rubber in nature, have been the subject of a recent review (Gilby, 1982).

BIBLIOGRAPHY

DeMarco, R. D. (1979). *Rubber Chem. Technol.*, **52**, 173.
Fram, P. (1964). Acrylic elastomers. In *Encyclopedia of Polymer Science and Technology*, Vol. 1. Interscience, New York, p. 226.
Gilby, G. W. (1982). In *Developments in Rubber Technology—3*, ed. A. Whelan & K. S. Lee. Applied Science, London, Chapter 4.
Lauretti, E., Mezzera, F., Santarelli, G. & Spelta, A. L. (1984). Paper presented at the Plastics and Rubber Institute Rubber Conference, Birmingham, UK, 12 March 1984.
Tucker, H. A. & Jorgensen, A. H. (1968). In *Polymer Chemistry of Synthetic Elastomers*, Part 1, ed. J. P. Kennedy & E. G. M. Törnqvist. Interscience, New York, Chapter 4D.
Vial, T. M. (1971). *Rubber Chem. Technol.*, **44**, 344.

REFERENCES

Bartl, H. (1972). *Kauch. u. Gummi Kunst.*, **25**, 452.
Gilby, G. W. (1982). In *Developments in Rubber Technology—3*, ed. A. Whelan & K. S. Lee. Applied Science, London, Chapter 4.
Holly, H. W., Mihal, F. F. & Starer, I. (1965). *Rubber Age (N.Y.)*, **96**, 565.
Lauretti, E., Mezzera, F., Santarelli, G. & Spelta, A. L. (1984). Paper presented at the Plastics and Rubber Institute Rubber Conference, Birmingham, 12 March 1984.
Tucker, H. A. & Jorgensen, A. H. (1968). In *Polymer Chemistry of Synthetic Polymers*, Part 1, ed. J. P. Kennedy & E. G. M. Törnqvist. Interscience, New York, Chapter 4D.

Chapter 12

Fluororubbers

12.1 INTRODUCTION

Where the highest levels of heat resistance possible in elastomers are required the choice lies between fluororubbers, silicone rubbers and the fluorosilicones. Members of the first class of materials, the subject of this chapter, may additionally possess other exceptional attributes such as outstanding resistance to weathering, burning, concentrated acids and swelling in hydrocarbons solvents.

The first fluorine-containing polymer to be produced was polychlorotrifluoroethylene (**I**) which was disclosed in patents in 1934, whilst the more successful polytetrafluoroethylene (**II**) was discovered in 1938. Both of these materials were thermoplastics and it was not until 1948 that the first elastomeric fluoropolymer, Fluoroprene (**III**), the fluorine-containing equivalent of CR, was introduced. This rubber, however, had properties that were far from outstanding and production was soon discontinued.

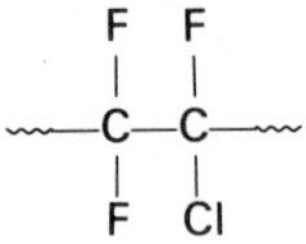

$$\sim\!\!\sim \overset{\displaystyle F}{\underset{\displaystyle F}{C}} - \overset{\displaystyle F}{\underset{\displaystyle Cl}{C}} \sim\!\!\sim$$

(I)

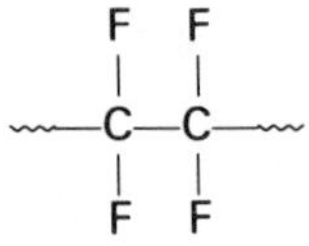

$$\sim\!\!\sim \overset{\displaystyle F}{\underset{\displaystyle F}{C}} - \overset{\displaystyle F}{\underset{\displaystyle F}{C}} \sim\!\!\sim$$

(II)

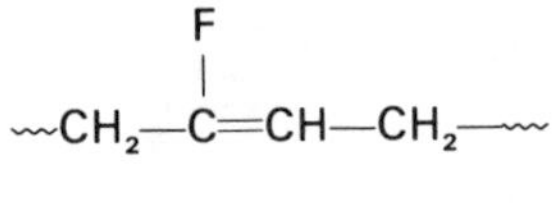

$$\sim\!\!\sim CH_2 - \overset{\displaystyle F}{C} = CH - CH_2 \sim\!\!\sim$$

(III)

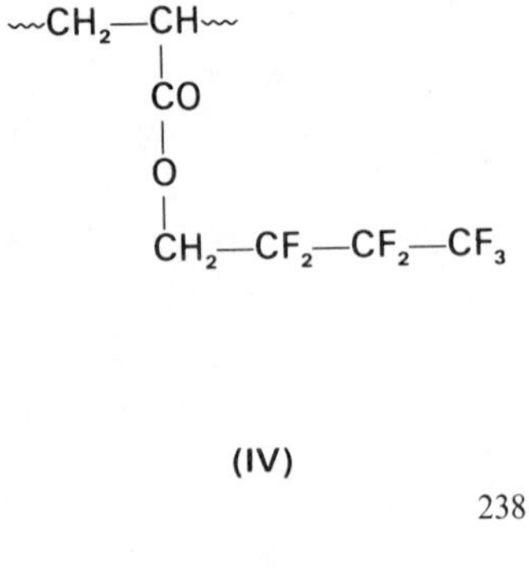

$$\sim\!\!\sim CH_2 - CH(-CO-O-CH_2-CF_2-CF_2-CF_3) \sim\!\!\sim$$

(IV)

$$\sim\!\!\sim CH_2 - CH(-CO-O-CH_2-CF_2-CF_2-O-CF_3) \sim\!\!\sim$$

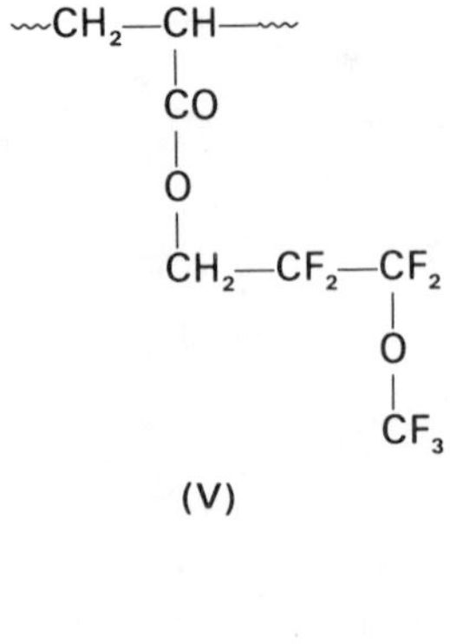

(V)

Shortly afterwards two polyfluoroacrylates (**IV** and **V**) were introduced but these are also no longer of any importance.

The fluororubbers only became established with the introduction of copolymers in which one of the monomers used was vinylidene fluoride (**VI**). Copolymers with chlorotrifluoroethylene (**VII**) were introduced in 1955 (as Kel-F Elastomer) and with hexafluoropropylene (**VIII**) as Viton and, subsequently, Fluorel, Tecnoflon and Dai-el. In the 1960s, largely to circumvent existing patents, rubbers were introduced in which one of the fluorine atoms in the hexafluoropropylene monomer molecules was replaced by a hydrogen atom to give 1-hydropentafluoropropylene (**IX**). With the expiry of the basic patents for vinylidene fluoride–hexafluoropropylene copolymers the raison d'être for such materials disappeared and they are no longer marketed. Subsequently terpolymers which also incorporated tetrafluoroethylene (**X**) were introduced, followed in the late 1970s by tetrapolymers containing a cure site monomer allowing cross-linking by means of peroxides. World capacity for fluororubbers is of the order of 10 000 tpa.

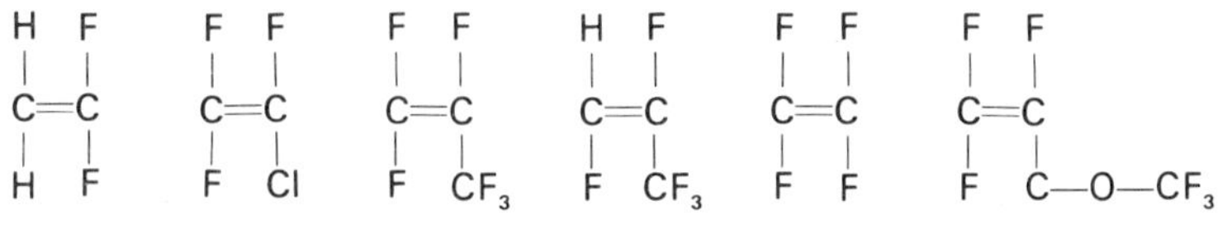

```
H  F     F  F     F  F      H  F      F  F     F  F
|  |     |  |     |  |      |  |      |  |     |  |
C==C     C==C     C==C      C==C      C==C     C==C
|  |     |  |     |  |      |  |      |  |     |  |
H  F     F  Cl    F  CF3    F  CF3    F  F     F  C—O—CF3

(VI)     (VII)    (VIII)    (IX)      (X)         (XI)
```

Copolymers and terpolymers based on vinylidene fluoride and either hexafluoropropylene or 1-hydropentafluoropropylene are given the ASTM designation FKM, whilst the copolymers and terpolymers based on vinylidene fluoride and chlorotrifluoroethylene are given the designation CFM.

Whilst the fluoroelastomer market is dominated by the vinylidene-fluoride-containing copolymers several other interesting fluororubbers have also become available. These include the copolymers of perfluoro(methyl vinyl ether) (**XI**) and tetrafluoroethylene (sometimes known as perfluorinated elastomer and marketed by Du Pont as Kalrez), the polyphosphazenes developed from an inorganic rubber first described in 1895 (Eypel), tetrafluoroethylene–propylene copolymers (Aflas) and the fluorosilicones. Also available on a developmental scale have been the carboxynitroso-rubbers, perfluoroalkylenetriazine elastomers and the poly(thiocarbonyl fluorides). These highly specialized materials have sometimes been accorded the classification of 'exotics'.

12.2 COPOLYMERS BASED ON VINYLIDENE FLUORIDE AND HEXAFLUOROPROPYLENE (FKM)

The FKM rubbers dominate the fluoroelastomer field and have been estimated to comprise about 95% of the market. They were originally introduced by Du Pont under the trade name Viton; there are today three further suppliers in Minnesota Mining and Manufacturing (Fluorel), Montefluos (Tecnoflon) and Daikin Kogyo (Dai-el).

The rubbers are prepared by free-radical emulsion polymerization, usually using peroxide initiators at temperatures from 80 to 125°C and pressures from 2 to 10 MPa. Molecular weight is controlled by varying monomer/initiator ratio or by the use of chain-transfer agents.

12.2.1 Grades

There are approximately 70 grades of FKM on the market. They may differ from each other in the following respects.

1. Presence or otherwise of tetrafluoroethylene (TFE) monomer units in the chain besides those of vinylidene fluoride (VDF) and hexafluoropropylene (HFP). About half of the FKM market comprises terpolymers of this type. Typically such materials have a 68% nominal fluorine content, compared with 65% for the copolymers, and as a result they have better long-term resistance to heat (particularly with respect to elongation and flexibility), resistance to swelling in oils and resistance to chemical degradation, particularly from oil additives. On the other hand the copolymers have a good balance of properties with better retention of tensile strength after high-temperature ageing, and some copolymer grades have outstanding compression set resistance.
2. Improved low temperature flexibility is obtained in at least one commercial material by incorporation of a small amount of perfluoromethyl vinyl ether in the polymerization recipe.
3. Replacement of hexafluoropropylene by 1-hydropentafluoropropylene. This was undertaken commercially largely as a patent circumvention exercise and is now of little importance.
4. Variation in molecular weight. Standard grades of typical copolymers have an average molecular weight of about 70 000 and those of VDF–TFE–HFP terpolymers, of about 120 000. Higher-molecular-weight grades are available with improved tensile and tear strength and low-molecular-weight grades for use in solutions, sealants and as a vulcanizable processing aid.

5. Inclusion of a bromotetrafluorobutene cure site monomer to allow peroxide curing. Such vulcanizates have particularly good resistance to aqueous fluids, steam, methanol and acids, and a lower brittleness temperature than VDF–HFP copolymers and VDF–HFP–TFE terpolymers.
6. A number of grades are supplied with proprietary vulcanization systems already included. These include the Viton E series, Viton B-910, the Fluorel 217 series, the Tecnoflon FOR series and Dai-el-G-701.

12.2.2 Compounding

With the exception of the materials described in the previous paragraph it is necessary to compound FKM rubbers with vulcanizing systems. A variety of approaches have been used of which the most important are amine derivatives, dihydroxy compounds, peroxides and high-energy irradiation.

Until the late 1960s the market was dominated by derivatives of aliphatic diamines. Aliphatic amines themselves are somewhat scorchy but this deficiency may be overcome by using blocked amines as their inner carbamates. The first such material to be developed was hexamethylene diamine carbamate (HMDA-C, marketed as Diak No. 1; **XII**) but this material tends to be scorchy with high-viscosity compounds and exhibits limited uncured storage stability. It is used at levels of 1–1·5 pphr. Rather more satisfactory is biscinnamylidene hexamethylene diamine (DCND, marketed as Diak No. 3; **XIII**), which provides the best balance of characteristics possible in an amine-based system. Blocked amine systems require the presence of a metal oxide in order to obtain commercially acceptable products with practical press cure times but even here a post-cure of up to 24 h at 200°C is required to develop optimum physical properties.

$^{+}H_3N—(CH_2)_6—N(CO_2^-)H$

(XII)

$[C_6H_5—CH{=}CH—CH—N(CH_2)_3]_2$

(XIII)

In practice about 15 pphr of magnesium oxide is used as the metal oxide although, in contrast to practice with polychloroprenes, a low-surface-activity oxide is preferred. Where resistance to water, aqueous chemicals and acids is required litharge may be used instead of magnesia. At the present time amine systems are largely restricted to food-contact applications. This restricted use arises from the fact that compound storage life is limited, particularly in humid conditions, because mould fouling occurs through interaction between amine and the metal mould surface and because in applications such as O-rings there is a loss in sealing force and ability to recover after compression.

Evidence has accumulated to show that amine cross-linking does not occur by a comparatively simple condensation with the polymer halogen group but by a more complex three-stage process involving:

1. Elimination of hydrogen halide to generate unsaturation;
2. reaction of difunctional agents at the double bonds formed; and
3. cross-linking (Bro, 1959; Paciorek *et al.*, 1960).

Much better results are obtained with aromatic polyhydroxy compounds used in conjunction with an 'onium' salt such as a quaternary ammonium compound or, more usually, a phosphonium compound. The polyhydroxy compounds used include bisphenols A, S and AF, the latter being most commonly employed.

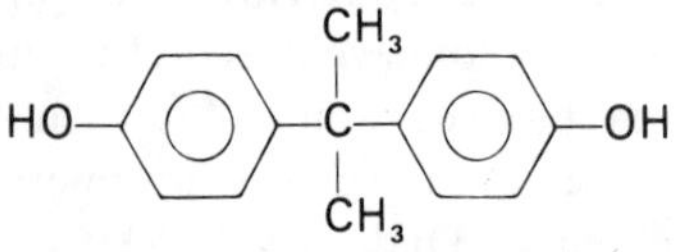

$$HO-C_6H_4-C(CH_3)_2-C_6H_4-OH$$

Bisphenol A

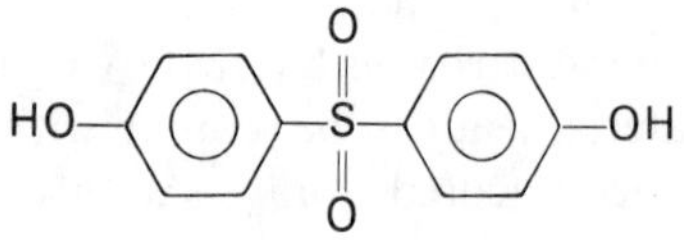

$$HO-C_6H_4-SO_2-C_6H_4-OH$$

Bisphenol S

$$HO-C_6H_4-C(CF_3)_2-C_6H_4-OH$$

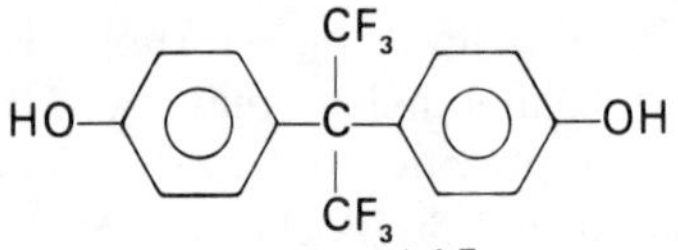

Bisphenol AF

Bisphenol cure systems also require the presence of a metal oxide, usually 3–5 pphr of *highly* active magnesium oxide. In addition 1·5–6 pphr of calcium hydroxide is recommended as a promoter, functioning possibly by the controlled release of water. Air-oven post-cure at 230–260°C is necessary for optimum properties.

A possible mechanism for the curing process is

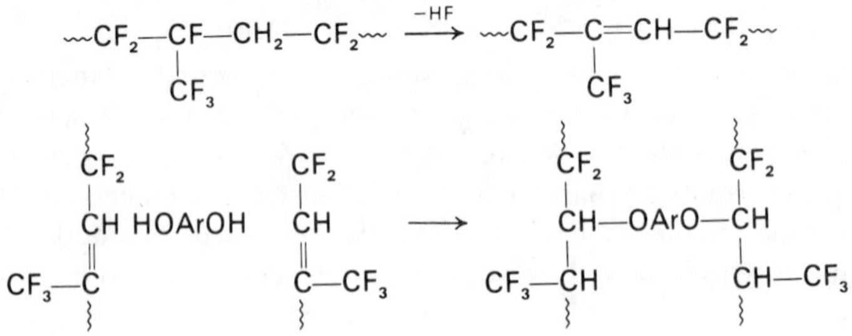

$$\sim CF_2-CF(CF_3)-CH_2-CF_2\sim \xrightarrow{-HF} \sim CF_2-C(CF_3)=CH-CF_2\sim$$

$$\begin{array}{c} \wr \\ CF_2 \\ | \\ CH \\ \| \\ CF_3-C \\ \wr \end{array} \quad HOArOH \quad \begin{array}{c} CF_2 \\ | \\ CH \\ \| \\ C-CF_3 \\ \wr \end{array} \longrightarrow \begin{array}{c} \wr \\ CF_2 \\ | \\ CH- \\ | \\ CF_3-CH \\ \wr \end{array} OArO \begin{array}{c} \wr \\ CF_2 \\ | \\ -CH \\ | \\ CH-CF_3 \\ \wr \end{array}$$

That this mechanism is not entirely correct has been implied by the observation that more than two cross-links can be produced per molecule of dihydroxy compound consumed.

FKM rubbers cured in this way have outstanding resistance to compression set up to 200°C, excellent compound storage stability and processing safety, and markedly reduced tendency to mould fouling.

Both amine and bisphenol systems generate water during cure and this can lead to problems such as splitting and degradation of mouldings. In order to overcome these problems the peroxide-curing rubbers were developed. Curing of these materials is effected by a curing system with three components, namely a peroxide, a co-agent and an acid acceptor as promoter. A favoured peroxide is 2,5-dimethyl-2,5-di(*t*-butylperoxy)-hexane, whilst triallyl isocyanurate or a proprietary triazine such as Diak No. 7 may be used as co-agent. Litharge and dibasic lead phosphite are preferred acid acceptors but low-activity magnesia may also be used (Brown *et al.*, 1977).

A variety of fillers may be used and amongst carbon blacks MT blacks are preferred for most applications whilst amongst non-black fillers blanc fixe, precipitated barium sulphate, is often used.

Most plasticizers are too volatile although liquid fluorosilicone rubbers may be used. Low-molecular-weight FKM rubbers may be useful as process aids, other materials being used for this purpose include pentaerythritol stearate and carnauba wax. A number of proprietary materials are also available from fluoroelastomer suppliers.

12.2.3 Processing

Mill mixing should be carried out with cool rolls, and after mixing stocks should be cooled as rapidly as possible. Shaping may be carried out by extrusion, calendering, compression moulding, transfer moulding and injection moulding. A typical injection moulding temperature would be in the range 177–200°C.

Post-curing in a ventilated oven is essential in order to realize optimum and stable properties. Post-curing times are usually 24 h at 200°C for amine cures and 230°C for dihydroxy and peroxide cure systems. For special compounds exhibiting minimum compression set, temperatures as high as 260°C may be used.

12.2.4 Properties

FKM rubbers find use because of their excellent heat resistance and their resistance to aliphatic and aromatic hydrocarbons, chlorinated solvents and petroleum fluids. They do however swell in ketones, monoesters, ethers

TABLE 12.1
Estimated continuous service life of FKM rubber in air at elevated temperatures (After Arnold *et al.*, 1973)

Temperature (°*C*)	*Service* (*h*)	*Temperature* (°*C*)	*Service* (*h*)
230	>3 000	290	240
260	1 000	315	48

and alkyl aryl phosphates. Unless compounded with litharge their resistance to hot water, steam and wet chlorine is rather poor. Ammonia and amines tend to cause embrittlement and they are also attacked by anhydrous HF and chlorosulphonic acid.

The continuous service life of FKM in air has been estimated and is summarized in Table 12.1.

The clear superiority in O-ring sealing force retention of FKM over other oil-heat-resistant rubbers is shown clearly in Fig. 12.1.

The low-temperature properties depend on the grade of FKM with Clash and Berg stiffness temperatures ranging from −13 to −30°C and

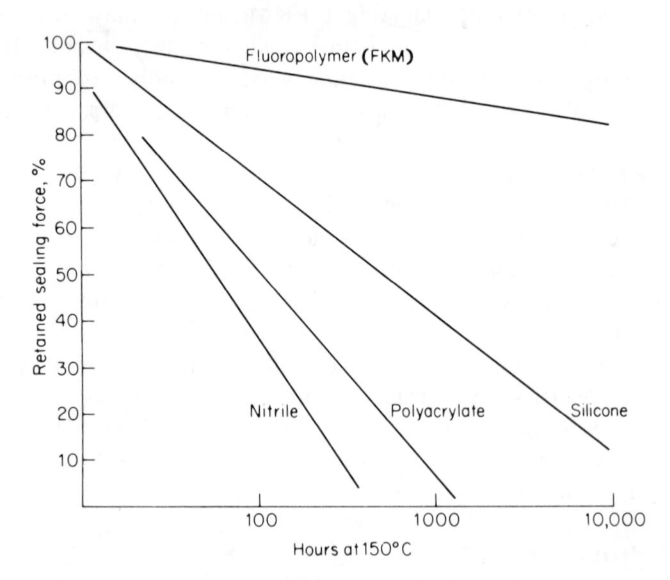

FIG. 12.1. Sealing-force retention of FKM fluorocarbon versus other elastomers used for seals. (From Sweet, 1979.)

TABLE 12.2
Typical mechanical properties of FKM rubbers

Raw polymer	
Specific gravity	1·72–1·86
Vulcanizate	
Hardness (Shore A)	60–95
Tensile strength, MPa	6·9–17·2
Elongation at break, %	100–300
Compression set,% (ASTM D395 Method B; 70 h at 200°C)	15–70

brittle points from −35 to −51°C. With many grades the function of the part may be impaired by stiffening at sub-zero temperatures.

As with other elastomers, the physical properties vary with grade and compound formulation. Table 12.2 indicates the range of values possible for some selected properties.

12.2.5 Applications

It has been estimated that about three-quarters of FKM consumption is for O-rings, packings and gaskets for the aerospace industry whilst automotive and other mechanical goods account for about 12%. Although the parts are expensive many motor manufacturers now appreciate the demand of their customers for reliability and increased service intervals. For this reason FKM is now used in valve-stem seals, heavy-duty automatic and pinion seals, crankshaft seals and cylinder liner O-rings for diesel engines. Other highlighted uses include seals for diesel engine glow plugs, seals for pilot-operated slide valves, protective suiting and flue duct expansion joints.

12.3 COPOLYMERS BASED ON VINYLIDENE FLUORIDE AND CHLOROTRIFLUOROETHYLENE (CFM)

CFM rubbers were first marketed in 1955, i.e. before the appearance of FKM rubbers, under the trade name Kel F Elastomer, by M. W. Kellogg. (The shorter trade name Kel F is used for the thermoplastic polychlorotrifluoroethylene homopolymer.) In 1957 manufacturing rights were given to the Minnesota Mining and Manufacturing Corporation, who marketed two grades, Kel F 3700 and Kel F 5500, which contained 70 and 50 mol% respectively of vinylidene fluoride. Subsequently rights were also given to

Ugine Kuhlmann in France who marketed corresponding materials as Voltalef 3700 and Voltalef 5500.

CFM rubbers are inferior to FKM types in respect of heat resistance but have better resistance to strongly oxidizing acids and to alkalies. The copolymer containing 70 mol% vinylidene fluoride has good low-temperature flexibility.

The preferred cross-linking systems for CFM elastomers are hexamethylene diamine carbamate, benzoyl peroxide or *p*-chlorobenzoyl peroxide.

The market for CFM rubbers is small compared with that for FKM.

12.4 PERFLUORINATED ELASTOMERS (FFKM)

In order to improve the thermal stability of fluorine-containing rubbers, chemists at Du Pont developed copolymers with no C—H groups (Barney, 1969). Copolymers of tetrafluoroethylene with 40 mol% of perfluoro(methyl vinyl ether) ($CF_2{=}CF(COCF_3)$) prepared by high-pressure emulsion polymerization have very good thermal stability and glass transition temperatures of about −10°C. This copolymer does not however provide a site for cross-linking.

By introduction of a third monomer cure sites may be introduced, and these have been described in the literature (Kalb *et al.*, 1973). In order to maintain the underlying excellent thermal, chemical and oxidation resistance of the basic copolymer, to avoid complications such as chain transfer during polymerization and to facilitate processing, the requirements of the cure site monomer become stringent. It has been found that in principle perfluorovinyl ethers of the type indicated below most nearly fulfil these requirements:

$$CF_2{=}CF{-}O{-}Rf{-}X$$

where Rf is perfluoroalkyl or perfluoroalkyl ether and X is —COOR, —CN or —O—ϕ. In particular, promising results were obtained using perfluoro(2-phenoxypropyl vinyl ether) (**XIV**) at about 2 mol%.

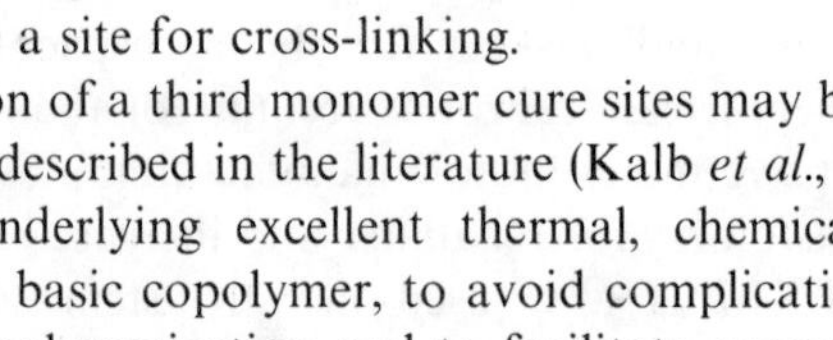

$$C_6F_5{-}O{-}CF(CF_3){-}CF_2{-}O{-}CF{=}CF_2$$

(XIV)

Vulcanization may be brought about by amines such as tetraethylene pentamine and hexamethylene diamine carbamate. It is believed that the process involves reaction of the diamine with the *para* fluorine atom on the aromatic ring.

$$2\ {\sim}O{-}C_6F_4{-}F + H_2N{-}R{-}NH_2 \longrightarrow {\sim}O{-}C_6F_4{-}NH{-}R{-}NH{-}C_6F_4{-}O{\sim} + 2HF$$

Magnesium oxide is present to react with the hydrogen fluoride and, as with other fluoroelastomers, a long post-cure process is required to expel the water formed. Other cross-linking agents quoted are the dipotassium salts of hydroquinone, of bisphenol A and of bisphenol AF. Another cross-linking agent quoted in the patent literature is tetraphenyltin (Du Pont, 1971).

Perfluororubbers of the FFKM type were introduced by Du Pont under the trade name Kalrez during the mid-1970s. However, because of processing difficulties, including high melt viscosities and very high post-curing temperatures, the material is only available from Du Pont as finished parts such as O-rings and sheet.

FFKM rubbers have a number of outstanding properties and are superior to the more conventional FKM rubbers in the following respects:

1. Continuous dry-heat resistance to 260°C and intermittent to 315°C, a level unmatched by any other elastomer;
2. they are virtually unaffected by both hydrocarbon and polar liquids;
3. resistance to the majority of chemicals that attack other elastomers such as acrylonitrile, amines, ketones, styrene, vinyl chloride, hot sodium hydroxide and fuming nitric acid;
4. resistance to oil-well sour gases.

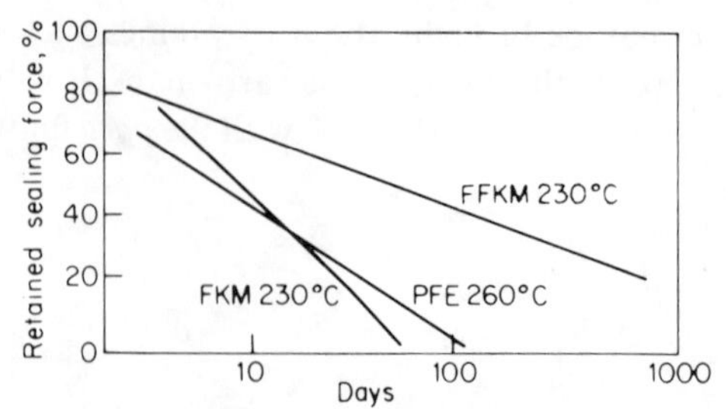

FIG. 12.2. FFKM versus FKM: effect of temperature on O-ring sealing-force retention. (From Sweet, 1979.)

5. resistance to high-temperature steam;
6. very good long-term high-temperature compression set resistance (Fig. 12.2);
7. resistance to outgassing under vacuum to 300°C.

The principal disadvantages of FFKM elastomers are:

1. The exceptionally high part cost, many times that for an identical part from FKM;
2. the severe limitation on size and complexity of finished parts because of the difficulties in processing;
3. the high coefficient of thermal expansion that may require redesign of seal housings when changing to FFKM;
4. slow recovery from compression at temperatures below 150°C;
5. tear resistance is only fair.

Typical data for dry heat ageing resistance are given in Table 12.3.

In passing it may be noted that the rubber compound supplied has a specific gravity of 2·01.

TABLE 12.3
Dry heat ageing resistance of an FFKM rubber of hardness 90 Shore A

Property	*Original before ageing*	*500 days at 230°C*	*112 days at 260°C*	*28 days at 290°C*	*7 days at 315°C*
Tensile strength, MPa	18·3	16·8	16·8	11·2	14·0
Elongation, %	120	220	240	320	230
Hardness, Shore A	90	87	87	86	87

In spite of their very high cost FFKM rubbers have found several uses, particularly for O-rings and seals in chemical plant, oil wells and refineries.

12.5 TETRAFLUOROETHYLENE–PROPYLENE COPOLYMERS

In 1976 the Asahi Glass company introduced under the name Aflas a copolymer of tetrafluoroethylene and propylene. It has found a niche in the marketplace largely on account of the broad range of chemical resistance exhibited and in particular where certain combinations of chemicals and/or environments are encountered. Early applications were largely found in oilfields where rubber parts may be exposed not only to oil but also to sour oil and gas, amine corrosion inhibitors, steam/hot water, water-base drilling and completion fluids, and high-pH completion fluids. Its ability to resist these environments has led to its use instead of FKM rubbers. Its ability to resist chemical combinations has subsequently led to additional uses in the chemical manufacturing and automotive industries.

The copolymers are available in range of molecular weights and are peroxide-cured. A recommended peroxide is α,α-bis(*t*-butylperoxy)di-isopropylbenzene with triallyl isocyanurate as co-agent. Typical cure conditions are 30 min at 165°C or 3 min at 195°C. Parts cured in open steam require wrapping in order to exclude air from the surface, which will otherwise become tacky. Post-curing, typically for 16 h at 200°C, is recommended. Carbon black fillers are usually used although mineral fillers may be employed providing that they also resist any chemicals with which the vulcanizates come into contact.

Whilst it is the chemical resistance that is of greatest importance, the following properties are also of interest.

1. Suitability for long-term service in air up to 200°C and for short terms up to 288°C.
2. Room-temperature tensile strengths of 20·8 MPa (3000 psi) and at 175°C of 6·9 MPa (1000 psi).
3. A T_g of -3°C.
4. Very good electrical insulation characteristics such as a volume resistivity at 21°C of $3{\cdot}0 \times 10^{16}$.

12.6 POLYPHOSPHAZENES

About 100 years ago (Stokes, 1895, 1897) a totally inorganic rubber, polyphosphonitrilic chloride, was prepared by the reaction of phosphorus pentachloride with ammonium chloride by the following reaction.

$$nPCl_5 + nNH_4Cl \xrightarrow[120^\circ C]{\text{Solvent}} n/3(NPCl_2)_3 + 4nHCl$$

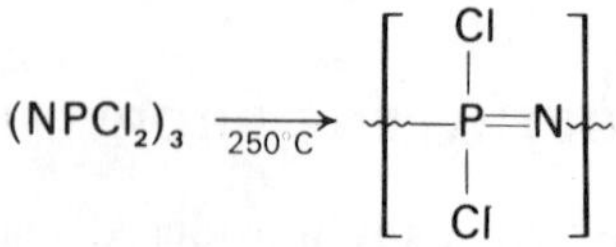

$$(NPCl_2)_3 \xrightarrow{250^\circ C} \left[-P(Cl)(Cl){=}N- \right]$$

This rubbery linear polymer is rapidly degraded by moisture and has remained little more than a laboratory curiosity. In order to improve the hydrolytic stability, chlorine groups have been replaced by such groups as alkoxy, aryloxy and amine. However, most of the early products were cross-linked and intractable and little progress was made until Allcock & Kugel (1965) disclosed methods of termination at low conversions to give a material that remained soluble. The first fluorophosphonitrilic polymers were produced three years later (Rose, 1968, patented 1970) by the addition of sodium trifluoroethoxide and heptafluorobutoxide to the solid polyphosphonitrilic chloride (also known as polydichlorophosphazene).

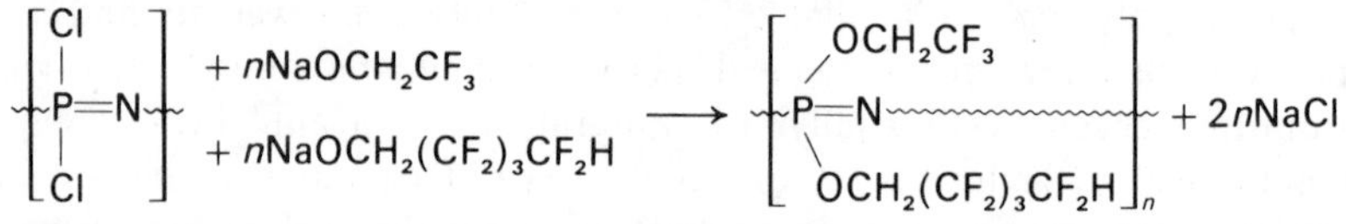

$$\left[-P(Cl)(Cl){=}N- \right] + nNaOCH_2CF_3 + nNaOCH_2(CF_2)_3CF_2H \longrightarrow \left[-P(OCH_2CF_3)(OCH_2(CF_2)_3CF_2H){=}N- \right]_n + 2nNaCl$$

This polymer was marketed in the mid-1970s by Firestone as PNF Rubber but in 1983 the Ethyl Corporation obtained exclusive rights to the Firestone patents and the polymer is now marketed as Eypel F.

A closely related polymer, although not a fluororubber, is the phenol-substituted polymer marketed by Firestone as APN Rubber and then by Ethyl Corporation as Eypel A.

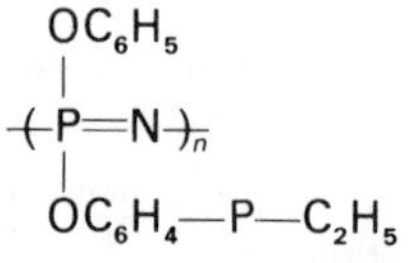

$$\left(-P(OC_6H_5)(OC_6H_4{-}P{-}C_2H_5){=}N- \right)_n$$

Both polymers are soft gum rubbers that may be processed on conventional equipment. In the case of the fluororubber a peroxide curing system is usually employed and compounds are normally supplied with the peroxide curative already added. Sulphur/accelerator and radiation curing have also been demonstrated. Mineral-filled compounds generally offer slightly improved thermal, compression set, and fuel and oil resistance compared with black-loaded compounds, although the latter confer superior mill processability and flex fatigue resistance (Penton, 1986).

The fluoro rubber is useful in the range -65 to $+175$°C and uses have stemmed from the value of such a wide operating range combined with

TABLE 12.4
Typical properties of three fluoropolyphosphazene rubbers

Property	*Eypel 6504 (GP grade)*	*Eypel 7003 (O-ring grade)*	*Eypel 7001 (Seal and hose grade)*
Hardness, shore A	65	70	70
100% modulus, MPa	4·1	9·0	6·7
Tensile strength, MPa	7·6	10·3	9·3
Elongation at break, %	200	120	140
Compression set, % (70 h at 149°C)	50	20	30
Density, mg m^{-3}	1·74	1·85	1·77
Mooney viscosity (ML 4 100°C)	66	45	45
Crescent tear Die B, kN m^{-1}	27	19·3	17·5

other desirable properties such as good damping characteristics over a broad temperature range, excellent resistance to hydrocarbon-based fluids and good flexural fatigue resistance.

Typical properties of three fluoropolyphosphazenes are given in Table 12.4.

The phenol-substituted polymer has, on the other hand, a narrower service range of −20 to +125°C and is mainly of interest because of its fire properties. An oxygen index of 28 for the raw polymer may be raised to 44 by appropriate compounding. Rather more importantly, the toxicity of the combustion products appears to be considerably less than for many other commonly used fire-resistant elastomers (Alarie *et al.*, 1981). Whilst there is some smoke evolution, the compounds char on burning rather than dripping or flowing. Current interest is in wire and cable insulation and for flame-resisting closed-cell insulating foams.

12.7 CARBOXY–NITROSO RUBBERS (AFMU)

Copolymerization of trifluoronitrosomethane and tetrafluoroethylene (Barr & Haszeldine, 1955) leads to the formation of an elastomer with the following features:

1. non-inflammability—even in pure oxygen;
2. excellent chemical resistance, including resistance to nitrogen tetroxide and chlorine trifluoride;
3. a low T_g of −51°C and an extremely low solubility parameter of 10·63 $MPa^{1/2}$.

The copolymer does however suffer from a number of disadvantages including poor heat resistance above 150°C, poor resistance to ionizing radiation, sensitivity to degradation by organic bases, highly toxic degradation products and high cost. A further disadvantage is the difficulty in obtaining vulcanizates of an acceptable tensile strength.

Cross-linking is facilitated if a third monomer, nitrosoperfluorobutyric acid, is used to provide a cure site, the products being referred to as carboxy–nitroso rubbers and given the ASTM designation AFMU:

$$\sim NO-CF_2-CF_2-NO\sim$$
$$\qquad | \qquad\qquad\qquad\qquad |$$
$$CF_3 \qquad\qquad\qquad (CF_2)_3COOH$$

AFMU rubbers can be cross-linked by metal oxides and amines but organometallic compounds such as chromium trifluoroacetate are claimed to provide the best balance of mechanical properties and chemical resistance. The tensile strength of a silica-filled vulcanizate is in the range 7–14 MPa whilst good compression set resistance may be achieved by an oven post-cure, typically 24 h at 175°C.

Carboxy–nitroso rubbers have been available on a developmental scale from the Thiokol Corporation and have been of interest in aerospace systems where flame resistance is mandatory. The properties, processing and applications of the rubbers have been reviewed (Levine, 1971).

BIBLIOGRAPHY

Arnold, R. G., Barney, A. L. & Thompson, D C. (1973). *Rubber Chem. Technol.*, **46**, 619.

Brown, J. H. (1976). *Annual Report on the Progress of Rubber Technology*. Plastics and Rubber Institute, London, p. 31.

Cooper, J. R. (1968). In *Polymer Chemistry of Synthetic Elastomers*, Part 1, ed. J. P. Kennedy & E. G. M. Törnqvist. Interscience, New York, Chapter 4E.

Montemerso, J. C. (1961). *Rubber Chem. Technol.*, **34**, 1521.

Wall, L. A. (ed.) (1972). *Fluoropolymers* Interscience, New York.

REFERENCES

Alarie, Y. C., Lieu, P. J. & Magill, J. H. (1981). *J. Combustion Technology*, **8**, 242.

Allcock, H. R. & Kugel, R. L. (1965). *J. Am. Chem. Soc.*, **87**, 4216.

Arnold, R. G., Barney, A. L. & Thompson, D. C. (1973). *Rubber Chem. Technol.*, **46**, 619.

Barney, A. L. (1969). *Polymer Preprints, Am. Chem. Soc. Div. Polym. Chem.*, **10**, 1483.
Barr, D. A. & Haszeldine, R. N. (1955). *J. Chem. Soc.*, 1881.
Bro, M. I. (1959). *J. Appl. Polym. Sci.*, **1**, 310.
Brown, J. H., Finlay, J. B., Hallenbeck, A., MacLachlan, J. D. & Pelosi, L. F. (1977). Paper presented to International Rubber Conference, Brighton, UK, May 1977 (Preprints produced by the Plastics and Rubber Institute.)
Du Pont Ltd. (1971). US Patent 3546186.
Kalb, G. H., Khan, A. A., Quarles, R. W. & Barney, A. L. (1973). *Advances in Chemistry Series*, **129**, 13.
Levine, N. B. (1971). *Rubb. Chem. Technol.*, **44**, 40.
Paciorek, K. L., Mitchell, L. C. & Lenk, C. T. (1960). *J. Polym. Sci.*, **45**, 405.
Penton, H. (1986). *Eur. Rubber J.*, **168**, 20.
Rose, S. H. (1968). *Polym. Letters*, **6**, 837.
Stokes, H. N. (1895). *Am. Chem. J.*, **17**, 275.
Sweet, G. C. (1979). In *Developments in Rubber Technology—1*, ed. Whelan, A. & Lee, K. S. Applied Science, London, Chapter 2.
Rose, S. H. (1970). US Patent 3515688.
Stokes, H. N. (1897). *Am. Chem. J.*, **19**, 782.

Chapter 13

Silicone Rubbers

13.1 INTRODUCTION

The silicone rubbers comprise an important group of the family of silicone polymers, the latter being distinctive in possessing an inorganic backbone of alternating silicon and oxygen atoms. In spite of their high price structure, such polymers have become of importance because of their good thermal stability, constancy of properties over a wide temperature range, good electrical insulation properties, water repellency and anti-adhesive properties. The polymers are available in a range of forms including fluids, greases, resins and elastomers, the last-named being the subject of this chapter.

The possibility of the existence of organosilicon compounds was first predicted by Dumas in 1840 and in 1872 Ladenburg produced the first silicone polymer, a very viscous oil, by reacting diethoxydiethylsilane with water in the presence of traces of acid.

$$C_2H_5{-}O{-}\underset{C_2H_5}{\overset{C_2H_5}{\underset{|}{\overset{|}{Si}}}}{-}O{-}C_2H_5 \xrightarrow[\text{acid}]{H_2O} \left({-}O{-}\underset{C_2H_5}{\overset{C_2H_5}{\underset{|}{\overset{|}{Si}}}}{-} \right) + C_2H_5OH$$

The theoretical basis of modern organosilicon chemistry was, however, laid down by F. S. Kipping of University College, Nottingham, who published over 50 papers on the subject between 1899 and 1944. However, as late as 1937 he appears to have foreseen no application for such materials when he concluded a lecture with the words '... the prospect of any immediate and important advance in this section of chemistry does not seem very hopeful'. This is quite surprising in view of the fact that in 1904 he introduced the use of Grignard reagents for the preparation of the important chlorosilane intermediates and later discovered the principle of intermolecular condensation of the silane diols, the basis of most polymerization practice. The term *silicone* was also designated by Kipping to the hydrolysis products of the disubstituted silicon chlorides because at the time he considered them to be silicon analogues of the ketones of organic chemistry.

The first commercial silicone resins resulted from the formation of the Dow Corning Corporation by Corning Glass and Dow Chemical in 1943 to manufacture silicone polymers. They were followed by the General Electric Company of Schenectady, New York, in 1946 and by Union Carbide in 1956. At the present time major suppliers include the above three companies together with Bayer, Rhone–Poulenc, Wacker and three Japanese companies—Toshiba Silicone, Toray Silicone and Shinetsu. There is also some Eastern bloc capacity. In the early 1980s world capacity, excluding the Eastern bloc, for all silicone polymers was estimated at about 270 000 tpa, the markets then being dominated by the USA (41%), Western Europe (33%) and Japan (17%). In the absence of firm data, current production of silicone rubbers is estimated to be of the order of 35 000 tpa.

13.2 NOMENCLATURE OF ORGANOSILICON COMPOUNDS AND ELASTOMERS

Before discussing the chemistry and technology of silicone rubbers it is useful to consider the nomenclature used for organosilicon compounds relevant to this chapter as adopted by the International Union of Pure and Applied Chemistry.

The structure used as the basis of the nomenclature is *silane*, SiH_4, corresponding to the organic compound methane, CH_4. Silicon hydrides of the type $SiH_3(SiH_2)_nSiH_3$ are referred to as disilane, trisilane, tetrasilane, etc., according to the number of silicon atoms present.

Alkyl-, aryl-, alkoxy- and halogen-substituted silanes are referred to by prefixing 'silane' with the specific group present. For example:

$(CH_3)_2SiH_2$	Dimethylsilane
CH_3SiCl_3	Trichloromethylsilane
$(C_6H_5)_3SiC_2H_5$	Ethyltriphenylsilane

Compounds having the formula $SiH_3(OSiH_2)_nOSiH_3$ are referred to as disiloxane, trisiloxane, etc., according to the number of silicon atoms. Polymers in which the main chain consists of repeating —Si—O— groups together with predominantly organic side groups attached to the silicon atoms are referred to as *polyorganosiloxanes* or more commonly as *silicones*.

Hydroxy derivatives of silanes in which the hydroxyl groups are attached to a silicon atom are named by adding the suffixes -ol, -diol, -triol, etc., to the name of the parent compound. Examples are:

H_3SiOH	Silanol
$H_2Si(OH)_2$	Silane diol
$(CH_3)_3SiOH$	Trimethylsilanol

The International Standards Organization (Recommendation R) and the American Society for Testing and Materials (ASTM D1418) use the following classification for silicone rubbers.

MQ	Silicone rubbers having only methyl substituent groups on the polymer chain (polydimethyl siloxanes)
VMQ	Silicone rubbers having both methyl and vinyl substituent groups on the polymer chain
PMQ	Silicone rubbers having both methyl and phenyl substituent groups on the polymer chain
PVMQ	Silicone rubbers having methyl, phenyl and vinyl substituent groups on the polymer chain
FVMQ	Silicone rubbers having fluoro, vinyl and methyl substituent groups on the polymer chain

VMQ materials may be considered as general-purpose silicone rubbers, the PMQ and PVMQ types for extreme low-temperature performance and the FVMQ materials where fuel, oil and solvent resistance are required. Few MQ and PMQ grades are now marketed.

Industrial practice today recognizes three distinct types of silicone elastomer with differing technologies:

1. Heat-curable or High-Temperature Vulcanizing (HTV) rubbers—these are usually semi-solid gum-like materials in the uncured form and require traditional rubber processing techniques for compounding and manufacture of finished parts.
2. Room-temperature vulcanizing (RTV) rubbers—usually flowable liquids supplied in a ready-to-use form in such applications as potting, encapsulation, sealants and mould making. These materials are not usually used in conventional rubber technology.
3. Heat-curable liquid materials (liquid silicone rubbers—LSR) processed on specially designed injection-moulding and extrusion equipment.

13.3 MANUFACTURE OF SILICONE ELASTOMERS

Polyorganosiloxanes are generally prepared by reacting chlorosilanes with water to give hydroxyl compounds which then condense to give a polymer structure, e.g.

$$\mathrm{Cl{-}\overset{\displaystyle R}{\underset{\displaystyle R_1}{Si}}{-}Cl} + H_2O \longrightarrow \mathrm{HO{-}\overset{\displaystyle R}{\underset{\displaystyle R_1}{Si}}{-}OH} \longrightarrow \left(\mathrm{{-}\overset{\displaystyle R}{\underset{\displaystyle R_1}{Si}}{-}O{-}} \right)$$

Similar reactions also take place with the alkoxysilanes but in commercial practice the chlorosilanes are favoured, those used for elastomers most commonly being prepared by the Direct Process involving the reaction of elementary silicon with an alkyl halide (usually methyl chloride):

$$Si + RX \longrightarrow R_nSiX_{4-n} \quad (\text{where } n = 0\text{–}4)$$

Vinyl-containing monomers may be produced by the reaction of acetylene with, for example, trichlorosilane, the latter being obtained by reacting silicon and hydrogen chloride:

$$CH{\equiv}CH + SiHCl_3 \longrightarrow CH_2{=}CH \cdot SiCl_3$$

A variety of types of polymer may be prepared according to the functionality of the chlorosilane. Thus reaction of chlorotrimethylsilane with water will lead only to the low-molecular-weight hexamethyldisiloxane:

$$2(CH_3)_3SiCl + 2H_2O \longrightarrow 2(CH_3)_3SiOH \longrightarrow (CH_3)_3Si{-}O{-}Si(CH_3)_3 + H_2O$$

Hydrolysis of dimethyldichlorosilane will yield a linear polymer:

$$\mathrm{Cl{-}\overset{\displaystyle CH_3}{\underset{\displaystyle CH_3}{Si}}{-}Cl} \xrightarrow{H_2O} \mathrm{HO{-}\overset{\displaystyle CH_3}{\underset{\displaystyle CH_3}{Si}}{-}OH} \longrightarrow \left(\mathrm{{-}\overset{\displaystyle CH_3}{\underset{\displaystyle CH_3}{Si}}{-}O{-}} \right)$$

whilst hydrolysis of trichloromethylsilane yields a network structure:

$$\mathrm{Cl{-}\overset{\displaystyle CH_3}{\underset{\displaystyle Cl}{Si}}{-}Cl} \xrightarrow{H_2O} \mathrm{HO{-}\overset{\displaystyle CH_3}{\underset{\displaystyle OH}{Si}}{-}OH} \longrightarrow \mathrm{{-}\overset{\displaystyle CH_3}{\underset{\displaystyle O \atop |}{Si}}{-}O{-}}$$

For use as silicone elastomers it is necessary to obtain high-molecular-weight polymers (*M c.* 500 000) free from branching. This requires very pure difunctional material free from mono- and tri-chlorosilanes (the former preventing chain growth, the latter leading to branching and cross-linking). In the case of pure dimethylsilicone rubbers this may be achieved by first producing the cyclic tetramer, purifying it and then equilibrating it, for example, with a trace of alkaline catalyst for several hours at 150–200°C, which causes rearrangement to give a long chain polymer. In recent years there has been interest in the ring-opening polymerization of cyclic trimers using a weak base such as lithium silanolate which gives high-molecular-weight products of narrow molecular weight distribution free of cyclic materials other than unreacted trimer.

13.4 STRUCTURE AND PROPERTIES OF SILICONE ELASTOMER POLYMERS

Silicon, like carbon, is found in Group 4 of the Periodic Table and is normally tetravalent. It is more electropositive than either carbon or hydrogen and has a marked tendency to oxidize. In spite of its normally tetravalent nature and because it is electropositive, it is not possible to produce silicone analogues of the multiplicity of carbon compounds that form the basis of organic chemistry. It is not even possible to prepare silanes higher than hexasilane because of the inherent instability of the silicon–silicon bond in the higher silanes. It is thus important to realize that simply including silicon in a polymer does not by any means ensure a high level of thermal stability.

However, the silicon–oxygen and silicon–methyl bonds *are* thermally stable and predictions that polydimethylsiloxane will also be thermally stable are found to be true in practice. On the other hand, the Si—O bond is partially ionic (51%) and it is easily broken by concentrated acids and bases at room temperature.

Of much significance is the large Si—O—Si bond angle of the order of 140–160°. This together with the favourable geometry which limits the steric attractions of groups attached to the silicon atoms results in a very free rotation of the molecules about the main chain Si—O bonds. Together with low intermolecular forces the low rotational energy of bonds leads to such features as low T_g, low T_m (of regular crystallizable polymers), low solubility parameter and low tensile strength of the rubbers.

The original silicone rubbers were essentially polydimethylsiloxanes (MQ rubbers). These materials have a very low T_g of $-120°C$. However, they stiffen up at higher temperatures because, being of regular structure, they are capable of crystallizing at $-60°C$.

In order to realize the potential low-temperature flexibility afforded by the low glass transition temperatures it is necessary to inhibit crystallization. This may be done, as with other polymers, by copolymerization. In practice the most convenient way of doing this is to co-hydrolyse dichlorodimethylsilane with dichloromethylphenylsilane (to give the PMQ rubbers). Only a small proportion of the phenyl derivative (*c.* 10%) is normally employed since, although this is sufficient to inhibit crystallization, the presence of the phenyl groups will cause the T_g to rise with increase in the dichloromethylphenylsilane content, and if used to excess the benefits of the phenyl derivative will be lost.

Neither the MQ nor the PMQ rubbers may be vulcanized with sulphur because of the absence of unsaturation in the chain. Such unsaturation may be incorporated into the polymer by incorporating dichloromethylvinylsilane into the reaction mixture to give the VMQ and PVMQ rubbers. In practice sulphur vulcanization has few attractions with silicone rubbers but the presence of vinyl groups has been found to be very beneficial in that their presence improves the cross-linking efficiency of peroxides for silicone rubbers. This is particularly valuable with alkyl hydroperoxides and dialkyl peroxides which, unlike many other peroxides, do not generate acidic by-products.

The silicone elastomers described above are not resistant to hydrocarbon oils and solvents. Oil-resistant silicone elastomers in which some of the methyl groups were replaced by groups containing nitrile or fluorine components first appeared in the 1950s:

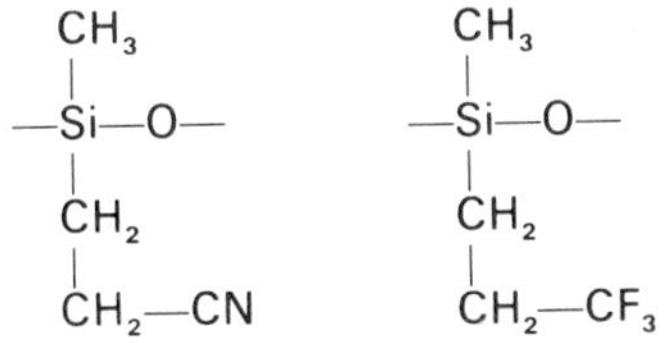

$$
\begin{array}{c} CH_3 \\ | \\ -Si-O- \\ | \\ CH_2 \\ | \\ CH_2-CN \end{array}
\qquad\qquad
\begin{array}{c} CH_3 \\ | \\ -Si-O- \\ | \\ CH_2 \\ | \\ CH_2-CF_3 \end{array}
$$

Although the nitrile-containing polymers failed to become commercially significant, the fluorine-containing polymers with their excellent resistance to oils, fuels and solvents have found important applications in spite of their high cost. Currently available materials also contain vinyl groups and thus are designated FVMQ. Interesting products may also be produced by introducing boron atoms into the chain, such as in the structure

$$\sim\!\!\sim\!\underset{\substack{|\\ R}}{\overset{\substack{R\\ |}}{Si}}-O-\underset{\substack{|\\ R'}}{B}-O-\underset{\substack{|\\ R}}{\overset{\substack{R\\ |}}{Si}}\!\sim\!\!\sim$$

The amount of boron used is usually small (B:Si 1:500 to 1:200) but its presence increases the self-adhesive tack of the rubber which is desirable where hand-building operations are involved.

The intriguing material 'bouncing putty' is also a silicone polymer with the occasional Si—O—B group in the chain, in this case with about one boron atom to every 3–100 silicon atoms. It is interesting in that if stressed slowly the material deforms like a viscous material flowing irreversibly; if it is dropped on a table the chain deformation is of the high-elastic type, with insufficient time for viscous flow, and the material behaves like a rubber whilst if it is subjected to a sudden blow with insufficient time for either high-elastic or viscous deformation, the sample may shatter. At one time it

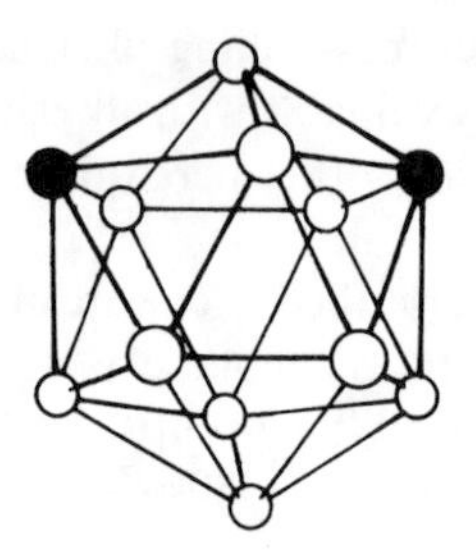

FIG. 13.1. Structure of *m*-carborane (●, C atom; ○, BH).

was widely available as a children's novelty and as a useful teaching aid but it is now difficult to obtain.

Boron is also incorporated into the *poly(carborane siloxane)* elastomers. These were first described in 1966 and introduced in 1971 by the Olin Corporation as Dexsil. The polymers have the essential structure

$$-Si(R)_2-CB_{10}H_{10}C\!\left[Si(R)_2-O\right]_x$$

where $CB_{10}H_{10}C$ represents a *m*-carborane group (Fig. 13.1).

13.5 GENERAL PROPERTIES

Silicone rubbers are notable for the following general characteristics:

1. Very wide service temperature range (−100 to +250°C);
2. excellent resistance to attack by oxygen, ozone and sunlight;
3. excellent non-stick, non-adhesive properties;
4. low toxicity (which may of course be altered by the presence of additives);
5. possibility of optical transparency;
6. good electrical insulation properties;
7. low chemical reactivity.

Possible limitations of the material are:

1. The low tensile strength of vulcanized rubbers;
2. poor hydrocarbon oil and solvent resistance (except with FVMQ types);
3. high gas permeability (a property which may also be used to advantage);
4. somewhat high cost.

Of these properties doubtless the most important is the high level of heat resistance, amongst rubbers only approached by the perfluorinated elastomers (see Chapter 12). Whilst the details of 'in-service' conditions will influence the 'lifetime' at any specific temperature, a useful rule-of-thumb guide is to note the time it takes, at a given temperature, before the elongation at break of the vulcanizate drops below 50%. On this basis Cush & Winnan (1981) have given the figures reproduced in Table 13.1 for the useful life for continuous operation of a silicone rubber specifically formulated for high-temperature use.

One problem that may be encountered at elevated temperatures in the

TABLE 13.1
Useful life of a silicone rubber formulated for high-temperature operation at selected elevated temperatures

Temperature (°C)	*Service life* (h)	*Temperature* (°C)	*Service life* (h)
150	15 000–30 000	316	100
200	7 500–10 000	370	1
260	2 000		

presence of traces of water is that of *reversion* due to hydrolysis of the main chain:

$$-\overset{|}{\underset{|}{Si}}-O-\overset{|}{\underset{|}{Si}}- + H_2O \longrightarrow -\overset{|}{\underset{|}{Si}}-OH + HO-\overset{|}{\underset{|}{Si}}-$$

This is more likely where the rubber is being used in an enclosed space and moisture is unable to escape from the system. Few problems occur where the rubber is not in close confinement. Additionally, appropriate choice of additives can be beneficial.

Whilst the tensile strengths of silicone rubbers are lower than for most organic rubbers at normal ambient temperatures, this property is less dependent on temperature with the silicones than with the other rubbers.

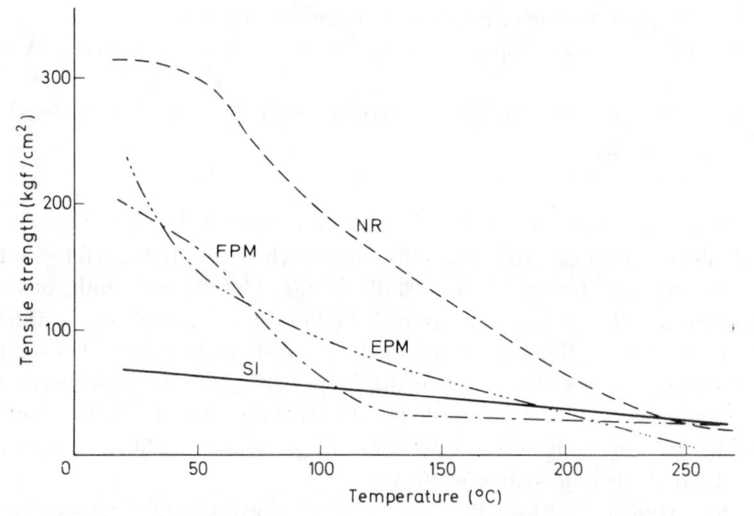

FIG. 13.2. Tensile strength of rubbers against temperature. (From Blow (1968), reproduced by permission of Butterworth Scientific Ltd.)

This is illustrated in Fig. 13.2 (Blow, 1968), which however does not do the silicones full justice, since higher-performance materials have been introduced since this diagram was first published.

Appropriately formulated silicone rubbers can also exhibit very low compression set over a broad range of temperature. This is indicated in Fig. 13.3 (Blow, 1964) but, as with Fig. 13.2, the data cannot reflect improved performance achieved with a number of polymers since 1964.

The low compression set achievable is one manifestation of the high level of elasticity with silicone rubbers. This is also reflected in the good dynamic properties of the polymer but which have not often been utilized because of the low tear resistance and rather low metal adhesion of the polymers.

An outstanding feature of the silicones is their very high permeability. At 25°C their permeability to oxygen is about 400 times greater than that of butyl rubber and 25 times greater than with natural rubber. Use has been

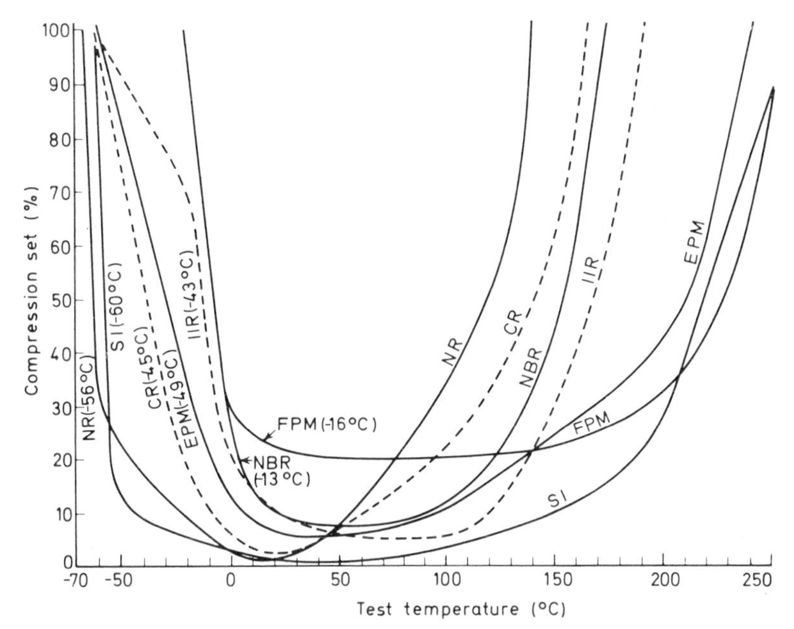

FIG. 13.3. Dependence of compression set of elastomers on temperature (the temperature at which the torsional modulus is 700 kgf cm^{-2} is stated on each curve). (From Blow, 1964.)

made of this property in the production of oxygen-permeable membranes for medical applications. Medical applications also benefit from the physiological inertness of correctly formulated and properly post-cured materials which are odourless, tasteless and generally accepted as non-toxic.

The silicone rubbers are naturally excellent electrical insulators with properties such as volume resistivity, dielectric strength and power factor little affected by temperature. They show excellent tracking resistance and exceptionally high arc and corona resistance which are surpassed only by those of mica. The materials have the additional advantage that if they are involved in a fire and burn, they leave behind non-conductive silica, thus providing additional safety in electrical equipment. Specially formulated conductive grades are also available with volume resistivities as low as 0·004 ohm cm.

The rubbers also exhibit very good resistance to gamma-rays and X-rays with standard grades withstanding doses of 40–50 Mrad at room temperature with specially formulated grades able to withstand doses of up to 90 Mrad.

13.6 COMPOUNDING

Traditionally silicone rubbers have been supplied to fabricators in fully compounded form by the polymer manufacturers, although in more recent years in-house compounding has become more common. Additives may be considered under the following categories:

1. Curing systems.
2. Fillers and structure control agents.
3. Other additives.

13.6.1 Curing Systems

Other than for the RTV and liquid curing systems, silicone rubbers are commonly cured using peroxides. Suitable peroxides include benzoyl peroxide, 2,4-dichlorobenzoyl peroxide, *t*-butyl perbenzoate and dicumyl peroxide. Alkyl hydroperoxides and dialkyl peroxides give rather poor cures with the non-vinyl silicones but their use is attractive because they do not generate by-products on cure. They also perform more satisfactorily with the vinyl-containing grades (VMQ, PVMQ and FVMQ).

An alternative approach to curing is that known as hydrosilylation (or hydrosilation) curing. In such a process a polymer containing vinyl groups is reacted with a reagent containing a number of hydrosilane (Si—H) groups, Pt(II) compounds being frequently used as catalysts for the addition reaction:

$$-\overset{|}{\underset{|}{\mathrm{Si}}}-\mathrm{H} + \mathrm{CH_2}{=}\mathrm{CH}-\overset{|}{\underset{|}{\mathrm{Si}}}- \longrightarrow -\overset{|}{\underset{|}{\mathrm{Si}}}-\mathrm{CH_2}-\mathrm{CH_2}-\overset{|}{\underset{|}{\mathrm{Si}}}-$$

A specific system used with liquid silicone rubbers is

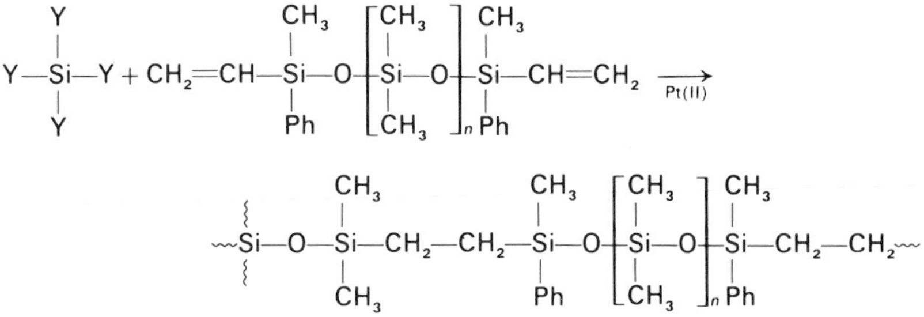

$$\mathrm{Y{-}\overset{\displaystyle Y}{\underset{\displaystyle Y}{Si}}{-}Y} + \mathrm{CH_2{=}CH{-}\overset{\displaystyle CH_3}{\underset{\displaystyle Ph}{Si}}{-}O{-}\left[\overset{\displaystyle CH_3}{\underset{\displaystyle CH_3}{Si}}{-}O\right]_n{-}\overset{\displaystyle CH_3}{\underset{\displaystyle Ph}{Si}}{-}CH{=}CH_2} \xrightarrow[\mathrm{Pt(II)}]{}$$

$$\sim\mathrm{\overset{\wr}{\underset{\wr}{Si}}{-}O{-}\overset{\displaystyle CH_3}{\underset{\displaystyle CH_3}{Si}}{-}CH_2{-}CH_2{-}\overset{\displaystyle CH_3}{\underset{\displaystyle Ph}{Si}}{-}O{-}\left[\overset{\displaystyle CH_3}{\underset{\displaystyle CH_3}{Si}}{-}O\right]_n{-}\overset{\displaystyle CH_3}{\underset{\displaystyle Ph}{Si}}{-}CH_2{-}CH_2}\sim$$

where

$$\mathrm{Y} = \sim\mathrm{O{-}\overset{\displaystyle CH_3}{\underset{\displaystyle CH_3}{Si}}{-}H}$$

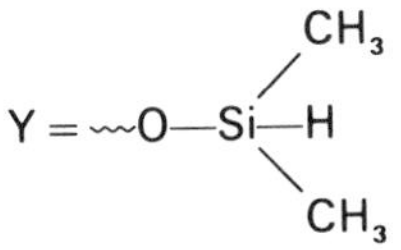

Such a system requires, in practice, a two-pack operation and once mixed together (for convenience in metering usually in a 1:1 ratio) the shelf life will be limited to a few days at room temperature.

Hydrosilylation curing has been used to produce heat-cured conventional silicone rubbers with high tear strength, in liquid silicone polymers and for two-pack RTV formulations. It is to be noted that this is an addition rather than a condensation cross-linking process so that volatiles are not produced during cure.

13.6.2 Fillers

Because of the very low inherent tensile strength of silicone rubbers it is important in most applications to incorporate a reinforcing filler. In practice fine silicas are more effective than carbon blacks and the latter are seldom used. There has been a considerable body of research work to optimize the performance of silica fillers (see, for example, the numerous references in Warrick *et al.*, 1979) and considerable advances have been made over the years. Fumed silicas in the 10–40nm diameter range are generally preferred, experimental work indicating that a high-filler structure is more important than particle size. Appropriate selection will lead to vulcanizates with tensile strengths in excess of 10·5 MPa and elongations at break of 500–800%.

One problem associated with the use of fillers is that there is interaction between filler and elastomer, leading to a form of *pseudo-vulcanization* which can occur during mixing on a mill (*crepe hardening*) or on storage (*bin ageing*). Whilst this structure may be broken down on milling, it is common to incorporate into the formulation certain siloxane-based materials known variously as *structure control additives* or *anti-structure agents* such as diphenylsilane diol and pinacoxydimethylsilane:

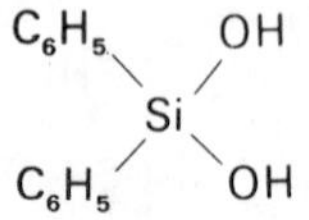

$$(C_6H_5)_2Si(OH)_2$$

Diphenylsilane diol

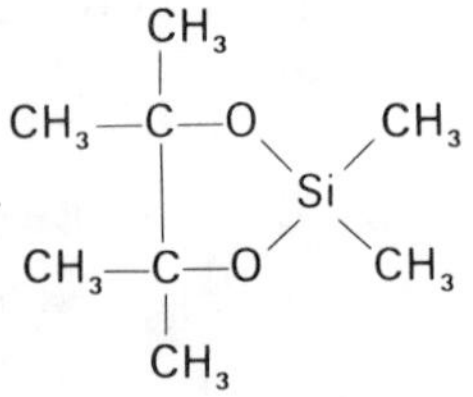

$$\begin{array}{c} CH_3 \\ | \\ CH_3-C-O \\ | \\ CH_3-C-O \\ | \\ CH_3 \end{array} \!\!> Si(CH_3)_2$$

Pinacoxydimethylsilane

13.6.3 Other Additives

Whilst silicone elastomers burn less easily than many organic rubbers, flame resistance may be further improved by use of additives. Traditional halogen-containing flame retardants used with other polymers interfere with cure and also produce a toxic smoke. In silicone rubbers flame-retardant additives include platinum compounds (sometimes in combination with titanium dioxide), carbon black, aluminium trihydrate, zinc or ceric compounds.

Other additives include carbon black to improve electrical conductivity, ferric oxide to improve heat stability, titanium dioxide as a white pigment and organometallic pigments for colouring.

13.7 PROCESSING

Silicone rubbers may be compounded in internal mixers or on mills but, because of the low viscosity of compounds, close-fitting cheek plates and nylon scraper blades need to be fitted with the latter. Shaping may be by standard techniques such as extrusion, compression moulding and injection moulding and, providing care is taken to avoid air trapping and allowance is made for the somewhat high curing shrinkage, the processes are quite easy. Curing is usually rapid and for most, but not all, grades this is followed by a post-cure in a forced-draught air oven at 200–250°C for 4–24 h. This may both improve physical properties and remove undesired peroxide decomposition products formed during cure.

13.8 LIQUID SILICONE RUBBERS

Considerable interest has been generated by the availability since the late 1970s of liquid silicone rubbers (LSRs). These are basically two-component systems in which the components are supplied in a paste form and which on mixing cure by an addition process such as hydrosilylation. Such materials exhibit the following desirable features.

1. The components are supplied deaerated and ready for use with appropriate multicomponent metering equipment and injection moulding machines.
2. The mixes are less scorchy than fast-curing peroxide systems but when cure commences it is very rapid, typical cures being 5–6 s per mm section at 200°C.
3. Only low injection pressures and consequently low clamping forces are required, thus allowing relatively inexpensive injection moulding machines to be used.
4. Post-curing is not generally necessary.

Physical properties are similar to general-purpose and medium to high strength peroxide cured elastomers. The LSRs are also self-extinguishing and inclusion of certain carbon blacks enables the materials to meet the Underwriters' Laboratory UL-94 test with a class of V-0. Conductive grades are also available.

Because of the low capital requirements and overheads, LSRs have competed not only with more conventional silicones but also with organic rubbers.

13.9 ROOM-TEMPERATURE VULCANIZING RUBBERS (RTV)

These rubbers have found considerable application outside of the field of conventional rubber technology, being used for mould-making and craft work, potting and encapsulation, sealants and a variety of other uses.

Two main types are distinguished:

1. One-pack systems (sometimes known as RTV-1 rubbers) which are moisture-cured;
2. two-pack systems (RTV-2 rubbers).

The single- or one-pack systems are prepared by first producing a polydialkylsiloxane with terminal hydroxyl groups. This is then reacted with a multifunctional organosilicon cross-linking agent to form a tetrafunctional structure stable under anhydrous conditions (Valles & Macosko, 1979; Macosko, 1985):

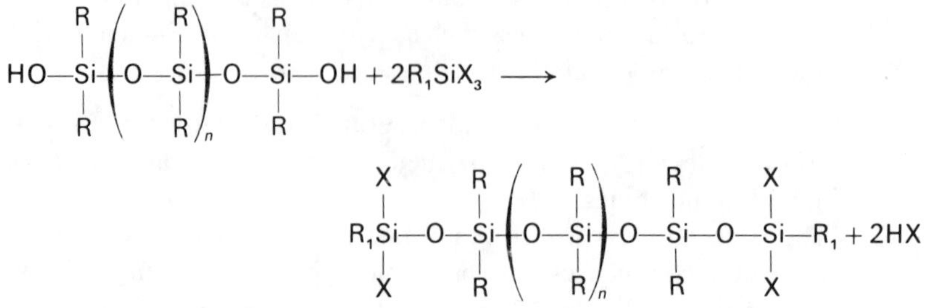

$$HO-\underset{R}{\overset{R}{|}}{Si}-\left(O-\underset{R}{\overset{R}{Si}}\right)_n-O-\underset{R}{\overset{R}{Si}}-OH + 2R_1SiX_3 \longrightarrow$$

$$R_1\underset{X}{\overset{X}{Si}}-O-\underset{R}{\overset{R}{Si}}-\left(O-\underset{R}{\overset{R}{Si}}\right)_n-O-\underset{R}{\overset{R}{Si}}-O-\underset{X}{\overset{X}{Si}}-R_1 + 2HX$$

where X may be —NH—R (amine), —O—CO—CH_3 (acetate) or —O—N=C(R_2) (oxime). The Si—X linkages react with water to form a Si—O—Si linkage with the liberation of HX. Typical catalysts include diarylalkyltin acrylates:

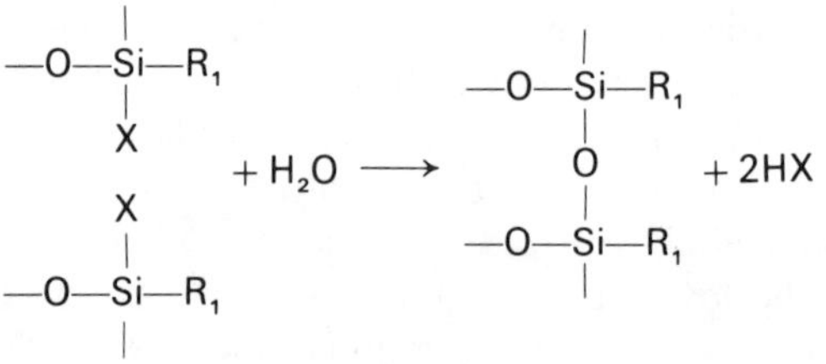

$$\begin{matrix} | \\ -O-Si-R_1 \\ | \\ X \\ \\ X \\ | \\ -O-Si-R_1 \\ | \end{matrix} + H_2O \longrightarrow \begin{matrix} | \\ -O-Si-R_1 \\ | \\ O \\ | \\ -O-Si-R_1 \\ | \end{matrix} + 2HX$$

Such a curing reaction may be brought about by atmospheric humidity and such rubbers are also known as moisture-curing silicones. Because of the need for diffusion of the water into the polymer, the rubber layer should not exceed 10 mm if the ingress is from one side only. For this thickness a typical cure at 23°C at 50% relative humidity would take about five days, whereas a sample 2 mm thick would cure in about one day.

One-pack systems are very widely used for sealing and caulking applications.

Two-pack systems, widely used for making flexible moulds, may also be divided into two types:

1. Condensation cross-linked materials;
2. addition cross-linked polymers.

A typical condensation system involves the reaction of a silanol-terminated polydimethylsiloxane with a multifunctional organosilicon cross-linking agent such as $Si(RO)_4$:

$$2\ \sim Si(CH_3)_2{-}OH + Si(OR)_4 + 2\ HO{-}Si(CH_3)_2\sim \;\longrightarrow\; Si\,(O{-}Si(CH_3)_2\sim)_4$$

Pot life will vary from a few minutes to several hours depending on the catalyst used and the ambient conditions.

Addition-cured materials are particularly suitable for use with polyurethane casting materials. Such addition-cured systems do, however, require absolute cleanliness when processing the rubbers, as cure may be affected by such materials as unsaturated hydrocarbon solvents, sulphur, metallo-organic compounds and materials such as plasticine and certain epoxide resins. Some catalysts for the condensation RTV rubbers may also interfere with cure.

13.10 APPLICATIONS

Silicone rubbers find use on account of their excellent thermal and electrical properties, their physiological inertness, their high permeability and their low compression set.

In mechanical engineering, uses include shaft sealing rings, spark plug caps, radiator hoses, O-rings, corona and embossing rollers and gaskets. In electrical engineering silicone rubbers are used for cables, corona-resistant insulating tubing, keyboards and contact mats, cable terminations and conductive profiled seals. In medicine, uses include transfusion and dialysis tubing, artificial respiration bellows, catheters and babies' dummies. In the building industry the rubbers are used for window and door profile seals and profiles for expansion joints. As previously mentioned, RTV rubbers are widely used for flexible mould making and for seals and caulking applications.

BIBLIOGRAPHY

Cush, R. J. & Winnan, H. W. (1981). In *Developments in Rubber Technology—2*, ed. A. Whelan & K. S. Lee. Applied Science, London, Chapter 8.

Lewis, F. M. (1969). In *The Polymer Chemistry of Synthetic Elastomers*, Part 2, ed. J. P. Kennedy & E. G. M. Törnqvist. Interscience, New York, Chapter 8B.

Noll, W. (1968). *Chemistry and Technology of the Silicones*. Academic Press, London.

Rochow, E. G. (1951). *An Introduction to the Chemistry of the Silicones*, 2nd edn. Wiley, New York.

Warrick, E. L., Pierce, O. R., Polmanteer, K. E. & Saam, J. C. (1979). *Rubber Chem. Technol.*, **52**, 437. (A most comprehensive review of developments during the period 1967–77 with 973 references.)

REFERENCES

Blow, C. M. (1964). *Aircr. Engng*, **36**, 208.

Blow, C. M. (1968). *Tribology*, **1** (March), 81.

Cush, R. J. & Winnan, H. W. (1981). In *Developments in Rubber Technology—2*, ed. A. Whelan & K. S. Lee. Applied Science, London, Chapter 8.

Macosko, C. W. (1985). *Rubber Chem. Technol.*, **58**, 436.

Valles, E. M. & Macosko, C. W. (1979). *Macromolecules*, **12**, 521.

Warrick, E. L., Pierce, O. R., Polmanteer, K. E. & Saam, J. C. (1979). *Rubber Chem. Technol.*, **52**, 437.

Chapter 14

Miscellaneous Rubbery Materials (Polysulphides, Polyethers and Halogenated Polyethylenes)

14.1 INTRODUCTION

The rubbery materials discussed in previous chapters use conventional rubber technology in their manufacture. Processing was carried out in the visco-elastic state with the rubber-like state being achieved after shaping by a chemical (usually covalent) cross-linking system. This chapter is intended as a survey of other rubbers which use such basic rubber technology before moving on to rubbers employing quite different technology. It is nevertheless to be noted that even with some of these materials processes have been developed which owe little to conventional methods.

14.2 POLYSULPHIDE RUBBERS (T)

In 1924 the American chemist J. C. Patrick was seeking applications for lightweight hydrocarbons then being flared-off as useless at oil refineries. In one series of experiments he hoped to devise an economic route to ethylene glycol from ethylene chloride which was in turn obtained from the refinery gases. One such attempt involved him in reacting ethylene dichloride (dichloroethane) with sodium tetrasulphide, but instead of obtaining ethylene glycol he obtained a rubbery material with an unpleasant odour.

Further work resulted in the first patent being taken out three years later (Patrick & Mnookin, 1927) and in 1929 commercial production of the first successful synthetic rubber to be invented in the United States began. The product was known as Thiokol A. With direct descendants of Thiokol A still on the market, this family of polysulphides may claim to have been in production for longer than any other synthetic rubber.

The polysulphides were of immediate interest because of their excellent resistance to swelling in a wide range of chemicals, particularly hydrocarbons. Even today they are superior to virtually every other rubber in this respect, with the exception of the FKM fluororubbers which are superior in some liquids. Certain undesirable characteristics were however very soon apparent. The rubbers had a most unpleasant odour, gas emission occurred on milling, the rubber was difficult to process and heat resistance was poor.

This led to a search for alternative intermediates and by 1940 bis-2-chloroethyl formal was being commonly employed in formulations. Odour problems were reduced and, when used in the absence of dichloroethane, gas emission on milling was much less. Also by the Second World War manufacturers in Germany and Britain had commenced production. One product made by the Thiokol Corporation was designed as GR–P (Government Rubber—Polysulphide) during this period.

In 1949 low-molecular-weight polymers were introduced, prepared by reductive cleavage of the higher-molecular-weight materials. The commercial grades of these materials were liquid.

By the mid-1980s capacity for polysulphide rubber production was estimated (IISRP, 1985) at 9100 tpa by the Thiokol Corporation in the USA, and 2000 tpa by Toray–Thiokol in Japan. There was also some Russian-produced material (SKE Rubber). Of this total the bulk is of the liquid rubber type. The solid rubber has lost a number of markets over the years, either due to the development of new alternatives or due to the raising of customer specifications, particularly in terms of heat resistance. It has however retained important applications in petrol- and oil-loading hose and for printers' rollers. The liquid rubbers are important for sealants, caulking compounds, rocket propellants, flexibilizers for epoxide resins and as castable rubbers. For some years after it was discarded as a rubber, Thiokol A continued to be used for the manufacture of sulphur cements.

Polysulphide rubbers are given the ASTM designation T.

14.2.1 Manufacture

The polysulphides are commonly prepared by a reaction of the following type:

$$\mathrm{Cl{-}R{-}Cl} + \mathrm{Na_2S}_x \longrightarrow \sim\!\mathrm{RS}_x\!\sim + 2\mathrm{NaCl} \quad (\text{where } x = 2\text{–}4)$$

Processes involving ring opening have also been developed.

The product may be a linear homopolymer (as in Thiokol A and B), a linear copolymer (as in GR–P) or branched as in Thiokol ST. Table 14.1 lists some of the polysulphide solid rubbers that have been available but

TABLE 14.1
Dihalides used in the production of some polysulphide rubbers that have been or are commercially available

Dihalide(s)	*Polymer designation*
Dichloroethane	Thiokol A
Bis-2-chloroethyl ether	Thiokol B
	Perduren G
	Novoplas
Bis-2-chloroethyl formal	Perduren H
(+ *c*. 2% 1,2,3-trichloropropane to provide a branch point)	Thiokol ST
Bis-2-chloroethyl formal/dichloroethane	Thiokol FA
1,3-glycerol dichlorohydrin	Vulcaplas
Dichloroethane/dichloropropane	Thiokol N (GR–P)

today only Thiokol FA and Thiokol ST are manufactured commercially outside the Soviet block.

Polymerization is commonly carried out in aqueous suspension. In one exemplified process (Bertozzi, 1968), 1·3 mol of 2M-aqueous $Na_2S_{2 \cdot 25}$ is taken to which is added a small quantity of colloidal magnesium hydroxide and a surface-active agent such as a sodium aryl sulphonate. This is then heated to 80°C with agitation. The bis-2-chloroethyl diformal is then added over a 1-h period, some cooling being necessary to maintain the reaction temperature at 80–90°C. At the end of the feed the temperature is raised to 100°C and the reaction mixture agitated for a further 30 min. The sulphonate surface-active agent helps to improve dispersion whilst the magnesium hydroxide helps to give easily washed and cleaned polymer particles.

Preparation of the polysulphide elastomers involves several interesting features which have been dealt with in detail elsewhere (Gobran & Berenbaum, 1969). They will however be touched on briefly here insofar as they have technological implications. These features are:

1. Interchange reactions;
2. ring formation;
3. sulphur ranking;
4. desulphurization and thionation;
5. reductive cleavage.

To achieve a high molecular weight in linear condensation polymerization reactions it is normally necessary to have an equivalence of monomers. If there is an excess of one monomer then all the chain ends on the growing polymer molecules become identical and reaction cannot proceed. This restriction does not apply entirely to polysulphides. In this case condensation takes place with an excess of sodium polysulphide and this yields terminal groups of the —S_xNa type:

$$\sim\!\sim RCl + Na_2S_x \longrightarrow \sim\!\sim RS_xNa + NaCl$$

Some hydroxyl end groups may also be produced by alkaline hydrolysis:

$$\sim\!\sim RCl + {}^{-}OH \longrightarrow \sim\!\sim R.OH + Cl^{-}$$

According to Gobran & Berenbaum (1969) the presence of S_2^{2-} ions causes cleavage of the polysulphide chain by the following equilibrium reaction:

$$\sim\!\sim SRS{-}SRS{-}SRS{-}SRS\sim\!\sim + S_2^{2-} \rightleftharpoons \sim\!\sim SRS{-}SRS{-}S^{-} + {}^{-}S{-}SRS{-}SRS\sim\!\sim$$

Where cleavage occurs well within the chain, the subsequent combination of the cleaved groups will give an average molecular weight similar to that of the original polymer. However, where cleavage occurs near the chain ends, small soluble fragments may be obtained that may be washed out of the system leaving the remaining large fragments to recombine and thus lead to an overall increase in molecular weight. Such an approach is not totally unique; a similar technique was used in the 1940s involving ester interchange with polyester rubbers.

More in common with other condensation polymerizations is the tendency to produce cyclic five- and six-membered rings rather than linear polymers where the dihalides contain 4–5 carbon (or carbon + oxygen) atoms. This is illustrated in Table 14.2. The value of x in the general polysulphide formula —RS_x—, known as the *sulphur rank*, depends primarily on the value of x in the original sodium polysulphide. This is an average value since both in the sodium polysulphide and in the polymer there is a range of values. In practice x is between 2 and 4. Whilst the original polysulphide elastomer (Thiokol A) had a sulphur rank of 4, current commercial materials have values of about 2–2·2.

At one time it was believed that where $x > 2$ the 'excess' sulphur was pendant as in **I** or **II**. However, a number of studies carried out about 1950 provided strong evidence that the sulphur atoms were linked linearly in the chain as in **III** and **IV**. (See Gobran & Berenbaum (1969) for references.)

(I)	(II)	(III)	(IV)
~~R—S(=S)—S(=S)~~	~~R—S(=S)—S~~	~~R—S—S—S—S~~	~~R—S—S—S~~

TABLE 14.2
Effect of dihalide structure on cyclization using 2M-Na_2S_2
(After Berenbaum, 1962)

Dihalide (Cl—R—Cl) —R—	*Cyclic sulphide (ml/mole of dihalide)*
$-(CH_2)_6-$	5
$-(CH_2)_5-$	39
$-(CH_2)_4-$	83
$-(CH_2)_3-$	0
$-(CH_2)_2-$	0
$-(CH_2)_2-O-(CH_2)_2-$	60
$-(CH_2)_2-O-CH_2-O-(CH_2)_2-$	0

Some of the sulphur from the sulphide linkages may be removed by a process known as *desulphurization.* Amongst effective agents for this are alkali hydroxides, sulphides and sulphites, of which the latter are usually preferred

$$\sim R-S-S-S-R\sim + SO_3^{2-} \rightleftharpoons R-S^- + {}^-O_3S-S-S-R\sim \rightleftharpoons \sim R-S-S-R\sim + S_2O_3^{2-}$$

The presence of excess sulphide during polymerization also allows some desulphurization to occur simultaneously by a similar mechanism.

$$\sim R-S-S-S-R\sim + S_2^{2-} \rightleftharpoons \sim R-S^- + {}^-S-S-S-S-R \rightleftharpoons \sim R-S-S-R\sim + S_3^{2-}$$

The reversing of this reaction, which can increase the sulphur rank to a value of about 5, is known as *thionation.*

Of commercial significance is the fact that the sulphide links are susceptible to *reductive cleavage.* One such process involves treatment of the polymer latex with sodium hydrosulphide and sodium sulphite followed by acidification (Patrick & Ferguson, 1949). It is this process which is used to make the liquid polymers of molecular weight 500–10 000 which today form the bulk of the polysulphide elastomer market.

$$\sim R-S-S-R\sim + SH^- \longrightarrow \sim SRS^- + HSSR\sim$$

$$\sim SRS^- \xrightarrow{H^+} \sim SRSH \qquad HSSR\sim \xrightarrow{Na_2SO_3} Na_2S_2O_3 + HSRS\sim$$

14.2.2 Structure and Properties

The key features of polysulphide rubbers are clearly a direct result of the presence of the main-chain sulphide groups. With a solubility parameter of 18·3–19·2 MPa$^{1/2}$ it would be expected that this material would have good resistance to swelling in hydrocarbon oils. In fact this resistance is very much better than even would be predicted in this way and is markedly superior to all but one or two of the highly specialist fluororubbers (see Chapter 12). Vapour permeability is also exceptionally low.

Hydrocarbon resistance is greatest where the percentage of sulphur is the highest. Thuş the original material based on dichloroethane (Thiokol A) has the greatest swelling resistance, that based on bis-2-chloroethyl formal has the least of the currently available rubbers, whilst the intermediate copolymer (Thiokol ST) has intermediate oil resistance (see Table 14.3).

TABLE 14.3
Comparison of solvent resistance of standard polysulphide compound vulcanizates based on three rubbers of different sulphide content

Solvent	*Volume swell[a] after 1 month at 25°C (%)*		
	Thiokol A	*Thiokol FA*	*Thiokol ST*
Benzene	30	96	114
Toluene	24	55	79
Xylene	17	31	39
Carbon tetrachloride	20	36	48
SR-6[b]	6	10	10
Iso-octane	−2	1	1
Glacial acetic acid	10	21	18
Butyl acetate	17	17	35
Dibutyl phthalate	6	7	8
Linseed oil	0	0	0
Ethyl alcohol	0	2	5
Ethylene glycol	0	1	3
Glycerol	0	2	1
Methyl ethyl ketone	34	28	49
Methyl isobutyl ketone	24	13	25

[a] Test based on ASTM D471-64T.
[b] 40% aromatic blended reference fuel (60% iso-octane, 20% xylene, 15% toluene, 5% benzene).

Whilst the sulphide groups impart hydrocarbon oil resistance, S–S bonds are not noted for great chemical stability and have low dissociation energies of about 268 kJ mol^{-1}. Hence the rubbers are attacked by concentrated mineral acids whilst disulphide interchange reactions occurring at elevated temperatures can lead to high creep and stress relaxation of stressed vulcanizates. Such undesirable interchange reactions are found to be accelerated by small amounts of sulphur, mercaptides and alkaline agents. Lowering of the sulphur rank also appears to reduce the reaction, as does selective choice of curing system. In particular the presence of manganese dioxide is said to reduce the tendency to stress-relaxation.

The high-temperature stability of the polysulphides is particularly limited by an acid-catalysed attack initiated by traces of water. Attempts to remove this water have however led to interference with curing reactions, whilst the use of antioxidants to counter subsequent free-radical processes has been of little value because of the overriding influence of the oxidizing agents commonly used for vulcanization. Curing with di-isocyanates and diepoxides avoids some of these problems and the polymers do have better heat resistance but they are less elastic. Curing with zinc chromate or lead dioxide also gives improved ageing resistance.

The low-temperature properties of the polysulphides are largely determined by the position of the T_g. Crystallization can occur with the homopolymers but not with copolymers such as Thiokol FA. Even with the virtual homopolymers such as Thiokol ST there is very little crystallization in the presence of fillers and curing agents.

The T_g tends to drop with an increase in the length of the chain of methylene groups between the sulphide linkage down to a minimum of about $-75°C$ with poly(hexamethylene sulphide) (Gobran & Berenbaum, 1969). The poly(ethyl formal disulphide) which is the basic polymer of current commercial formulations has a reasonably low T_g of $-59°C$, a figure slightly higher than for the corresponding oxygen-free polymer. Such a low T_g is lower than for many oil-resisting rubbers and is adequate for most purposes.

14.2.3 Vulcanization and Compounding

The two commercially available solid elastomers differ structurally in several ways which affect vulcanization and processing and which are summarized in Table 14.4.

The high-molecular-weight linear copolymer (Thiokol FA) has first to be treated with peptizing agents (chemical plasticizers) such as DPG or MBTS to reduce the molecular weight and enable the material to be processed. The

TABLE 14.4
Structural differences between commercial polysulphide elastomers

Feature	*Thiokol FA*	*Thiokol ST*
General polymer type	Copolymer	Homopolymer with cure site monomer
Disulphide(s) used	Bis-2-chloroethyl formal/dichloroethane	Bis-2-chloroethyl formal + 2% trichloropropane
Polymer shape	Linear	Branched
Molecular weight	Very high	80 000
Terminal groups used in vulcanization	Hydroxyl	Mercaptan

reaction mechanism is of the general form

$$\underset{\text{Polymer}}{(\sim\text{SS—R}'\text{—SS—R}''\text{—SS}\sim)} + \underset{\text{Peptizer}}{\text{R}'''\text{—SS—R}'''} \longrightarrow (\sim\text{SS—R}'\text{—SS—R}''') + \underset{\text{Lower-mol.-wt polymer}}{(\text{R}'''\text{—SS—R}''\text{—SS}\sim)}$$

The mechanism of vulcanization is somewhat obscure but appears to be primarily a chain extension process involving *hydroxyl* end groups and an oxide such as zinc oxide, typically used at levels of 10 pphr, is usually employed for this purpose.

The branched 'homopolymer' Thiokol ST is obtained by reductive cleavage of higher-molecular-weight material during manufacture. It is immediately more processable and in addition possesses mercaptan (—SH) end groups. Since there are more than two of these to each polymer molecule, chain extension on reaction with 'curing agents' leads to the formation of a three-dimensional network.

Zinc, lead and calcium peroxides are widely used as coupling agents (curing agents), the reaction in the last case being described by the equation:

$$2(\sim\text{RSH}) + \text{CaO}_2 \longrightarrow \sim\text{RSSR}\sim + \text{Ca(OH)}_2$$

An alternative system is provided by the use of *p*-quinone dioxime in conjunction with diphenylguanidine. Some typical curing systems based on 100 parts of Thiokol ST are:

1. *p*-Quinone dioxime, 1·0; zinc oxide, 0·5; stearic acid, 0·5–3·0;
2. *p*-quinone dioxime, 1; zinc chromate, 10; stearic acid, 1;
3. zinc peroxide, 6; stearic acid, 1;
4. zinc peroxide, 4·0; calcium hydroxide, 1·0; stearic acid, 1·0.

TABLE 14.5
Effect of carbon black on physical properties of polysulphide vulcanizate (After Bertozzi, 1968)
Formulation: Thiokol FA, 100; ZnO, 10; stearic acid, 0·5; MBTS, 0·3; DPG, 0·1; SRF Black, *x*.

Carbon black loading	*Tensile strength*		*Elongation at break (%)*	*Hardness (Shore A)*
	(psi)	*(MPa)*		
0	155	1·06	450	40
10	300	2·06	550	45
20	600	4·12	700	49
40	1 000	6·89	600	58
60	1 200	8·27	380	68
80	1 200	8·27	230	78
100	100	0·69	210	82

The polysulphides are extremely weak unless reinforced by carbon black. For this purpose SRF (N770) types are usually used. As seen from Table 14.5, a loading of about 60 pphr is most practical and the one commonly used.

In the case of mercaptan-terminated rubber (Thiokol ST), up to 5% of a liquid polysulphide may be added as a co-curing softening agent.

Typical vulcanizate properties are given in Table 14·6.

Although Thiokol FA has superior hydrocarbon resistance, Table 14.6 clearly shows the very poor compression set resistance of this polymer which makes it generally unsuitable for O-rings and seals. It also exhibits inferior low-temperature flexibility.

14.2.4 Liquid Polysulphides

These mercaptan-terminated low-molecular-weight materials now form the bulk of the polysulphide market. Marketed as Thiokol LP polymers, they may be cured using similar materials to those used with Thiokol ST. They are mainly used for sealants, encapsulating and impregnation. They may be blended with epoxide resins in any ratio to give the resins varying degrees of flexibility. In these circumstances reaction proceeds between terminal epoxy groups and mercaptans. It is however necessary to include a curing agent for the epoxide resin.

14.2.5 Applications

The uses for polysulphide elastomers arise from their excellent resistance to aliphatic hydrocarbons. Use has however been restricted by their poor

TABLE 14.6
Typical properties of polysulphide vulcanizates
(After Cooper, 1971)

	Thiokol FA	*Thiokol ST*
Formulation (pphr)		
Polymer	100	100
Zinc oxide	10	0·5
Stearic acid	0·5	3
p-Quinone dioxime	—	1·5
MBTS	0·3	—
DPG	0·1	—
SRF Black	60	60
Cure	50 min at 150°C	30 min at 145°C
Tensile strength, MPa	8·1	8·6
Elongation at break, %	380	310
Hardness (IRHD)	70	70
Low-temperature flexibility, °C	−45	−53
Compression set (22 h at 70°C), %	100	37

temperature resistance, high compression set, poor mechanical properties and cost. In spite of these deficiencies they continue to be used for fuel hose and tubing and printers' rollers, and to some extent for gaskets, diaphragms, cork binders, putties and sealants.

14.3 EPICHLORHYDRIN RUBBERS (CO, ECO, AECO)

High-molecular-weight polymers of epichlorhydrin that were rubbery in nature were prepared by E. J. Vandenberg in the laboratories of the Hercules Powder Co. in 1957 using aluminium alkyl catalyst systems somewhat related to the Ziegler–Natta catalysts. This was an example of a ring-opening polymerization.

$$\underset{\diagdown O \diagup}{CH_2—CH}—CH_2—Cl \longrightarrow —CH_2—\underset{\underset{CH_2Cl}{|}}{CH}—O—$$

CO Rubber

The product is today given the ISO designation CO rubber, the CO being derived from the alternative name of the monomer, chloromethyl oxirane.

It was found to have excellent ozone, oil and petrol resistance, and very low gas permeability together with very good heat resistance. One limiting factor was the high brittle point of about −17°C.

In the mid-1960s copolymers (ECO rubbers) were prepared by copolymerization of epichlorydrin with ethylene oxide.

$$-\!\!\left[CH_2-\underset{\underset{CH_2Cl}{|}}{CH}-O\right]\!\!\left[CH_2-CH_2-O\right]\!\!-$$

Following the discovery of these copolymers Hercules licensed manufacturing rights to Goodrich, who have since marketed the material as Hydrin rubber whilst Hercules also commenced manufacture using the trade name Herclor. In 1977 sulphur-curable grades using allyl glycidyl ether as a third monomer were introduced (AECO rubbers) which partly overcame the problems caused by the softening effect of hydroperoxide-containing gasoline (the sour gas problem). In the early 1980s two Japanese companies, Nippon Zeon and Osaka Soda, commenced manufacture, marketing their products respectively under the names Gechron and Epichlomer.

By 1986 world nameplate capacity was estimated at 17 000 tpa with the Hercules plant at Hattiesburg alone having a capacity of 12 000 tpa. The market, about 75% of which was in the automotive industry, was however only about one-third of capacity and in 1976 the Hercules operations were sold to Goodrich.

14.3.1 Manufacture

The epichlorhydrin rubbers are prepared by ring-opening polymerization with aluminium alkyls such as tributylaluminium used as catalysts in conjunction with water and in the presence of chelating agents such as acetylacetone and zinc acetylacetonate. The highest polymerization rates are said to occur when the aluminium alkyl and water are used in equimolar proportions (Dimonie & Gavăt, 1968). Typical aluminium alkyl concentrations are of the order of 5×10^{-2} mol litre^{-1}. The detailed mechanisms of the reaction appear complex and it has been suggested that in the relatively simple aluminium alkyl–water system a cationic mechanism is operative but that when chelates are present a co-ordination polymerization occurs.

14.3.2 Structure and Properties

The epichlorhydrin rubbers are in essence halogenated linear aliphatic polyethers. The presence of the ether oxygen group has two opposing effects:

1. An increase in chain flexibility which depresses T_g (and T_m if the polymer is crystallizable);
2. an increase in polarity and hence in interchain attraction that will increase T_g and T_m.

The relative effects of these two factors depend on the ether group concentration, i.e. on the value of R in a polymer of type

$$—(CH_2)_R—O—$$

with the T_m being at a minimum with the CH_2/O ratio of 3–4. This is shown clearly in Fig. 14.1.

Copolymerization of two ethers will usually inhibit crystallization so that rubberiness will depend on the (much lower) T_g rather than the T_m.

Introduction of chlorine atoms onto the main chain would improve oil resistance but by increasing interchain attraction would raise T_g and, in the case of crystalline polymers, the T_m. If however the chlorine is incorporated into a side chain, such as in a $—CH_2Cl$ group, the chain-separating effect of the side group will more than offset the interchain attraction effect of the halogen, leading to a decrease in the transition points. If the polymer is substantially atactic there will be no crystallization and the T_g may be so low that the polymer is rubbery at room temperature. This is the case with the polyepichlorhydrin homopolymer (CO rubber).

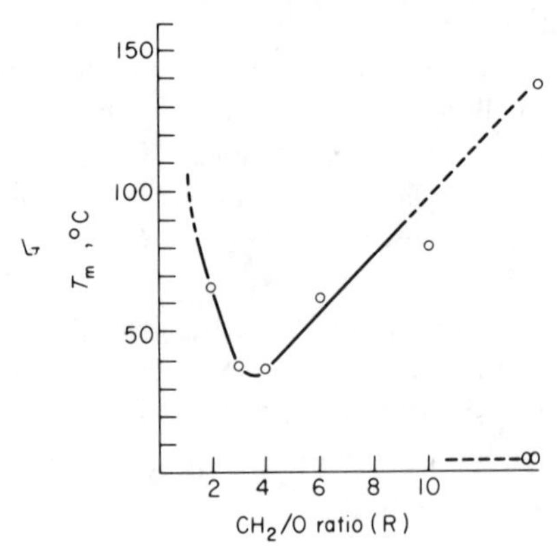

FIG. 14.1. Crystalline melting points of linear polyethers $—[(CH_2)_R—O—]_n$. (From Ledwith & Fitzsimmonds (1968), produced by permission of John Wiley & Sons Inc.)

Copolymerization of epichlorhydrin with the less polar ethylene oxide markedly depresses the T_g and thus enables the rubber to be used at much lower temperatures.

The somewhat high T_g of the homopolymer influences several other properties. These include resilience, which is very low at normal room temperatures, and low permeability which is much less than for either butyl or nitrile rubber, both of which are generally considered outstanding in this respect.

The chlorine content (38·4% for the homopolymer; 25% for the normal copolymer) has a notable effect in improving oil resistance and resistance to burning. The absence of double bonds results in the rubbers having excellent resistance to ozone and weathering and good heat resistance.

14.3.3 General Properties

The principal features of the polyepichlorhydrin homopolymer (CO rubber) in approximate order of importance are:

1. Very good resistance to hydrocarbon oils, bettered only by the polysulphides, fluororubbers and high-acrylonitrile nitrile rubbers;
2. good heat resistance with a maximum continuous service temperature of 150°C, a figure bettered only by the acrylic, fluoro- and silicone rubbers;
3. excellent resistance to ozone and weathering;
4. lowest permeability of all commercial synthetic rubbers (less than half that for butyl rubber with most gases and vapours);
5. good resistance to burning;
6. good mechanical properties of black-filled grades;
7. need to plasticize to achieve good low-temperature properties;
8. susceptibility to sour gas attack;
9. poor electrical insulation resistance and radiation resistance;
10. low resilience at normal ambient temperatures.

The ECO copolymers have 1:1 molar ratios and differ mainly from the homopolymers in respect of:

1. Lower brittle point;
2. higher resilience at normal ambient temperatures;
3. not self-extinguishing;
4. higher permeability to gases similar to a medium-acrylonitrile NBR.

The main differences between the two types are illustrated in Table 14.7.

TABLE 14.7
Comparison of typical CO and ECO rubbers

Property	*Homopolymer, CO*	*Copolymer, ECO*
Raw polymer		
Specific gravity	1·36	1·27
Chlorine content (%)	38·4	25
Compound[a]		
Specific gravity	1·49	1·42
Mooney scorch at 250°F (121°C), min	10	7
Mooney cure at 250°F (121°C), min	15	14
Modulus at 200% elong., psi (MPa)	2 250 (15·5)	1 650 (11·4)
Tensile strength, psi (MPa)	2 305 (15·9)	2 180 (15·0)
Elongation at break, %	230	320
Hardness (Shore A)	82	74
Compression set (ASTM Method B; 70 h at 100°C), %	20	20
Low-temp. brittleness (ASTM D746), °C	−18	−40
Resilience (Lupke; room temp.), %	26	67
Self-extinguishing	Yes	No

[a] Formulation: Polymer, 100; FEF Black, 50; Red lead, 5; nickel dibutyl dithiocarbamate, 1; 2-mercaptoimidazoline, 1·5; zinc stearate, 1 (for CO), 1·5 (for ECO) pphr. Cure: 30 min at 160°C.

14.3.4 Processing

Processing of epichlorhydrin rubbers is conventional. In particular they are characterized by low mill shrinkage, good building tack and easy flow properties with low die swell. Typical mill mixing is carried out with the front roll at 70°C and the back roll at 90°C. Mould fouling can be a problem requiring care in the choice of mould release agents and compound formulation. The sulphur-curing grades are less prone to mould fouling.

14.3.5 Compounding

Formulation of epichlorhydrins is concerned mainly with the selection of curing system.

Vulcanization is readily accomplished by reagents that act difunctionally with the chloromethyl groups and include diamines, ureas, thioureas and ammonium salts. One widely quoted curing agent is 2-mercaptoimidazoline (ethylene thiourea) which is most effective in the presence of an acid acceptor such as red lead. As may be expected, the degree of cross-linking

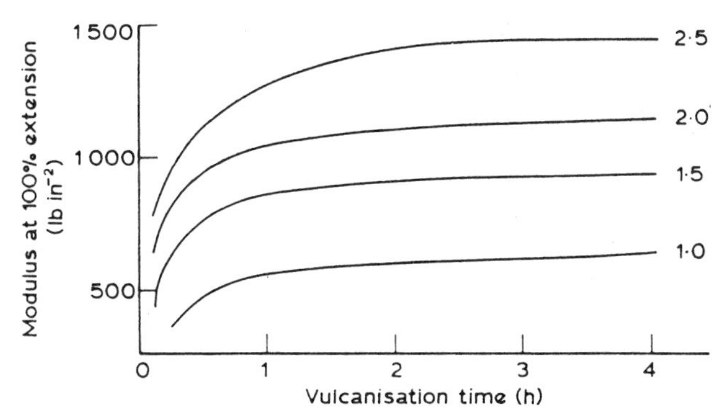

FIG. 14.2. Vulcanization behaviour of an epichlorhydrin rubber. Formulation: epichlorhydrin–ethylene oxide rubber, 100; FEF carbon black, 50; red lead, 5; zinc stearate, 0·75; nickel dibutyldithiocarbamate, 1; 2-mercaptoimidazoline, variable (levels indicated by numbers appended to curves). Vulcanization carried out at 160°C. (From Blackley, 1983.)

will be affected by the amount of the curing agent used, as is shown in Fig. 14.2, which shows the influence of loading on tensile modulus.

As already mentioned in Chapter 10, there are some specific hazards associated with the use of 2-mercaptoimidazoline and this has led to the development of alternative proprietary systems. One such system based on hexamethylene diamine carbamate gives low compression set and long flex life but poorer ageing. A retarder may also be required to reduce scorching.

Most formulations use carbon black as a filler and the N550 (FEF) blacks are widely used. Typical mixes also contain some zinc stearate and some nickel dibutyldithiocarbamate as antioxidant. A typical formulation would be

	pphr
Polymer	100
2-Mercaptoimidazoline	1–2·5
Red lead	5
Nickel dibutyldithiocarbamate	1
FEF black	40–60
Zinc stearate	0·75–1·0
Plasticizer	0–25

Because of the polar nature of the polymer, it is necessary to use polar plasticizers such as esters and ethers rather than petroleum oils. They are very useful in extending downwards the minimum service temperatures for which the polymers may be used.

The influence of plasticizer on the brittle temperature can be seen clearly in Table 14.8.

The plasticizers mentioned in Table 14.8 are however subject to loss through volatilization at elevated temperatures (of the order of 90% loss after 70 h at 150°C). Where this is a problem, higher-molecular-weight plasticizers may be considered although they are less effective in depressing the brittle point.

TABLE 14.8
Effect of plasticizer on low-temperature flexibility ($T_{10\,000}$)
Basic formulation as for Table 14.7 with varying amounts of plasticizer

Plasticizer	*Amount*	$T_{10\,000}$ (°C)	
		CO	*ECO*
DOP	0	−18	−38
	15	−26	−46
	25	−31	−47
TP 90B[a]	15	−33	−49
	25	−42	−53

[a] TP90B is a high-molecular-weight polyether supplied by the Thiokol Corporation.

Pollution control regulations of many governments have led to modifications in fuel delivery systems, and in turn this led to many failures using traditional fuel lines. Problems have been encountered in the softening of ECO rubber in contact with hydroperoxide-containing gasoline ('sour gas').

In this connection, allyl-group-containing epichlorhydrin rubbers were introduced to reduce the problem (AECO rubbers). Where allyl glycidyl ether is used as the third monomer at a level of 2·6 mol/100 units, sour gas causes deterioration by softening whilst at the 8 pphr level deterioration by hardening is observed (Mori & Nakamura, 1984). Best results are obtained using the allyl chemical at the 5 mol/100 unit level with the following cure system: polymer, 100; 1,3,5-triazine-2,4,6-trithiol, 1; MgO, 3; $CaCO_3$, 5; stearic acid, 1 pphr.

A reduction in the sour gas problem is also claimed where nickel dibutyldithiocarbamate is replaced by bis(4-hydroxy-3,5,di-*t*-butylhydrocinnamic acid) hydrazide in CO homopolymer (Macarthur, 1981).

14.3.6 Applications

About three-quarters of polyepichlorhydrin rubbers produced are used by the automotive industry for such uses as hose, tubing, diaphragms and a variety of moulded products.

It is understood that about three-quarters of the market is taken up by the copolymer and the larger part of the rest by the homopolymer. The sulphur-curing terpolymers usage is limited by their lower heat resistance. It is also understood that for many applications the homopolymers and copolymers are used together in blends.

14.4 PROPYLENE OXIDE RUBBERS (GPO)

Copolymers of propylene oxide with a cure site monomer (usually allyl glycidyl ether used to a level of about 10% of the total monomer) were first described in 1963 (Grüber *et al.*, 1963, 1964). A commercial product was first marketed by General Tire and Rubber as Dynagen XP-139 and then by Hercules as Parel. In 1986 Hercules sold their interest in this rubber to Goodrich along with their interest in epichlorhydrin rubbers (see Section 14.3). Goodrich have therefore become the sole manufacturers of both epichlorhydrin rubbers (outside Japan) and polypropylene oxide rubbers as well as being major producers of both acrylic and nitrile rubbers. Production capacity for propylene oxide rubbers, designated as GPO rubber by ASTM, has not been disclosed by the manufacturers but it has been estimated at 6000 tpa by the International Institute of Synthetic Rubber Manufacturers.

14.4.1 Structure and Properties

The chemical structure of the commercial polypropylene oxide rubber may be represented as

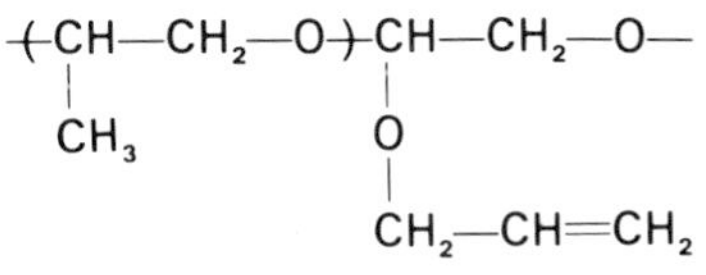

$$\left(\underset{\substack{|\\ CH_3}}{CH}-CH_2-O\right)\underset{\substack{|\\ O\\ |\\ CH_2-CH=CH_2}}{CH}-CH_2-O-$$

It thus has a strong resemblance to the epichlorhydrin terpolymers which have the structure

$$\left(CH_2-\underset{\underset{CH_2Cl}{|}}{CH}-O\right)\left(CH_2-CH_2-O\right)\underset{\underset{\underset{CH_2-CH=CH_2}{|}}{\underset{O}{|}}}{CH}-CH_2-O-$$

The most distinctive difference is the absence of a chlorine atom in GPO; although this reduces resistance to hydrocarbon oils and solvents and to burning it does give a very low T_g. Although sulphur-cured, the rubbers have a very good heat resistance and as a result an operating range from $-60°$ to $+150°C$.

GPO rubbers are similar to natural rubber in exhibiting high resilience and excellent flex life, but in addition show exceptional low-temperature properties together with good heat resistance, good ozone resistance and moderate oil resistance. Like NR they may be vulcanized with sulphur.

A typical compound for the commercial material Parel 58 is as follows.

	pphr
Parel 58 elastomer	100
HAF carbon black	50
Zinc oxide	5
Stearic acid	1
Nickel dibutyldithiocarbamate	1
Tetramethylthiuram monosulphide	1·5
Mercaptobenzothiazole	1·5
Sulphur	1·25

This compound has a Mooney ML $(1+4)$ at 100°C of 86, a Mooney scorch MS at 121°C of 9 min to a three-point rise and requires curing for 30 min at 160°C.

Typical properties of the vulcanizate are given in Table 14.9.

14.4.2 Compounding and processing

GPO rubbers may be processed using conventional techniques. Lightly loaded compounds however require 'upside-down' internal mixing techniques in order to ensure satisfactory incorporation and dispersion of filler. High filler levels are also required for satisfactory extrusion. Compounds also suffer from low green strength and where this is a problem it may be advisable to blend with about 10 pphr of an EPDM rubber.

With reinforcing grades of carbon black, tensile strength increases with

TABLE 14.9
Typical properties of polypropylene oxide rubber vulcanizates

Physical property	*Tested after cure*	*After air ageing for 70 h at 150°C*
Tensile strength, MPa	13·3	10·9
Modulus at 200%, MPa	8·2	10·0
Elongation at break, %	370	210
Hardness Shore A	70	73
Low-temp. stiffness		
(ASTM D1053 $T_{10\,000}$), °C	−58	—
De Mattia flex at 110°C, kc to first crack	3 000	—
Volume swell (70 h at 23°C), %		
Water	13	—
ASTM Fuel A	69·7	—
ASTM Fuel B	104	—

the level of black up to at least 80 pphr with at the same time a loss in resilience and compression set resistance.

Because of the inherently very good low-temperature properties, plasticizers are not added to improve low-temperature flexibility but may be used to lower hardness or compound viscosity. For this purpose naphthenic or aromatic oils may be used.

The formulation given in Section 14.4.1 is somewhat scorchy and has a bin storage life of the order of only 2–3 weeks. Replacement of the MBT by a CBS delayed-action accelerator reduces scorch and increases bin storage life.

14.4.3 Applications

The polypropylene oxide rubbers are best looked on as highly resilient and flexible materials with excellent low-temperature flexibility coupled with good ozone resistance, heat resistance and moderately good hydrocarbon oil and solvent resistance.

In practice the principal use for this material is for automotive hoses.

14.5 CHLORINATED AND CHLOROSULPHONATED POLYETHYLENES (CM AND CSM)

In Chapter 7 it was seen that copolymerization of ethylene and propylene, which separately can homopolymerize to give crystalline plastics materials, may yield rubbery materials if used in the appropriate ratios. This is because copolymerization introduces irregularities in the structure, severely curtailing crystallization, and since chain flexibility and low interchain attraction also lead to a low T_g, a rubbery polymer results.

An alternative to copolymerization as a means of introducing structural irregularities is achieved by random substitution on to preformed polyethylene molecules. Being alkanes, polyethylenes are of very limited chemical reactivity and in practice substitution is restricted to halogenation and certain related processes. Of the various halogens only chlorine is used commercially.

The introduction of chlorine groups into a polyethylene molecule has two opposing effects:

1. It reduces chain regularity, thus restricting and eventually eliminating the ability of the polymer to crystallize. In view of the low T_g of polyethylene this tends to increase rubberiness.
2. It increases interchain attractions causing the T_g to be raised, eventually to the point where the amorphous polymer is no longer rubbery.

These effects are clearly indicated in Table 14.10.

This approach towards making rubbery polymers has led to the commercial appearance of two classes of elastomers, CSM and CM.

The CSM chlorosulphonated polyethylenes were first introduced by Du Pont in 1951 under the trade name Hypalon. Although they continue to

TABLE 14.10
Effect of chlorine content on the properties of chlorinated polyethylenes
(After Oakes & Richards, 1946)

Cl content (%)	*Structure at room temp.*	*Physical form at room temp.*	*Brittleness temp. (°C)*	*Softening pt (°C)*
2	Mainly crystalline	Flexible	—	82
8	Crystalline	More flexible	<-70	69
25	Mainly amorphous	Rubbery	<-70	*c.* 20
40	Amorphous	Soft and flexible	<-70	20
45	Amorphous	Flexible, leathery	-20	30
54	Amorphous	Rigid	$+20$	52
60	Amorphous	Rigid, brittle	$+40$	67

Although traces of crystallinity are discernible up to 35% chlorine content, the polymer is essentially elastomeric in the range 25–40%.

dominate the market, Toyo Soda and Denko in Japan also produce CSM rubbers and manufacture is reported to have commenced in China. Manufacturing capacity is of the order of 50 000 tpa. With very good chemical, flame, heat, oil, ozone and weathering resistance these rubbers have been widely used for coated fabrics, wire and cable covering and hose applications.

Chlorinated polyethylenes have been available for many years and were originally used mainly as additives for PVC. In the late 1960s rubbery grades were introduced and given the ASTM designation CM. Manufacturers include Bayer and Osaka Soda.

14.5.1 Structure and Properties of CSM Rubbers

The chlorosulphonated polyethylene rubbers contain both chlorine and sulphonyl chloride ($—SO_2Cl$) substituents. Chlorosulphonation is brought about by the Reed–Horn reaction that involves co-reaction of sulphur dioxide and chlorine in the presence of light according to the following reaction sequence.

$$
\begin{aligned}
Cl_2 &\longrightarrow 2Cl\cdot \\
RH + Cl\cdot &\longrightarrow R\cdot + HCl \\
R\cdot + SO_2 &\rightleftharpoons RSO_2\cdot \\
RSO_2\cdot + Cl_2 &\longrightarrow RSO_2Cl + Cl\cdot
\end{aligned}
$$

Typical polymers contain about 1% sulphur (in sulphonyl chloride units) and the structure of a typical rubber of this type may be represented by:

$$
\sim\!\!\left[\left(CH_2{-}CH_2{-}CH_2{-}CH_2{-}CH_2{-}CH_2{-}CH_2{-}\underset{Cl}{\underset{|}{CH}}\right)_2\underset{SO_2Cl}{\underset{|}{CH}}\!-\right]_{17}\!\!\sim
$$

The sulphonyl chloride group acts as a cure site.

CSM polymers have the following characteristics.

1. The balance of chain flexibility and chlorine content is such that the material exists in the rubbery range.
2. The polymer is saturated and thus not susceptible to oxidation, ozone and weathering in the manner of diene rubbers. It is also a rubber of generally limited reactivity and has therefore good chemical resistance.
3. The presence of the chlorine confers a measure of hydrocarbon oil resistance and also of flame resistance.

Commercial grades vary in the following respects.

1. Level of branching in the original polyethylene used. Low-density polyethylenes contain substantial amounts of ethyl and butyl side chains together with some long chain branching; high-density polyethylenes are more linear in structure and have little branching. The low-density branched grades are now used mainly for solution coating applications with high-density polymers now preferred for other applications.
2. Molecular weight of original polyethylene. For solution applications these are quite low and of the order of 20 000. Somewhat higher molecular weights giving rise to Mooney ML 1 + 4 (100°C) viscosities in the range 40–85 are used elsewhere.
3. Chlorine content. This is usually in the range 25–43%, the higher values giving improved oil and flame resistance but at some loss of low-temperature flexibility.
4. Sulphur content. Although early grades had values up to 1·5%, most of the more recent grades have settled on a figure of 1%.

14.5.2 Compounding and Vulcanization of CSM Rubbers

Because of the high reactivity of the sulphonyl chloride cure site, a wide variety of curing systems are practical. A selection of such systems is shown in Table 14.11. Each system has its own niche according to the properties required, as indicated in the table.

Conventional fillers such as carbon blacks may be used but for optimum heat resistance plasticizers should, as with other heat-resisting rubbers, have a low volatility.

14.5.3 Applications of CSM Rubbers

Although several rubbers have better heat resistance (fluororubbers, silicones, acrylics, EVA and EPDM) and many better resistance to hydrocarbon oils, the chlorosulphonated polyethylene rubbers have become well-established in the market place. Major applicational areas are coated fabrics and sheeting, and wire and cable covering, which between them take up about three-quarters of the market. Other important uses include hose and automotive applications.

Coated fabrics have been a particularly successful application, especially for inflatable boats but also for many other leisure products. Some 20 years ago the writer purchased a CSM groundsheet for camping and other outdoor activities which today after much service appears as good as new.

TABLE 14.11
Vulcanization systems for CSM
(After Brown, 1976)

	Cure system							
	Metallic oxide + rubber accelerator			*Metallic oxide + polyfunctional alcohol + rubber accelerator*	*Epoxy resin + rubber accelerator*		*Metallic oxide + bismaleimide*	*Metallic oxide + peroxide co-agent*
Magnesia	20	10	—	4	3	—	—	20
Litharge	—	20	25–40	—	—	—	30	—
Tetrone® A	2	0·75	1	2	2	1·5	—	—
MBTS	—	0·5	1	—	—	0·5	—	—
DOTG	—	—	—	—	—	0·25	—	—
Ni BD, NBC	—	3	3	—	—	—	1·5	—
HVA-2®	—	1	1	—	—	—	1·5	—
Pentaerythritol	—	—	—	3	—	—	—	—
Epoxy resin (Epikote 828®)	—	—	—	—	4	15	—	—
40% Dicumyl peroxide	—	—	—	—	—	—	—	8
Triallyl cyanurate	—	—	—	—	—	—	—	4
Hydrogenated wood resin	—	—	—	—	—	—	3	—
Type of cross-links formed	Ionic and covalent			Ionic, ester and covalent	Epoxy and covalent		Complex	Complex
Outstanding property	Colour	Heat resistance	Water resistance	General-purpose	Water resistance		Colour Good electricals	Compression set

® HVA-2-(*N*,*N*′-phenylene dimaleimide) is a registered trade name of Du Pont (UK) Ltd.
Epikote 828 is a registered trade name of Shell Chemicals.

A more recent development has been the use of high-green-strength unvulcanized sheet for roofing and industrial effluent pit liners, applications where good weathering and, in the latter case, chemical resistance are required.

Cable applications arise from the heat and oil resistance of the material. Amongst hose applications, attention has recently been drawn to the fact that radiator and heater hoses fitted to London buses about 1970 were still operating quite satisfactorily in the late 1980s. The hoses are of particular value in applications where difficulty of accessibility would cause replacement problems.

14.5.4 Properties of CM Rubbers

(Whilst the ASTM designation is CM, these materials are also widely referred to as CPE rubbers.)

The CM rubbers are structurally very similar to the chlorosulphonated polyethylenes, the obvious main difference being the absence of the sulphonyl chloride cure site. Vulcanizate properties are therefore frequently similar, one particular advantage claimed for CM being the lower compression set at elevated temperatures (e.g. at 125°C).

A further advantage of CM over CSM rubbers is the generally lower price structure. The main disadvantage has been the constraints imposed by available curing systems. The usual curing systems involve peroxides and these are unsuitable for hot air curing, as used for example with rubberized fabrics. Whilst non-peroxide curing systems have been developed, the vulcanizate properties are inferior in terms of hot air resistance and high-temperature compression set.

The grades available have chlorine contents in the range 25–42, which is very similar to that available for CSM rubbers. Increasing the chlorine content improves oil, fuel and flame resistance but lowers the heat resistance. Mooney ML $1+4$ (120°C) viscosities range from about 35 to about 130 whilst crystallinity levels, as rated by the suppliers, range from 'very slight' to 'appreciable'. The market is dominated by a grade with a chlorine content of 36%, a Mooney ML $1+4$ (120°C) of 85 and very slight crystallinity.

14.5.5 Compounding and Vulcanization of CM Rubbers

As mentioned above, peroxides are usually used for curing CM rubbers. The most commonly used peroxides are:

1,1-Di-*t*-butylperoxy-3,3,5-trimethylcyclohexane;
4,4-di-*t*-butylperoxy-*n*-butyl valerate;

dicumyl peroxide;
dis(*t*-butylperoxyisopropyl)benzene;
2,5-dimethyl-2,5-di(*t*-butylperoxy)hexane.

The peroxides are usually used in conjunction with a promoter such as triallyl cyanurate.

CM rubbers require a heat stabilizer. Magnesium oxide is commonly used but unsuitable where resistance to swelling in water is an important factor. In these circumstances lead stabilizers may be considered. Care must be taken in the selection of plasticizers in order that they do not interfere with cure, and rather expensive ester plasticizers such as dioctyl sebacate rather than the cheaper mineral oils may be required.

There has been some recent interest in non-peroxide cures involving thiadiazole derivatives in conjunction with aniline–butyraldehyde derivatives. As mentioned above these vulcanizates have inferior heat resistance but do exhibit significantly better tear resistance.

CM rubbers are supplied as fine powders which lend themselves to compounding in internal mixers when upside-down mixing procedures are generally recommended.

14.5.6 Applications of CM Rubbers

The major interest in chlorinated polyethylene rubbers has been for wire and cable applications. A second major area of interest is for technical hoses where their heat, ozone, oil and weathering resistance make them particularly suitable as cover materials. Whilst car radiator hoses are usually prepared from EPDM the oil resistance requirements with car diesel engines are often more demanding than can be met by that rubber, and this has led to developments of EPDM/CM blends. Amongst moulded goods, car axle boots set close to the engine and spark plug covers make use of the good heat and ozone resistance combined with medium oil resistance of the material.

BIBLIOGRAPHY

Polysulphides

Berenbaum, M. B. (1962). In *Polyethers*, Part III, ed. N. G. Gaylord. Interscience, New York, Chapter 13.

Bertozzi, E. R. (1968). *Rubber Chem. Technol.*, **41**, 114.

Cooper, W. (1971). In *Rubber Technology and Manufacture*, ed. C. M. Blow. Butterworth, London, Chapter 4D.

Gobran, R. H. & Berenbaum, M. B. (1969). In *The Polymer Chemistry of Synthetic Elastomers*, Part 2, ed. J. P. Kennedy & E. G. M. Törnqvist. Interscience, New York, Chapter 8C.

Polyethers

Ledwith, A. & Fitzsimmonds, C. (1968). In *The Polymer Chemistry of Synthetic Elastomers*, Part 1, ed. J. P. Kennedy & E. G. M. Törnqvist. Interscience, New York, Chapter 5C.

Chlorosulphonated Polyethylenes

Brown, J. H. (1976). *Annual Report on the Progress of Rubber Technology—1977*, ed. M. J. Quinlan. Plastics & Rubber Institute, London, **39**, p. 31.

Maynard, J. & Johnson, P. (1963). *Rubber Chem. Technol.*, **36**(1), 963.

REFERENCES

Berenbaum, M. B. (1962). In *Polyethers*, Part III, ed. N. G. Gaylord. Interscience, New York, Chapter 13.

Bertozzi, E. R. (1968). *Rubber Chem. Technol.*, **41**, 114.

Blackley, D. C. (1983). *Synthetic Rubbers: Their Chemistry and Technology*. Applied Science Publishers, London.

Brown, J. H. (1976). *Annual Report on the Progress of Rubber Technology—1977*, ed. M. J. Quinlan. Plastics & Rubber Institute, London, **39**, p. 31.

Cooper, W. (1971). In *Rubber Technology and Manufacture*, ed. C. M. Blow. Butterworth, London, Chapter 4D.

Dimonie, M. & Gavăt, G. (1968). *Eur. Polymer J.*, **4**, 541.

Gobran, R. H. & Berenbaum, M. B. (1969). In *The Polymer Chemistry of Synthetic Elastomers*, Part 2, ed. J. P. Kennedy & E. G. M. Törnqvist. Interscience, New York, Chapter 8C.

Grüber, E. E., Briggs, R. A. & Mayer, D. A. (1963). *Atti. Congr. int. Plast. e Elastomer*, Turin, p. 315.

Grüber, E. E., Meyer, D. A., Swart, G. H. & Weinstock, K. V. (1964). *Ind. Engng Chem., Prod. Res. Dev.*, **3**, 194.

IISRP (1988). *Worldwide Rubber Statistics—January 1988*. International Institute of Synthetic Rubber Producers Inc., Houston.

Ledwith, A. & Fitzsimmonds, C. (1968). In *The Polymer Chemistry of Synthetic Elastomers*, Part 1, ed. J. P. Kennedy & E. G. M. Törnqvist. Interscience, New York, Chapter 5C.

MacArthur, N. C. (1981). Paper presented at the PRI London Section Jubilee Conference, April 1981.

Mori, K. & Nakamura, Y. (1984). *Rubber Chem. Technol.*, **57**, 665.

Oakes, W. G. & Richards, R. B. (1946). *Trans. Faraday Soc.*, **42A**, 197.

Patrick, J. C. & Ferguson, H. R. (1949). US Patent 2 446 963.

Patrick, J. C. & Mnookin, N. M. (1927). British Patent 302 270.

Chapter 15

Polyurethane Rubbers

15.1 INTRODUCTION

Reaction of an isocyanate and an alcohol results in the formation of a urethane:

$$R \cdot NCO + HO \cdot R_1 \longrightarrow R \cdot NH \cdot COOR_1$$

In a corresponding reaction, di-isocyanates will react with diols (dialcohols; glycols) to form a long-chain *polyurethane*:

$$OCN \cdot R \cdot NCO + HO \cdot R_1 \cdot OH + OCN \cdot R \cdot NCO \longrightarrow$$
$$\sim\sim R \cdot NH \cdot COOR_1OOC \cdot NH \cdot R \cdot NH \cdot COO \sim\sim$$

In this reaction no small molecule is split out as in condensation polymerization and the process is referred to as a rearrangement polymerization (and occasionally as a polyaddition reaction). Similar reactions also occur when triols such as glycerol and tetrafunctional alcohols such as pentaerythritol are used; in such cases branched or cross-linked structures will be obtained.

The first commercial polyurethanes were fibres produced by reacting 1,4-butanediol and hexamethylene di-isocyanate, but subsequently a wide spectrum of polymers has been prepared including thermoplastics, rigid foams, flexible foams, adhesives, surface coatings, vulcanizable rubbers and thermoplastic rubbers. Although all are known as polyurethanes, the urethane group comprises only a small portion of the total polymer molecule with other groups such as ether, ester and urea groups often at least as important. Crucial to an understanding of polyurethanes is an understanding of isocyanate chemistry and this will be outlined below in Section 15.2.

Probably the best-known polyurethane rubbers are the flexible polyurethane foams (often illogically classified as plastic foams), but in addition solid vulcanizable materials have become of use as a result of their oil resistance and excellent abrasion resistance, whilst the thermoplastic polyurethane rubbers exhibit these properties together with the ability to be processed as thermoplastics. These last-named materials are dealt with separately in Chapter 16, which deals with thermoplastic rubbers. This chapter will be concerned with reaction-moulded solid rubbers (cast polyurethane rubbers), microcellular reaction-moulded polyurethanes and millable gums. Flexible polyurethane foams will be dealt with only very briefly.

15.2 ISOCYANATES

Isocyanates are reactive chemicals containing the —NCO group. Their most important reactions are with groups containing active hydrogen and are of general form

$$\text{—NCO} + \text{HX—} \longrightarrow \text{—NHCOX—}$$

Those of greatest importance are:

1. With water:

$$\sim\text{NCO} + H_2O \longrightarrow \underset{\text{An unstable carbamic acid}}{\sim\text{NH}\cdot\text{CO}\cdot\text{OH}} \longrightarrow \sim NH_2 + CO_2$$

2. With hydroxyl groups:

$$\sim\text{NCO} + \text{HO}\sim \longrightarrow \underset{\text{Urethane link}}{\sim\text{NH}\cdot\text{COO}\sim}$$

3. With primary amines:

$$\sim\text{NCO} + H_2N\sim \longrightarrow \underset{\text{Urea link}}{\sim\text{NH}\cdot\text{CO}\cdot\text{NH}\sim}$$

4. With urea derivatives:

$$\sim\text{NCO} + \sim\text{NH}\cdot\text{CO}\cdot\text{NH}\sim \longrightarrow \underset{\text{Biuret link}}{\sim\text{NHCO}\cdot\overset{\wr}{\text{N}}\cdot\text{CONH}\sim}$$

5. With urethane links:

$$\sim\text{NCO} + \sim\text{NH}\cdot\text{COO}\sim \longrightarrow \underset{\text{Allophanate link}}{\sim\text{NH}\cdot\text{CO}\cdot\overset{\wr}{\text{N}}\cdot\text{COO}\sim}$$

These reactions are all used to make polyurethane rubbers.

In addition, isocyanates may react with themselves to give dimers, trimers and carboimides:

RNCO → $RN{=}C{=}NR + CO_2$

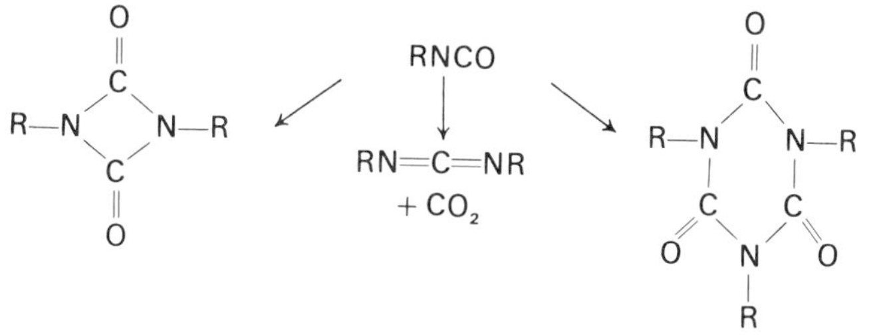

(The dimers have been obtained only from aromatic isocyanates and usually break down to monomer above 150°C. The trimers and carboimides have also been obtained from aliphatic isocyanates and both these and the aromatic derivatives are stable, often to 200°C and above. All of these reactions are usually undesirable.)

The influence of isocyanate type was studied at an early stage in the development of polyurethane elastomers (Bayer *et al.*, 1950*a,b*). It was established that aliphatic di-isocyanates gave unsatisfactory products but that good properties were obtainable using 2,4-tolylene di-isocyanate (2,4-TDI) and 1,5-naphthalene di-isocyanate (NDI). The latter material led to particularly tough rubbers with a high tear resistance marketed as Vulkollan by Bayer. Subsequently diphenylmethane di-isocyanate (MDI) has become widely used, one attractive feature being the low vapour pressure and relative ease of safe handling.

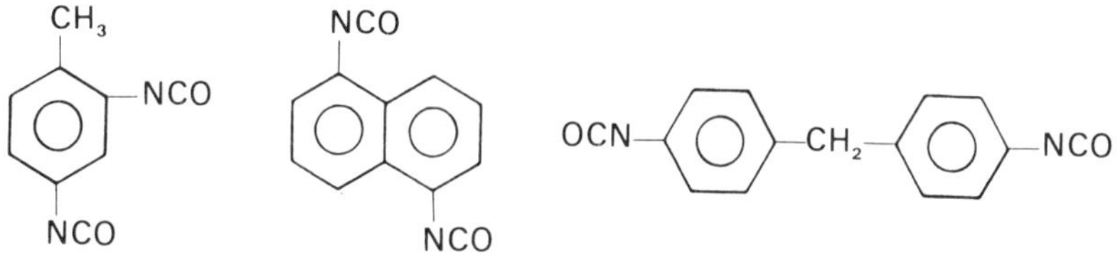

2,4-TDI NDI MDI

15.3 PREPARATION CHEMISTRY OF A POLYURETHANE RUBBER

Before discussing manufacturing variables that lead to such a diverse range of products it is instructive to consider the preparation of one type of polyurethane rubber (Vulkollan).

The first stage of the reaction is to prepare a polyol. This is a low-molecular-weight polymer with at least two hydroxyl end groups. Whilst polyols may be either polyethers or polyesters, in the case of Vulkollan rubbers polyesters are used, made for example by condensing ethylene or propylene glycol with adipic acid.

This polyol is then reacted with an excess of 1,5-naphthalene di-isocyanate to give a polymer with isocyanate end groups. The molar excess is about 30% so that the number of polyester molecules joined together is only about 2–3 on average. A typical structure for the *prepolymer* with P for polyester groups, U for urethane and I for isocyanate groups may be represented as IPUPUPI. This prepolymer, with its active end groups, is unstable and must be prepared just prior to the next stage of the process. This next stage is that of *chain extension*, which may be carried out by the use of water, a diol or a diamine to join the prepolymers by urea, urethane or double urea links respectively.

$$
\begin{array}{l} \sim\mathrm{NCO} \\ \\ \sim\mathrm{NCO} \end{array}
\; + \mathrm{H_2O} \longrightarrow
\begin{array}{c} \sim\mathrm{NH} \\ | \\ \mathrm{C{=}O} \\ | \\ \sim\mathrm{NH} \end{array} + \mathrm{CO_2}
$$

Urea link

$$
\begin{array}{l} \sim\mathrm{NCO} \\ \\ \sim\mathrm{NCO} \end{array}
\quad
\begin{array}{l} \mathrm{HO} \\ \quad \mathrm{R} \\ \mathrm{HO} \end{array}
\longrightarrow
\begin{array}{l} \sim\mathrm{NH \cdot COO} \\ \qquad\qquad \mathrm{R} \\ \sim\mathrm{NH \cdot COO} \end{array}
$$

Urethane link

$$
\begin{array}{l} \sim\mathrm{NCO} \\ \\ \sim\mathrm{NCO} \end{array}
\quad
\begin{array}{l} \mathrm{H_2N} \\ \quad \mathrm{R} \\ \mathrm{H_2N} \end{array}
\longrightarrow
\begin{array}{l} \sim\mathrm{NH \cdot CO \cdot NH} \\ \qquad\qquad\quad \mathrm{R} \\ \sim\mathrm{NH \cdot CO \cdot NH} \end{array}
$$

Double urea link

Where water is used, carbon dioxide is evolved and this reaction may be used to produce cellular products (although much of the expansion is nowadays brought about by the addition of a volatile blowing agent such as a fluorocarbon). For solid materials, aromatic amines tend to give the highest modulus and tear strength whilst extenders with flexible linkages such as thiodiethylene glycol give the product the greatest elasticity.

It will be noted that the reaction between isocyanate and glycol (diol) is of the same form as isocyanate and polyol. Where, however, the *number* of glycol molecules is severalfold higher than the number of the much larger polyol molecules, blocks consisting of sequences of the glycol–isocyanate reaction product will be formed. As a result the polymer chain will be composed of flexible (or 'soft') polyol segments separated by rather rigid ('hard') polyurethane segments:

$$\sim\!\!\left[\underbrace{\text{O—polyol—O}}_{\text{Soft segment}}\right]\!\!\left[\underbrace{\text{OCNHRNHCOO}(\text{R}_1\text{OOCNHRNHCO})_n}_{\text{Hard segment}}\right]\!\!\sim$$

The presence of such soft and hard blocks has an important bearing on the properties of all polyurethane rubbers, in particular the thermoplastic rubbers.

In the case of amine chain extenders, the reaction will lead analogously to polyurea blocks linked to the flexible polyol by means of urethane linkages:

$$\sim\![\text{O—polyol—O}][\text{OCNH}(\text{RNHCONHR}_1\text{NHCONH})_n\text{RNHCO}]\!\sim$$

Provided that the initial polyol has only two reactive hydroxyl groups, i.e. it is difunctional, then the product of the above sequence of reactions will be a long-chain molecule. Two further reactions may occur, however, leading to cross-linking. They are:

1. The reaction of isocyanate with an in-chain urea link to yield a biuret structure;
2. the reaction of isocyanate with an in-chain urethane link to yield an allophanate structure.

```
Urea      ~~NH·CO·NH~~                        ~~NH·CO·N~~     Biuret
                                                      |
                        OCN                          OCNH
                          \        ——→                 \
                           NCO                          NH·CO
                                                          |
Urethane        ~~NHCOO~~                               ~~N·COO~~
                                                     Allophanate
```

Uncatalysed, these reactions are quite slow between 80 and 100°C and may be reversed in the range 150–200°C. Urea groups are usually more reactive towards isocyanates than are urethane groups so that biuret cross-linking will be more favoured than allophanate cross-linking.

Polyvalent metal cations may be very selective in catalytic activity. For example, using the model system of the reaction of phenyl isocyanate with methanol in dibutyl ether at 30°C, the rate of reaction was increased by a factor of 37 000 in the presence of dibutyltin dilaurate. It was also found that this compound promoted the alcohol–isocyanate reaction more strongly than the isocyanate–water reaction and this in turn more strongly than the isocyanate–urea reaction. It is particularly important to control the relative rates of these reactions in polyurethane foam manufacture.

15.4 POLYOLS

Polyols are of two basic types: polyesters and polyethers.

The polyesters were used in the early work by Bayer and continue to be used by that company. The Vulkollan castable materials are based either on polyethylene adipate (Desmophen 2000) or on a copolymer obtained by reacting adipic acid with a mixture of ethylene glycol and 1,4-butanediol (Desmophen 2001). The copolymer is better for lower-temperature applications but physical properties at normal ambient temperatures are somewhat inferior. Similar systems are used by Bayer for their millable rubbers including a more hydrolysis-resistant adipic acid–hexanediol polymer. There is also some interest in polycaprolactones, polyesters obtained by ring-opening of caprolactone.

$$2n\,(CH_2)_5\begin{matrix} CO \\ | \\ O \end{matrix} + HOROH \longrightarrow HO[(CH_2)_5COO]_nR[OOC(CH_2)_5]_nOH$$

The most commonly used polyethers are polypropylene and polytetramethylene glycol, whose chemical structures are given in Fig. 16.6.

15.5 CAST POLYURETHANE RUBBERS

Cast solid polyurethanes first became available in the early 1950s under the trade name Vulkollan (Bayer). Since then many other materials have come onto the market and cast polyurethane elastomers have become well established and accepted as a consequence of their outstanding abrasion and oil resistance coupled in many cases with very high values for tensile strength.

During this period many types of cast polyurethanes have been marketed and it is convenient to divide them into the following classes.

1. Unstable, usually polyester, prepolymer systems using NDI.
2. Stable prepolymer systems.
3. Polyether or polyester quasi-prepolymer systems.
4. One-shot materials.

15.5.1 Unstable Prepolymer Systems

The unstable prepolymer systems are dominated by the Vulkollan materials, which are of importance because of their excellent load-bearing and, for a polyurethane, excellent heat-resistant characteristics. Their preparative chemistry was described in Section 15.3.

Mixing of the prepolymer and chain extender may be by either batch mixing or continuous processing in a mixing head at about 80–120°C. Fabrication involves pouring the reacting mix into moulds although such techniques as rotational and centrifugal casting, spraying and compression moulding may be used. Cast articles may be demoulded after 4–120 min, depending on the grade of elastomer, but several grades may require post-curing for 6–24 h at 110–130°C to obtain optimum properties.

Some typical formulations and properties of rubbers cast from them are given in Table 15.1.

One inherent weakness of Vulkollan-type materials is their susceptibility to hydrolysis. Life in a humid environment may, however, be doubled by the incorporation of a carbodi-imide at approximately 2 pphr.

15.5.2 Stable Prepolymer Systems

One serious disadvantage of the Vulkollan-type materials is the necessity to use the prepolymer within 30 min of manufacture. In 1958 DuPont introduced the Adiprene L range of prepolymers (now marketed by Uniroyal) which were storage-stable. In this case the polyol used was the polyether polytetramethylene glycol and the isocyanate, tolylene di-isocyanate, the latter ingredient providing the key to the storage stability.

The use of diols as chain extenders for these materials leads to generally inferior products of the unstable prepolymer systems but, using suitable amines, good phase separation into hard and soft blocks may be achieved, giving the products good physical properties. Technically the preferred diamine has been 3,3′-dichloro-4,4′-diaminodiphenylmethane (MOCA) but concern about the toxicity of this additive has led to the search for alternative materials.

Most of the more recently developed cast elastomer systems have been based on the use of MDI. MDI-based prepolymers are storage-stable, have the advantage of being available from several manufacturers and present less of a health hazard than the TDI/MOCA-based systems. Both polyesters and polyethers are used, diols being the usual chain extenders.

Both TDI/MOCA and MDI/diol prepolymer systems are of importance for harder grades of rollers (Shore A 55–95), mainly for the printing industry, and also for wear-resisting applications, particularly in mining

TABLE 15.1
Properties of typical Vulkollan materials
(After Wright & Cumming, 1969)

	Formulation (pphr)					
	A	*B*	*C*	*D*	*E*	*F*
Desmophen 2000[a]	100	100	—	100	—	100
Desmophen 2001[b]	—	—	100	—	100	—
Desmodur 15[c]	18	18	18	30	30	60
1,4-Butanediol	1·38	2	2	7	7	—
2,3-Butanediol	—	—	—	—	—	16
Trimethylolpropane	0·92	—	—	—	—	3
Pot life, min	5	4	4	1	1	1·5
Demoulding time, min	45	25	25	10	10	10
Specific gravity	1·26	1·26	1·26	1·26	1·26	1·26
Hardness (Shore A)	65	80	85	94	96	98
Tensile strength, psi	4 267	4 267	3 556	3 982	3 414	5 405
Modulus						
At 20% extension	100	213	213	1 000	900	1 990
At 300% extension	711	996	924	2 500	2 100	4 500
Elongation at break, %	600	650	650	450	500	300
Tear resistance (Graves), pli	140	310	250	390	310	750
Abrasion loss, mm³	50	40	65	55	61	42
Compression set, %						
70 h at 20°C	12	7	9·5	5	6	23
24 h at 70°C	22	17	22	14	12	41
24 h at 100°C	55	43	47	27	25	56
Rebound elasticity (DIN 53512), %	47	50	55	45	53	33

[a] Desmophen 2000, polyethylene adipate.
[b] Desmophen 2001, a copolymer from adipic acid, 1,4-butanediol and ethylene glycol.
[c] Desmodur 15, 1,5-naphthalene di-isocyanate.

and quarrying, such as linings for pumps, pipes and impellers. In the latter area the MDI/diol systems are claimed to be superior, with polyethers being preferred where hydrolytic stability is important.

15.5.3 Quasi-Prepolymer Systems

A quasi-prepolymer system is one in which a prepolymer is prepared with a substantial excess of isocyanate to give a low-molecular-weight prepolymer which is isocyanate-terminated. This is then reacted with a blend of unreacted polyol, chain extender, catalysts, surfactant and (for microcellular applications) water. The advantage of this system is that the component streams to the mixing head may be of both a similar volume and a similar viscosity which helps to simplify and control metering and mixing.

Such quasi-prepolymer systems are of particular importance with polyester-based microcellular polyurethanes, with diols being the preferred chain extenders.

15.5.4 One-Shot Systems

One-shot systems were developed to overcome the problems of having to make prepolymers which require many precautions in their preparation if consistent high-quality products are to be obtained. In one-shot processes polyol and chain extender are mixed together prior to addition of the isocyanate. Success in the process depends on selection of catalysts which balance the various reaction rates at the desired level. Such one-shot processes are all important with polyurethane foams; their use in cast solid polyurethane technology is more restricted.

One particular example is for the manufacture of soft rollers (Shore A 15–55) for use mainly in the printing industry, usually being based on polyesters and TDI. Cold-curing systems based on a one-shot technology have also been developed based on MDI and some of its variants. They find use for cable jointing and potting compounds, commercial vehicle mats, coatings for oil-discharge hoses and for moulds for precast concrete.

15.5.5 Markets for Cast Polyurethane Elastomers

In spite of their special processing requirements, which do not easily fit in with the standard techniques of rubber technology, the cast polyurethane elastomers have established a range of applications which utilize the excellent wear-resisting properties of the material, in some cases coupled with the good oil resistance.

Examples where the polyurethanes are used in what is sometimes described as 'wear parts' are separation screens and liners for pipes, pumps, cyclones and impellers for handling abrasive ores and slurries by the mining and quarrying industries; hydraulic seals, sleeves, couplings, timing belts, shock absorbers and O-rings amongst mechanical goods; doorstops, lock parts, slide and joint bushings and door stops for the automotive industry.

Cast polyurethanes have also become widely used for rollers, as mentioned in earlier sections. Soft rollers made by the one-shot technique are used in printing whilst harder, prepolymer-based rollers are used in mechanical handling operations by the steel, textile and paper industries.

A long-established use has been for solid tyres for slow-moving vehicles such as forklift trucks, in this case the market being dominated by the polyester/NDI systems. Cold-cure elastomers are also used for cable jointing, potting compounds, sealants and as moulds for precast concrete.

15.6 MILLABLE GUMS

The introduction of millable gums based on polyurethanes resulted from the demands of the rubber industry for a polyurethane-type material which could be processed by the conventional processes of rubber technology. However, these materials are of limited value today, for two reasons:

1. In broad terms their properties are not as good as those of the cast polyurethane rubbers;
2. the development of thermoplastic polyurethane rubbers (see Chapter 16) has led to such materials making inroads into markets for products not easily made by cast systems such as diaphragms and thin-walled gaiters.

The millable gums are best classified according to the curing system used. These are of three types:

1. Isocyanate curing systems;
2. peroxide curing systems;
3. sulphur curing systems.

A typical isocyanate-cured millable gum is Urepan 600 (Bayer) which was formerly known as Desmophen A. In the production of this material a polyester is first prepared by reacting diethylene glycol and adipic acid. This is then reacted with a slight deficiency of tolylene di-isocyanate so that the chain ends are hydroxyl rather than isocyanate and are reasonably stable on storage. The average molecular weight of this prepolymer is 'less than 20 000'. The prepolymer is then compounded with a di-isocyanate, usually dimerized 2,4-tolylene di-isocyanate.

The isocyanate reacts with the terminal groups of the prepolymer to cause chain extension and also with urethane groups within the prepolymer chain to produce allophanate cross-links. There will also be some reaction with traces of water, which should not be in excess of 0·7 pphr, which will also lead to some biuret cross-links. These considerations lead to a di-isocyanate level of about 8–10 pph prepolymer.

It is also possible to enhance the hardness of the product by introducing 'hard blocks' into the polymer chain by the use of a short-chain glycol such as hydroquinone diethylol ether and additional isocyanate. Under cool dry conditions the compounds may be stored for several days with typical curing being carried out for 30 min at 130°C. The use of certain organic lead salts as accelerators can lead to cure which may be faster or at lower temperatures, with room-temperature curing possible.

Where greater hydrolysis resistance is required, a closely related product (Urepan 601) which is based on a hexane diol–adipic acid polyester is offered by Bayer. Some typical properties of vulcanizates of these materials are given in Table 15.2.

The isocyanate-cured gums are rather hard and have a higher compression set than acceptable for some applications. Softer rubbers with lower set may be obtained using diphenylmethane di-isocyanate as the isocyanate. Materials of this type include Vibrathane 5004 (Uniroyal) and Urepan 640 (Bayer).

In order to enable sulphur curing, it is necessary to incorporate double bonds into the polymer structure by the use of a material such as glyceryl monoallyl ether. Whilst such materials may give vulcanizates of good initial properties, the highly active accelerator system necessary is detrimental to

TABLE 15.2
Typical properties of isocyanate-cured millable gums
(After Ellegast *et al.*, 1966)

	Formulation (pphr)				
	1	*2*	*3*	*4*	*5*
Urepan 600	100	100	100	100	—
Urepan 601[a]	—	—	—	—	100
Stearic acid	0·5	0·5	0·5	0·5	0·5
HAF Black	20	—	—	20	—
Hydroquinone diethylol ether	—	12·5	23·9	—	12·5
Desmodur TT[b]	10	30	50	10	30
Accelerator	0·5	0·5	0·3	—	—
Polycarbodi-imide	—	—	—	3	—
Cure schedule					
Temperature, °C	150	130	130	130	—
Cure time, min	5	10	10	10	—
Post-cure, h at 110°C	15	15	15	15	—
Physical properties					
Hardness (Shore A)	84	92	96	84	95
Tensile strength, psi	3 400	3 400	4 000	3 100	6 000
Elongation at break, %	650	450	310	680	400
Tear strength, pli	280	250	450	250	500
Rebound resilience, %	45	37	36	45	37
Abrasion loss, mm^3	45	50	60	60	55

[a] Cure details for the Urepan 601 compound not supplied in source reference.
[b] Desmodur TT, dimerized 2,4-tolylene di-isocyanate.

TABLE 15.3
Comparison of sulphur and peroxide cures for Adiprene C

	Formulation (pphr)	
	1	*2*
Adiprene C	100	100
HAF Black	30	30
Coumarone–indene resin	10	—
Butyl oleate	—	10
Sulphur	0·75	—
MBTS	4	—
MBT	1	—
Activator RCD 2098	0·35	—
DiCup 40C	—	2·5
Physical properties after cure for 45 min at 153°C		
Hardness (Shore A)	64	62
Tensile strength, psi	5 150	3 150
Modulus at 300%, psi	2 475	1 750
Elongation at break, %	540	450
Tear resistance (Graves), pli	380	260
Compression set, %		
22 h at 70°C	20	21
70 h at 100°C	86	59
Resilience (Yerzley), %	57	56

such properties as hydrolysis resistance. Examples of sulphur-curable gums that have been available are Elastothane 455 (Thiokol) and Adiprene C (Uniroyal, formerly DuPont). This latter material may be cured by either sulphur or peroxide (see Table 15.3).

15.7 POLYURETHANE FOAMS AND MICROCELLULAR REACTION-MOULDED POLYURETHANES

Early on in the development of polyurethane rubbers it was recognized that traces of water reacted with isocyanate to form a urea bridge which led to chain extension and also to the evolution of carbon dioxide which caused bubbles in the product. Originally a nuisance, this was turned to good effect with the appearance of the *polyurethane foams* which are today of considerable commercial importance. These materials have their own technology and will only be treated briefly here, partly to demonstrate their relationship to the solid polyurethane elastomers and also to deal with some intermediate products.

The basic chemistry for the foams is the same as for solid materials and commercially both polyesters and, more importantly, polyethers are used. TDI and MDI are the most important isocyanates. The polyols usually have more than two hydroxyl groups so that chain bridging also leads to cross-linking. The intensity of cross-linking will depend to a great extent on the distance between hydroxyl groups on the chain. Where the distance is short rigid foams may be produced, but where it is long products will be more flexible.

Three polyol structures are indicated schematically as (a) to (c):

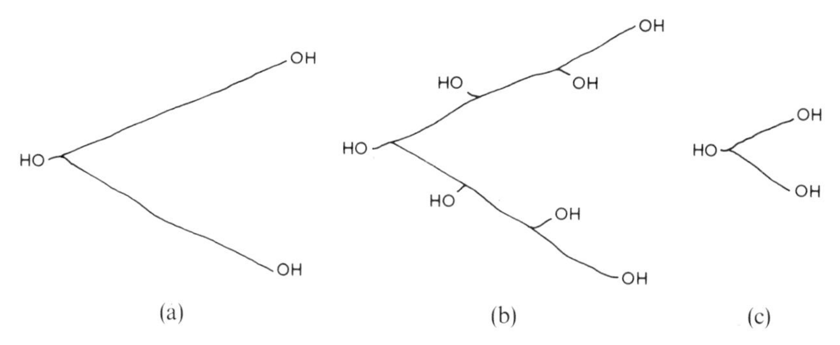

Structures (a) and (b) have similar molecular weights but (b) has more hydroxyl groups and thus (b) will be expected to yield a more rigid product than (a). Structure (c) has the same number of hydroxyl groups as (a) but a much lower molecular weight, with the ratio molecular weight/number of hydroxyl groups similar to that of (b). At a first approximation (b) and (c) would therefore be expected to give foams of similar rigidity. In practice the presence of excess isocyanate will also affect the rigidity because of its ability to introduce additional cross-links of the biuret and allophanate types.

Whilst the foaming can arise totally from the reaction of isocyanate with water, this is a somewhat expensive way of making a blowing agent. As an alternative, blowing may be brought about by incorporating a volatile material such as a fluorocarbon into the formulation. The isocyanate–water reaction also introduces additional cross-links which do not occur where the system is fluorocarbon-blown so that, everything else being equal, the water-blown foams may be somewhat stiffer.

The best known process for making foams is that in which a mixing head oscillates horizontally across the width of a large trough which is slowly moving at right angles to the movement of the mixing head. The ingredients of the foam formulation are fed to the head and the emerging reacting stream of mixed components is thus spread evenly over the surface of the trough. As the mix reacts, foaming and cross-linking occur simultaneously giving rise to the foamed product.

In recent years there has been particular interest in the *reaction injection moulding* process (RIM process). In this process the ingredients are mixed by impingement in a small chamber at the side of a mould and then injected into the mould cavity. The process has a number of advantages, particularly over conventional injection moulding, such as

low plant investment;
low process energy demands;
excellent surface quality;
the possibility of making large parts without the need for high mould locking pressures;
large variations in wall thickness without sink marks;
the ability to produce a high-density skin with a low-density, foamed core.

In practice it is possible to obtain foams with densities from as low as $0{\cdot}02\,\mathrm{g\,cm^{-3}}$ to as near as required to the solid material density of about $1{\cdot}2\,\mathrm{g\,cm^{-3}}$. Materials in the density range $0{\cdot}5$–$0{\cdot}8\,\mathrm{g\,cm^{-3}}$ produced by reaction moulding are somewhat intermediate between the traditional foams and the traditional solid rubbers. Today they are of considerable importance and the market for them may well be larger than that for the truly solid materials. The main markets are flexible automobile parts prepared by RIM methods and microcellular shoe soling, for which lower-pressure metering and mixing machinery are employed.

Microcellular polyurethanes for shoe soling may be based on either polyesters or polyethers. In the case of polyesters, quasi-prepolymer systems using MDI are usual. Polyether systems also tend to use MDI and, whilst polyester systems tend to be water-blown, fluorocarbons are used with polyethers.

15.8 OTHER USES OF POLYURETHANES

It will be appreciated that a very wide spectrum of polyurethane materials exists, some of which are not rubbery and which may be either solid or cellular. These include fibres, paints and adhesives. Brief mention should be made of flexible/rubbery materials being used as leather finishes and for fabric coatings. Although these have in the past been applied from solution, fabric-coating processes have also been developed to avoid the use of solvents. This may be achieved by handling a solventless liquid reacting mixture and using this to coat the fabric, or by using a hot-melt calendering process.

BIBLIOGRAPHY

Frisch, K. C. (1972). Recent advances in the chemistry of polyurethanes. *Rubber Chem. Technol.*, **45**, 1442–66.

Redman, R. P. (1978). Developments in polyurethane elastomers. In *Developments in Polyurethanes—1*, ed. J. M. Buist. Applied Science, London, Chapter 3.

Saunders, J. H. (1969). In *Polymer Chemistry of Synthetic Elastomers*, Part 2, ed. J. P. Kennedy & E. G. M. Törnqvist. Interscience, New York, Chapter 8A.

Saunders, J. H. & Frisch, K. C. (1962, 1964). *Polyurethanes—Chemistry and Technology*, Parts 1 and 2. Interscience, New York.

Wright, P. & Cumming, A. P. C. (1969). *Solid Polyurethane Elastomers*. Maclaren, London.

REFERENCES

Bayer, O., Müller, E., Petersen, S., Piepenbrink, H. F. & Windemuth, E. (1950*a*). *Angew. Chem.*, **62**, 57.

Bayer, O., Müller, E., Petersen, S., Piepenbrink, H. F. & Windemuth, E. (1950*b*). *Rubber Chem. Technol.*, **23**, 812.

Ellegast, K., Kallert, W., Peter, J. & Schultheis, H. (1966). Preparation and properties of polyurethane elastomers. In *Kunstoff-Handbuch*, Vol. 7, ed. R. Viewig & A. Hochten. Carl Hanser Verlag, München.

Wright, P. & Cumming, A. P. C. (1969). *Solid Polyurethane Elastomers*. Maclaren, London.

Chapter 16

Thermoplastic Rubbers

16.1 INTRODUCTION

In the years after World War II natural and synthetic rubbers were replaced by thermoplastics for a number of applications where flexibility rather than high elasticity was the main requirement. Such applications included belting, cables, flooring, protective sheeting and shoe soling. In most of these instances the use of a thermoplastics material resulted in a superior end-product and additionally this was often also accompanied by production economies. In large measure this was due to the fact that, after shaping the polymer, setting the shape with a thermoplastics material simply involved cooling whereas with conventional rubbers setting involved cross-linking (vulcanization). Furthermore, whilst it is possible to reprocess thermoplastics scrap, albeit with care, it is not possible to do this with vulcanized rubber.

From the above considerations it may be seen that the concept of a rubber which behaved as if it were cross-linked at room temperature but which flowed like a thermoplastics material at processing temperatures was a very attractive one. In consequence, over the years a number of materials have been marketed with such characteristics. For reasons which will be explained later this has not led to the general replacement of conventional vulcanized rubbers but rather to the creation of new markets.

16.2 THE HEAT-FUGITIVE CROSS-LINK

The common feature of these so-called *thermoplastic rubbers* is that the chains are held together at normal ambient temperatures by a bond which becomes ineffective at elevated temperature to give what is known as a heat-fugitive cross-link (Fig. 16.1). Such cross-links may be obtained in a number of ways, of which the most important are:

1. Ionic cross-linking;
2. cross-linking by hydrogen bonding;

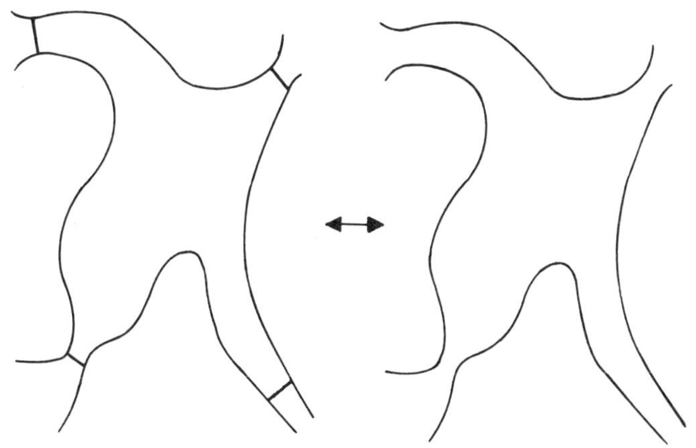

FIG. 16.1. Heat-fugitive cross-links; links which break down on heating and form on cooling.

3. cross-linking by thermally unstable covalent bonds which re-form on cooling;
4. linking of molecules via small crystalline structures;
5. the use of block copolymers (the most important route).

Ionically bonded polymers have been known for many years; the best known are the ionomers introduced by Du Pont as Surlyn in the 1960s. These are copolymers predominantly composed of ethylene monomer units but with a small amount of a second monomer that contains carboxylic acid groups. When this material is blended with metallic salts, ionic cross-links are formed which are strong at room temperature but which disappear on heating. Whilst the Surlyn materials are more plastic than elastic, the approach has also been used experimentally with natural rubber and commercially with carboxylated nitrile rubbers cross-linked with zinc oxide. It is, however, difficult to argue that these are genuine thermoplastic rubbers.

Cross-linking by hydrogen bonding probably occurs with polyurethane rubbers but is masked by other effects. It is also the belief of the writer that much of the rubberiness of plasticized PVC is due to hydrogen bonding between chains via ester plasticizer molecules which by spacing the polymer molecules also depress the T_g.

Thermally unstable covalent bonds are also encountered. There is a belief that during the manufacture of many polyurethane products allophanate groups are formed which break down on heating and re-form on cooling. Once again this effect, the existence of which is disputed by some workers, is masked by other influences and believed to be unimportant. The high-green-strength SBRs produced by Polysar in the 1970s may, however, be considered as a variation on this theme.

Low-density polyethylenes are to some extent rubbery, and this is even more evident when ethylene is copolymerized with small amounts of vinyl acetate. In such circumstances crystallization is reduced to a very low level but the remaining crystal structures effectively link the chains together at normal ambient temperatures. Such EVA polymers are sometimes considered as thermoplastic rubbers but will not be discussed further here; the interested reader is referred to other texts (Streib *et al.*, 1977; Gilby, 1982). The thermoplastic polyolefin rubbers, dealt with in Section 16.7, may, however, be regarded as functioning by a crystalline cross-linking mechanism.

The breakthrough in thermoplastic rubbers took place with the development of certain block copolymers. In these materials blocks of rubbery polymer of low glass transition alternate with polymer blocks with a T_g or T_m well above normal ambient temperatures. For convenience these blocks are usually referred to as soft blocks and hard blocks, the soft blocks providing rubberiness, the hard blocks holding the chains together like cross-links.

Important thermoplastic rubbers based on block copolymers are:

1. Styrene–butadiene–styrene triblock copolymers and related materials;
2. polyether–polyester rubbers of the Hytrel type;
3. some thermoplastic polyurethane rubbers;
4. some thermoplastic polyamide rubbers.

16.3 STYRENE–BUTADIENE–STYRENE TRIBLOCKS AND RELATED MATERIALS

Because of the many interesting aspects of their preparative chemistry and their physical properties, these materials have proved a great attraction to academics and there is a very substantial literature concerned with them. Commercially they have found application more on the fringes of the rubber industry than in the mainstream, being particularly established in footwear, adhesives and as bitumen additives.

Styrenic block copolymers were introduced by Shell in 1963 (Cariflex TR, Kraton) followed by Phillips in 1968 (Solprene) and Anic (now part of

Enoxy SpA) in 1972. A number of other companies are now in the market, some of whom operate the Phillips process under licence. It is difficult to estimate capacity as much of this is interchangeable with that for solution SBR. Production is now understood to be well in excess of 100 000 tpa.

16.3.1 Preparation

The styrenic block copolymers are prepared using the process of sequential anionic polymerization, of which there are three main variants. In what is in principle the simplest system, polymerization of styrene is initiated, in a typical example, by *sec*-butyllithium in a solvent such as cyclohexane. This is a reaction of general type

$$R^-M^+ + mA \longrightarrow RA_{m-1}A^-M^+$$

In this process it is possible to select initiators and solvents so that initiation is instantaneous and all of the chains start to grow at about the same time (as distinct from free-radically initiated polymerization) and because growth continues until all of the styrene is consumed, all of the chains at this stage will be of approximately the same length. The degree of polymerization will be given simply by [styrene]/[initiator] (the squared brackets indicating molecular concentrations). Under carefully controlled conditions it is possible to prevent termination reactions and the chain ends remain active (the term *living polymers* has been used to describe this situation).

If additional monomer is then added to the reactor, polymerization recommences and the chains increase in length. This additional monomer may be of a different species, for example butadiene, and this will add to the chains until all is consumed, also giving blocks of a narrow molecular weight distribution. In turn another monomer species, or possibly additional styrene, may then be added. These reactions may be represented by the equations

$$R^-M^+ + mA \longrightarrow RA_{m-1}A^-M^+ \xrightarrow{nB} RA_mB_{n-1}B^-M^+$$

$$\xrightarrow{mA} RA_mB_nA_{m-1}A^-M^+$$

If alternatively we represent the styrene blocks by S and the butadiene blocks by B then we may represent the triblock copolymer produced as S—B—S. Alternatively it is possible to initiate using a difunctional initiator that allows chain growth in two directions. In this case it is possible first to grow a polybutadiene block with two 'live ends' and then when all of the butadiene has been consumed to add styrene. In this way a triblock may be obtained in two stages. The process may be formally represented by the following equation:

$$M^{+-}R^{-}M^{+} \xrightarrow{2mA} M^{+}A^{-}A_{m-1}RA_{m-1}A^{-}M^{+}$$

$$\xrightarrow{2nB} M^{+}B^{-}B_{n-1}A_{m}RA_{m}B_{n-1}B^{-}M^{+}$$

Commercially, greatest interest centres in a two-stage sequential process followed by chain coupling. In this method the styrene is first polymerized followed by the butadiene (the first two stages of the first method described above) to give a product of formal structure

$$RS_{m}B_{n-1}B^{-}M^{+}$$

The living polymer that exists at this stage may then be coupled by difunctional coupling agents to give linear polymers, trifunctional agents to give T-shaped polymers, tetrafunctional agents to give X-shaped polymers and with agents of higher functionality to give star-shaped polymers. Some examples of coupling agents are given in Table 16.1. Whilst the original Shell materials were linear, other manufacturers concentrated on branched types. The Shell Company now also offer branched polymers.

16.3.2 Structure

Somewhat over 30 grades of rubbery styrenic block copolymers are marketed (as well as several oil-extended grades). The main structural variables between the grades are:

1. Selection of monomer;
2. ratio of monomers used;
3. molecular weight of each block;
4. polymer shape (linear or branched);
5. chemical modification, if any, of polymer, e.g. by hydrogenation.

Before discussing the effect of changing these variables it is first instructive to consider the nature of a fairly typical material. As an example we will take a linear styrene–butadiene–styrene triblock of molecular weight of about 80 000 and relative block sizes (by weight) of 14:72:14. Although the solubility parameters .of the butadiene and the styrene blocks will be similar, the blocks will try to separate although in an individual chain they will be linked. In the above ratio the styrene blocks will tend to aggregate into spherical (or perhaps cylindrical) domains which will be embedded in a

TABLE 16.1
Effect of type of coupling agent on shape of SBS block copolymer
(After Brydson, 1982)

Coupling agent	*Functionality*	*Polymer shape*
 $ClCH_2-C_6H_4-CH_2Cl$	2	Linear 
PI_3	3	T-shaped 
CH_3SiCl_3	3	T-shaped 
$SiCl_4$	4	X-shaped 
Divinylbenzene	7–14	Star-shaped 

continuous matrix of polybutadiene blocks. This is illustrated schematically in Fig. 16.2.

The continuous butadiene phase is rubbery down to very low temperatures which are related to the T_g of the polybutadiene. The polystyrene domains are glassy and remain so up to the T_g of the polystyrene blocks (which because of their low molecular weight will be somewhat lower than for conventional commercial polystyrene). In effect these domains act like end-of-chain cross-links forming a polymer network at normal ambient temperatures. (Because the cross-links are at the chain ends it is interesting to note that unreacted chain ends do not have to be considered and it is not necessary to use a high-molecular-weight polymer to obtain a reasonable modulus.) If, however, the polymer mass is heated

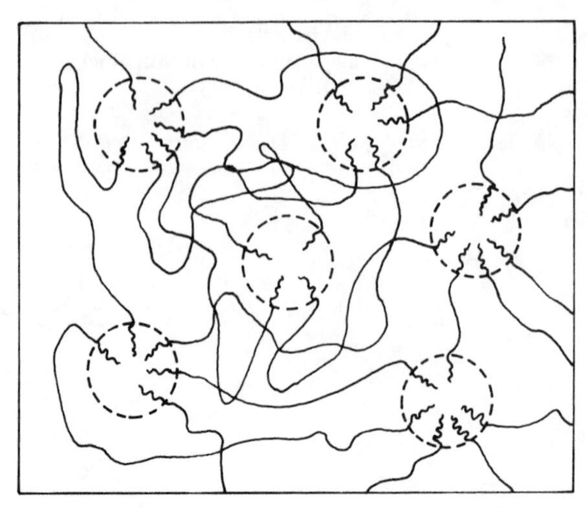

FIG. 16.2. Schematic representation of the polystyrene domain structure in styrene–butadiene–styrene triblock copolymers. (From Holden *et al.* (1969), reproduced by permission of John Wiley & Sons, Inc.)

well above the T_g of the styrene blocks, these too become rubbery and the chains are able to move relative to each other and, on application of a stress, the polymer mass becomes capable of flow. (It does appear that domains tend to persist to temperatures well above the styrenic block glass transition temperature and, as with amorphous thermoplastics, processing takes place some 100°C above this temperature.) In effect this means that these materials are effectively cross-linked at room temperatures but behave like thermoplastics at processing temperatures.

The polystyrene domains also have a further important role. They act like particulate reinforcing fillers so that materials such as carbon black are unnecessary and tough transparent products may be obtained. This effect is illustrated by the data in Table 16.2.

The most important of the styrenic block copolymers are of the SBS type. Styrene–isoprene–styrene (SIS) materials are, however, important for adhesives. Whilst there has been some interest in α-methylstyrene materials with their higher glass transitions they have not, at the time of writing, become significant. Chemically modified SBS materials will be discussed below.

In a number of ways SBS and SIS rubbers reflect the properties of BR and NR (or IR). For example, the SBS materials are quite amorphous, do not break down on mastication and on ageing the rubbery phase tends to

TABLE 16.2
Comparative tensile strengths of SBS and polybutadiene rubbers

Rubber	*Tensile strength (MPa)*
Uncompounded polybutadiene	0·3–0·5
Polybutadiene (sulphur-vulcanized)	3–5
Polybutadiene + carbon black (sulphur-vulcanized)	15–20
SBS (unvulcanized)	30–35

harden due to oxidative cross-linking. On the other hand, the SIS materials exhibit limited crystallinity, tend to suffer chain scission during mechanical working and soften on ageing.

For SBS materials the bound styrene content is usually in the range 30–40% although for oil-extended grades bound styrene levels as high as 55% are used commercially. For SIS materials 14% bound styrene is more common although up to 21% may be used.

Molecular weights should be such that the styrene blocks are long enough to be firmly embedded in the domains. Both theory and practice indicate a block molecular weight of the order of at least 7000 is required. At such a low molecular weight the glass transition of the styrene block is only about 40°C. There is, however, evidence (Chung *et al.*, 1976, 1978) that microphase separation persists up to at least 150°C and that as a result melt viscosities are higher than for related homopolymers or random copolymers of similar molecular weight.

Branched SBS materials have lower melt and solution viscosities than unbranched triblock polymers of similar molecular weight. This effect is quite common amongst polymer melts where the long chain branches result in a more compact molecule less likely to become entangled with other molecules than is the case with an unbranched molecule. A corollary to this is that it is possible to have a branched polymer with a higher molecular weight, and hence higher tensile strength for a given melt viscosity, that is possible for linear polymers.

Hydrogenated SBS materials have been marketed by at least one company (Shell—Kraton G and Elexar). Because the butadiene units in the soft block are joined together by both 1,4- and 1,2-addition, the hydrogenated structure of the block is that of a random ethylene–butene copolymer—hence the abbreviation SEBS which is commonly used for this type of polymer. The SEBS materials, free of double bonds, have better ageing and weathering resistance.

16.3.3 Properties

Besides the obvious thermoplastic-type behaviour, the SBS polymers show many features of a hydrocarbon rubber. Thus they lack hydrocarbon oil resistance, are soluble in many solvents of similar solubility parameter, and have good electrical insulation properties and good resistance to chemicals that do not attack double bonds. Unlike natural rubber and the bulk of the major synthetics, finished products can be virtually glass-clear since it is not necessary to add fillers or other particulate additives.

Compared with natural rubber vulcanizates, SBS materials show less instantly recoverable elasticity. Thus not only do the materials show high compression and tension set but the tensile stress–strain curve on a second deformation is quite different from that of a first deformation (strain softening) (Fig. 16.3).

Further storage of the sample for, say, a week may, however, result in a hardening on storage, after which a further deformation will have a similar pattern to the first (Fig. 16.4).

This is a consequence of a material having a morphology very sensitive to deformation, fascinating to the academic but less so to the manufacturer. Some indication of the potential complexity is given in Table 16.3, which

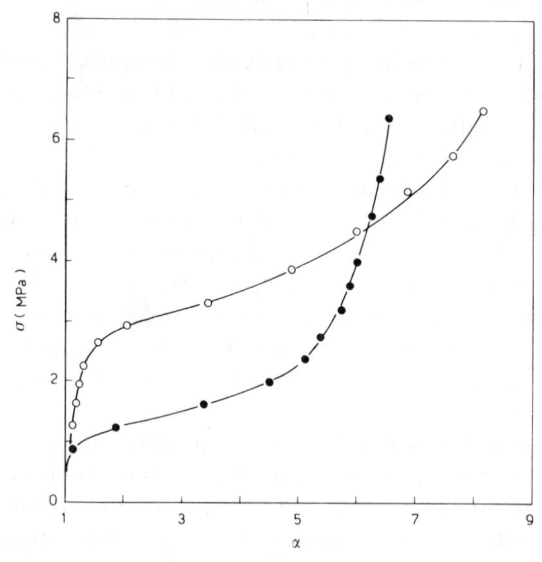

FIG. 16.3. Simple extension stress–strain behaviour of extruded and annealed linear SBS: ○, first deformation; ●, second deformation. (From Pedemonte *et al.*, 1974.)

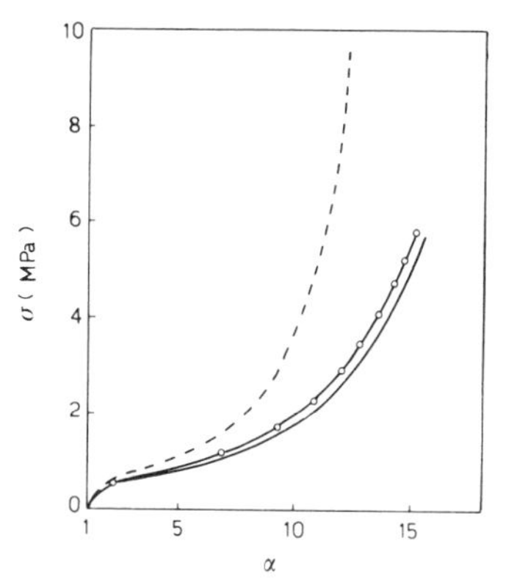

FIG. 16.4. Simple extension stress–strain behaviour of cyclohexane-cast SIS as a function of storage time at zero load between the two following deformations: —, first deformation; - - -, second deformation (storage time 10 min); —O—O—, second deformation (storage time 1 week). (From Pedemonte *et al.* (1977), reproduced by permission of Butterworth & Co. Ltd.)

shows the effect of processing conditions on the properties of a commercial branched SBS polymer.

16.3.4 Compounding and Processing

Whilst the SBS materials do not require a vulcanization system and can be quite adequate for many purposes without the need for fillers, there is some scope for compounding. The effect of the additives depends on whether they are eventually located in the styrenic hard block phase or in the rubbery matrix.

Paraffinic and naphthenic oils are taken up by the rubbery phase and lower melt viscosity, and reduce hardness, modulus, tensile strength and abrasion resistance. Aromatic oils tend to solvate the hard block and lower its glass transition temperature.

The addition of polystyrene resin increases modulus but has little influence on tensile strength. Blends with EVA may be used to improve ozone resistance, with polyethylene to improve abrasion resistance and polypropylene to improve hardness without reducing flexibility.

Non-reinforcing fillers may be used to reduce cost, whilst reinforcing fillers are reported to improve hot tear resistance. Antioxidants are required for many applications, one particular problem being the need to protect against UV attack using such additives as nickel dibutyl dithiocarbamate, benzophenones and benzotriazoles. The frequent

TABLE 16.3
Effect of processing conditions on the properties of a star-branched styrene–butadiene thermoelastomer (Enoxy Europrene SOL T163) (After Arcozzi *et al.*, 1975)

Sample	*Young's modulus*[a] ($kg\,cm^{-2}$)		*Anisotropy*	*Morphology*	
	I	*E*		*Domains shape*[b]	*Domains arrangement*
Cyclohexane-cast film	65	40	None	SC	Random bundles
Methyl ethyl ketone-cast film	450	92	None	BC and S domains in B matrix	Random bundles
Compression moulded (200°C, 40 min)	10–100	46–60	Very high	S, B lamellae	Regular alternation
Calendered					
Calender direction	120	80	Very low	Fibrillae	Practically randomness
Normal to calender direction	115	80			
Injection moulded	110	75	None	Fibrillae	Randomness
Extruded (screw speed 20 rpm)					
Extrusion direction	230	92	High	Fibrillae	High statistical orientation
Normal to extrusion direction	66	38			
Extruded (screw speed 40 rpm)					
Extrusion direction	110	58	Low	Fibrillae	Low statistical orientation
Normal to extrusion direction	68	54			

[a] I, specimen just shaped; E, specimen strained into equilibrium.
[b] S = polystyrene; B = polybutadiene; SC = polystyrene cylinders; BC = polybutadiene cylinders.

reworking of scrap may also place additional demands, not met in conventional vulcanized rubbers, on the antioxidant system.

Additives are usually compounded by dry blending although conventional internal mixers and screw compounders may be used. Injection moulding is the most common process for making dry rubber products.

16.3.5 Applications

The principal application for SBS materials is currently as a bitumen additive where it improves many properties, including the heat sensitivity of the bitumen, and finds extensive use both for road surfaces and for roofing compounds. The materials are also of importance for footwear soling, varying ratios of SBS, polystyrene and oil being used to achieve appropriate properties. Other uses of importance are pressure-sensitive and contact adhesives, and for plastics modification.

16.4 POLYETHER–ESTER THERMOPLASTIC RUBBERS

In 1972 the Du Pont company introduced a polyether–ester block copolymer rubber under the trade name Hytrel. For reasons that will be explained later this material may be used over a wider temperature range than the SBS materials and is also resistant to hydrocarbon oils. The material has consequently become regarded as an important speciality engineering rubber and has probably realized the original objectives for a thermoplastic rubber better than any of the other materials discussed in this chapter. Similar materials have also been marketed by AKZO as Arnitel, General Electric (Lomod) and the Japanese company Toyobo as Pelprene. Total capacity is believed to be of the order of 30 000 tpa. The materials are designated YBPO.

16.4.1 Synthesis and Structure

The polyether–ester thermoplastic rubbers are prepared by a melt transesterification process involving a phthalate ester, usually dimethyl terephthalate, a low-molecular-weight glycol, usually 1,4-butanediol, and a linear low-molecular-weight polyether, usually polytetramethylene ether glycol of molecular weight 600–3000. The resulting condensation polymer will consist of blocks consisting largely of flexible, low-T_g polyether (soft blocks) separated by blocks of short polyester units derived from the low-molecular-weight glycol and the terephthalic acid derivative which are rigid and crystalline (hard blocks). By varying component ratios, the ratio of hard to soft blocks may be altered as desired to give a family of materials which at one extreme are quite soft and rubbery and at the other more like a leather-like thermoplastics material. A typical structure of such a polymer may be represented as

$$\sim\!\!\left[[(CH_2)_4O]_{n\sim14}\!-\!OOC\!-\!\langle\bigcirc\rangle\!-\!COO\right]\!\left[(CH_2)_4OOC\!-\!\langle\bigcirc\rangle\!-\!COO\right]\!\!-$$

Soft block (PTMEG/T) Hard block (4GT)

The softest commercial grades (40 Shore D; 92 Shore A) consist of about 30% of hard segments (designated as 4GT—tetramethylene glycol terephthalate—in the above formula) and the hardest grade (72 Shore D hardness) about 82%. The overall molecular weight is low, with values in the range of 25 000–30 000 giving optimum physical properties.

Electron microscopy shows that the elastomers have a two-phase morphology with phase separation taking place below the melting point. A schematic diagram of this structure is given in Fig. 16.5, where A represents crystalline domains and B an area of crystallites within the crystalline domain. Interspersed with the continuous crystalline domain network is a continuous amorphous domain largely consisting of soft segments C but with some hard segments D which have not crystallized. It is the interpenetrating networks which confer high strength to the rubbers.

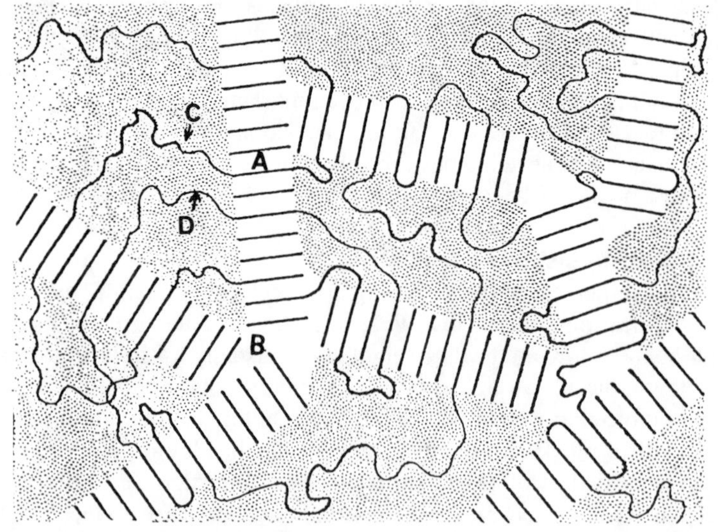

FIG. 16.5. Schematic diagram of the morphological structure of segmented polyesters. (From Lloyd, 1982.)

Also of great significance is the wide operating temperature range for these materials. This is because the soft blocks remain flexible down to temperatures approaching the T_g of the soft block amorphous phase of about -50°C whilst the hard blocks continue to exist up to the T_m of the crystalline hard block phase of about 200°C. This situation is quite different from that of the amorphous SBS-type materials where the upper service temperature is fixed by the T_g (which in crystalline polymers is about two-thirds of the T_m when expressed in Kelvin units).

Also of importance is the fact that the polyether–ester rubbers, being polar, have very good resistance to hydrocarbons.

16.4.2 Properties

As inferred above, the key features of these rubbers are:

1. Good low-temperature flexibility;
2. excellent hydrocarbon oil and solvent resistance;
3. high toughness and flexibility;
4. ability to be processed on typical thermoplastics equipment.

Some typical mechanical and thermal properties of polyether–ester thermoplastic rubbers are given in Table 16.4.

Although environmental resistance may be considered to be good, the rubber does have some limitations. The heat resistance is more limited by a tendency to degradation rather than the crystalline melting point. This is a problem above 100°C although the use of hindered amine antioxidants can be beneficial. Ultraviolet resistance may also be improved by the use of UV stabilizers, particularly of the benzotriazole type, in conjunction with a phenolic antioxidant. Hydrolysis resistance against water is good up to 70°C but at higher temperatures the incorporation of a polycarbodi-imide stabilizer may be required. As may be expected of a polar polyester rubber, the elastomers are affected by phenols, low-molecular-weight organic acids and some chlorinated hydrocarbons.

16.4.3 Processing

The main features to note are:

1. The need to ensure that granules and reground material are thoroughly dry before processing;
2. the need to process 5–30°C above the T_m;
3. the need to avoid degradation by processing below 250°C;
4. standard grades have a low melt viscosity and melt strength so that for blow moulding specially formulated grades should be employed.

TABLE 16.4
Typical mechanical properties of Hytrel polyester resins
(After Lloyd, 1982)

Property[a]	*ASTM method*	*Polyester elastomer hardness (durometer) grades*				
		40D	*47D*	*55D*	*63D*	*72D*
Tensile strength, MPa	D-638	25·5	21·0	37·9	39·3	39·3
Ultimate elongation, %	D-638	450	500	450	350	350
25% Modulus or yield point, MPa	D-638	7·6 (M25)	—	13·8 (M25)	17·2 (Yield)	26·2 (Yield)
Flexural modulus, MPa	D-790	65	109	207	345	585
Resistance to flex cut growth, cycles to failure, Ross (pierced)	D-1052	3×10^5	$1{\cdot}5 \times 10^5$	3×10^5	$2{\cdot}8 \times 10^5$	—
Notched impact (Izod)						
At 24°C, $J\,cm^{-1}$	D-256A	No break	No break	No break	No break	2·1
At −40°C, $J\,cm^{-1}$	D-256A	10·6 No break	—	10·6 No break	0·3	0·4
Taber abrasion CS-17 wheel (1 000 g load), mg/1 000 cycles	D-1044	3	—	5	8	13
Softening point (Vicat), °C	D-1525	112	159	180	184	203
Brittleness temperature, °C	D-746	−70	−70	−70	−70	−70
Coefficient of linear expansion, $mm\,mm^{-1}\,°C^{-1}$	D-696	20×10^{-5}	20×10^{-5}	18×10^{-5}	17×10^{-5}	21×10^{-5}
Specific gravity	D-720	1·17	1·20	1·20	1·22	1·25

[a] Properties were measured on injection-moulded test specimens.

16.4.4 Applications

The polyether–ester rubbers have found application as an engineering material, particularly where properties somewhat intermediate between that of a rubber and a rigid plastics material are required and where there is a demand for toughness, low-temperature flexibility and hydrocarbon resistance. In some applications it is possible to achieve the desired stiffness without the fabric reinforcement used with conventional rubbers.

Amongst applications are flexible diaphragms, automotive convoluted bellows, segmented tracks for snow vehicles, ski boot soles, rollers, slow-running tyres, quiet-running gear wheels and high-pressure hose linings.

16.5 THERMOPLASTIC POLYURETHANE RUBBERS

The thermoplastic polyurethane rubbers have much in common with the polyether–ester thermoplastic rubbers. Both are somewhat polar and thus have good hydrocarbon oil resistance and both appear to comprise interpenetrating domains of soft blocks with low glass transition temperatures and hard blocks which retain their rigidity up to the crystalline melting point.

In general the thermoplastic polyurethanes are outstanding in their abrasion resistance but possess neither the temperature range nor the hydrolysis resistance of the polyether–ester rubbers. However, within the classification of thermoplastic polyurethane rubbers it is possible to obtain a spectrum of properties.

There are several suppliers of thermoplastic polyurethane elastomers, including American Cyanamid (Cytor), Elastogran BASF (Elastollan), Bayer (Desmopan), Davathane (Davathane), Goodrich (Estane), Thiokol (Plastothane), Uniroyal (Roylar) and Upjohn (Pellethane). Production for the USA and Western Europe is believed to be of the order of 25 000 tpa.

16.5.1 Preparation and Structure

SBS thermoplastic rubbers are produced by double-bond polymerization and the polyether–esters by condensation polymerization. The polyurethanes are prepared by rearrangement polymerization, sometimes known as polyaddition. The general chemistry of polyurethanes was discussed in Chapter 15 and the rest of this section will be restricted to aspects relevant to the thermoplastic rubbers.

If a polyol, a di-isocyanate and a glycol (diol) are reacted together the reaction may be schematically represented as follows:

$$\underset{\text{Polyol}}{HO\sim P\sim OH} + OCNRNCO + HOR_1OH$$
$$\downarrow$$
$$\underbrace{\sim P\sim O}_{\text{Soft segment}}\underbrace{OCNHRNHCOO[R_1OOCNHRNHCOO]_x\sim}_{\text{Hard segment}}$$

Since there will be many more glycol than polyol molecules, the structure in the right-hand bracket will probably be repeated many times to form a block. If R and R_1 are small and regular units, the polyurethane segment will show high intersegment attraction including hydrogen bonding and may also be able to crystallize. As the polyol segments will try to agglomerate separately from the polyurethane segments, soft domains and hard domains will be formed due to the polyol and polyurethane segments respectively. Whilst there may be differences of detail, the morphology of the thermoplastic polyurethanes may be considered as broadly similar to that of the polyether–esters.

As with the thermoplastic polyether–esters, it is possible to vary the ratios of the three main components, and commercial polyurethane materials are available with an even broader range of hardness.

More generally the properties of the thermoplastic polyurethanes may be varied considerably by use of the following formulation variables:

1. Choice of polyol (sometimes referred to as macroglycol);
2. choice of chain extender;
3. choice of isocyanate;
4. ratio of ingredients.

There are two basic types of polyol: polyester and polyether. Polyesters confer to the rubber the better oil resistance, mechanical properties and resistance to ultraviolet light, whilst polyethers have better hydrolysis resistance. The most commonly used polyethers are polytetramethylene glycol (as used with the polyether–ester rubbers described in Section 16.4) and polypropylene glycol. Polyethylene adipate is commonly used as the polyester, although polycaprolactone glycol ester is also used. The structures of these four polyols are shown in Fig. 16.6 and comparative properties of three thermoplastic polyurethanes based on a polyether, polyethylene adipate and a polycaprolactone glycol ester are shown in Table 16.5.

Whilst the chain extender is usually a glycol such as 1,4-butanediol (also

TABLE 16.5
Comparative physical properties of various Elastogran extrusion-grade thermoplastic polyurethanes
(After Harrop, 1982)

Property	*Standard test method*	*Elastollan T88 AKO, polyethylene adipate-type polyester*	*Caprolan T88 AKO, polycaprolactone glycol ester*	*Elastollan P85 A-10, polyether*
Hardness				
(IRHD)	(ISO S1400–75)	88	88	87
(Shore D)	(ISO S868–78 (E))	40	40	40
Density, $g\,cm^{-3}$		1·25	1·16	1·09
Tensile strength, $MN\,m^{-2}$	(ISO S37–77 (E))	35	35	30
100% Modulus, $MN\,m^{-2}$	(ISO S37–77 (E))	6·4	6·6	6·6
300% Modulus, $MN\,m^{-2}$	(ISO S37–77 (E))	11	12	13
Elongation at break, %	(ISO S37–77 (E))	650	600	630
Tear strength				
(Crescent) N standard thickness specimen	(ISO R34–57 (E))	185	195	145
(Angle) $kN\,m^{-1}$	(ASTM D624–54)	95	105	88
Abrasion (volume loss), mm^3	(DIN 53 516)	38	25	50
Compression set				
At RT, %	(ISO S815–72 (E))	15	14	20
At 70°C, %	(ISO S815–72 (E))	35	37	35
Rebound resilience, %	(ISO S4662–78 (E))	32	35	42
Low-temperature flexibility (T10), °C	(ISO S1432–76)	−25	−37	−59

The test samples were cut from 2-mm injection-moulded sheets and tempered at 100°C for 20 h. The values are the typical values from a series of individual measurements.

TABLE 16.6
Typical properties of Elastollan C range of polyester-based thermoplastic polyurethanes
(After Harrop, 1982)

	C 78 A	*C 80 A*	*C 85 A*	*C 90 A*	*C 95 A*	*C 59 D*	*C 60 D*	*C 64 D*	*C 74 D*
Hardness (Shore A)	80 ± 2	$83 \pm ^{2}_{1}$	87 ± 2	92 ± 2	$94 \pm ^{3}_{0}$	(98)	(98)	(98)	(98)
Hardness (Shore D)	27 ± 2	30 ± 2	34 ± 2	42 ± 2	47 ± 3	57 ± 2	60 ± 3	$64 \pm ^{3}_{2}$	$74 \pm ^{2}_{4}$
Density, $g\,cm^{-3}$	1·18	1·19	1·19	1·21	1·22	1·24	1·24	1·24	1·24
Tensile strength measured with test specimen S2 with a strain rate of 100 $mm\,min^{-1}$, $N\,mm^{-2}$	35	35	40	45	45	45	45	40	35
Elongation at break, %	600	550	500	450	450	450	400	350	300
20% Tensile modulus, $MN\,m^{-2}$	2·0	2·7	4	5·6	8·5	12	16	17	28
100% Tensile modulus, $MN\,m^{-2}$	3·5	5	8	10	12	17	20	24	30
300% Tensile modulus, $MN\,m^{-2}$	7·5	10	13	15	18	30	35	35	35
Tear strength, $N\,mm^{-1}$	35	40	40	60	65	70	80	80	90
(Graves) $N\,mm^{-1}$	55	60	70	90	110	140	160	160	200
Abrasion loss, mm^3	50	50	50	50	55	60	60	65	70
Compression set									
At RT, %	25	20	20	20	25	30	40	40	40
At 70°C, %	40	40	35	35	40	50	50	55	60

Rebound resilience, %	44	40	35	35	35	37	38	38	41
Freezing temperature, °C	−45	−45	−45	−45	−45	−45	−45	−35	−25
Hydrolysis resistance									
Tensile strength after storage in water for 21 days at 80°C, N mm^{-2}	25	25	33	35	40	40	38	35	32
Elongation at rupture after 21 days storage at 80°C in water, %	650	600	600	500	500	500	400	400	300

The tests were carried out according to the following standards:

Shore hardness: DIN 53 505, in accordance with ASTM 2240–68.
Density: DIN 53 479, in accordance with ASTM D 792–66.
Tensile strength, elongation at break and tensile strain: DIN 53 504, in accordance with ASTM D 638–71a.
Tear strength (single tear strips): DIN 53 507, approx. in accordance with ASTM D 1938.
Tear strength (Graves): DIN 53 515, in accordance with ASTM D 1004–66.
Abrasion loss: DIN 53 516.
Compression set: DIN 53 517, in accordance with ASTM D 395, method B.
Rebound resilience: DIN 53 512, in accordance with ASTM D 1054–66.
Glass transition temperature: DIN 53 513.

The test samples were punched out of 3- or 6-mm injection-moulded sheets and tempered at 80°C for 24 h. The values are the minimum values from a series of individual measurements.

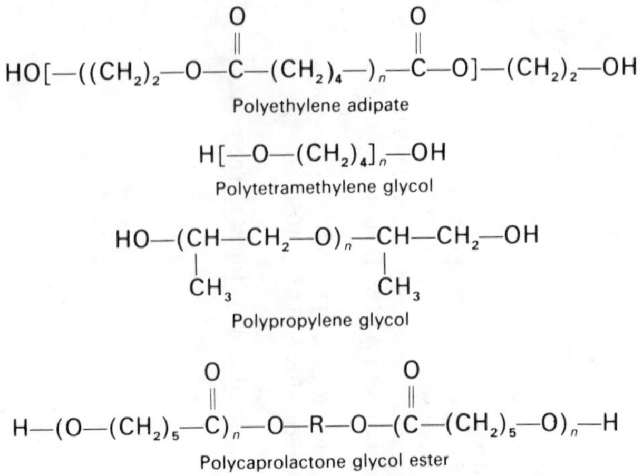

$$HO[-((CH_2)_2-O-\overset{\overset{O}{\|}}{C}-(CH_2)_4-)_n-\overset{\overset{O}{\|}}{C}-O]-(CH_2)_2-OH$$

Polyethylene adipate

$$H[-O-(CH_2)_4]_n-OH$$

Polytetramethylene glycol

$$HO-(\underset{\underset{CH_3}{|}}{CH}-CH_2-O)_n-\underset{\underset{CH_3}{|}}{CH}-CH_2-OH$$

Polypropylene glycol

$$H-(O-(CH_2)_5-\overset{\overset{O}{\|}}{C})_n-O-R-O-(\overset{\overset{O}{\|}}{C}-(CH_2)_5-O)_n-H$$

Polycaprolactone glycol ester

FIG. 16.6. Structure of polyols commonly used in production of thermoplastic polyurethane rubbers.

used with the polyether–ester rubbers) amines may be used. In this case the reaction will be of general form:

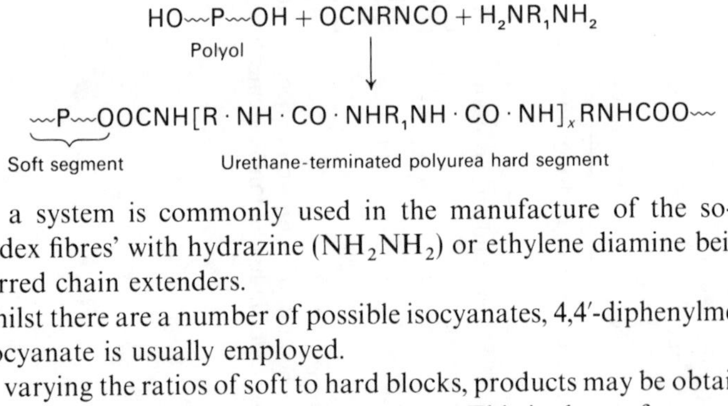

$$\underset{\text{Polyol}}{HO\sim P\sim OH} + OCNRNCO + H_2NR_1NH_2$$

$$\downarrow$$

$$\underbrace{\sim P\sim}_{\text{Soft segment}}\underbrace{OOCNH[R\cdot NH\cdot CO\cdot NHR_1NH\cdot CO\cdot NH]_xRNHCOO\sim}_{\text{Urethane-terminated polyurea hard segment}}$$

Such a system is commonly used in the manufacture of the so-called 'spandex fibres' with hydrazine (NH_2NH_2) or ethylene diamine being the preferred chain extenders.

Whilst there are a number of possible isocyanates, 4,4′-diphenylmethane di-isocyanate is usually employed.

By varying the ratios of soft to hard blocks, products may be obtained in a range of properties, particularly hardness. This is shown for a range of commercial polyester-based polyurethanes in Table 16.6.

16.5.2 Properties

Typical mechanical and thermal properties have been shown in Tables 16.5 and 16.6. In addition, the following further points should be made.

1. The high abrasion resistance indicated in laboratory wear tests has been confirmed by service tests such as actual shoe wear trials.

2. The rubbers show a high resistance to tear propagation as well as to tear initiation.
3. Chemical and solvent resistance is largely as to be expected of a polar material subject to hydrolysis.
4. Whilst the rubbers are resistant to ozone, breakdown may occur on weathering, particularly in hot and tropical climates due to UV attack in the case of polyether-based materials and hydrolytic or microbiological attack in the case of polyester-based polymers.

Hydrolysis resistance of polyester-based rubbers may be improved by the incorporation of up to 2% of polycarbodi-imide. Ultraviolet resistance may be improved by incorporating a UV stabilizer together with an antioxidant.

16.5.3 Processing

As with the polyether–ester rubbers, the main concern in processing is in ensuring that the granules are thoroughly dry before being fed to the injection-moulding machine or extruder. Traces of moisture will cause rapid degradation by hydrolysis. Also to be noted is that the melt viscosity is usually more temperature-sensitive than with many conventional thermoplastics and it is important that good control of temperature is achieved when processing.

16.5.4 Applications

The thermoplastic polyurethane rubbers, like several other thermoplastic rubbers, are frequently used in the hardness range intermediate between that of conventional rubbers and plastics, and particularly where high abrasion resistance coupled with good oil resistance is required. The main limitations are the limited temperature range, the susceptibility to either UV or microbiological attack and the poor resistance of some grades to synthetic lubricants such as phosphates and diesters.

Important uses include ski boot shells, shoe heel top pieces, soling of studded sports footwear, soling of high-quality tennis shoes, grommets, seals, washers and rollers for use in the automotive, mining and mechanical handling industries and many other general industrial mouldings.

16.6 POLYETHER–AMIDE BLOCK COPOLYMERS

These materials first appeared at the end of the 1970s and are related to polyamides of the nylon type as the polyether–esters are related to linear polyesters such as polyethylene terephthalate. Amongst the suppliers are Atochem (Pebax), EMS-Chemie AG (Grilamid ELY 60) and Hüls. The Monsanto Nyrim materials processed by reaction injection moulding (RIM) techniques are similar.

A typical polymer may be prepared by condensation of a polyether (usually polytetramethylene ether glycol) with laurin lactam and decane-1,10-dicarboxylic acid. By appropriate adjustment of the conditions block copolymers are obtained with soft (polyether) and hard (polyamide) blocks. The materials are therefore similar to the polyether–ester and thermoplastic polyurethane rubbers in being polar rubbers consisting of hard and soft blocks. Some comparative data for the three materials (comparing grades of similar hardness) are given in Table 16.7.

TABLE 16.7
Comparison of three polar thermoplastic rubbers

Property	*Polyether–amide*	*Polyether–ester*	*Polyester–urethane*
Hardness (Shore D)	63	63	64
Density, g cm^{-3}	1·01	1·22	1·24
Tensile strength, MPa	35	39·3	40
Elongation at break, %	300	350	350
Melting point, °C	160	*c.* 200	*c.* 120
Vicat temperature, °C	150	184	—

As with the other two classes of polar thermoplastic rubber, at least one supplier (Atochem) offers a range of materials. These vary in the choice of polyether, in the nature of the amide block (it may be homopolymeric, i.e. crystallizing, or copolymeric with limited crystallization) and in the ratio of hard to soft blocks.

Currently the hardness range is from Shore A 60 to Shore D 63, which is broader than for the polyether–esters and the polyurethanes. Flexural modulus may range from 20 to 410 MPa, density from 1·01 to 1·15 g cm^{-3}, melting points from 120 to 210°C and moisture absorption from 1·2 to as high as 120%. Hardness, flexibility and impact strength are little changed over the range −40 to +80°C. The flexing resistance of some grades is particularly noteworthy; they have withstood 36 million cycles in a De Mattia flexing test. Chemical resistance is broadly similar to that of the polyether–esters. Whereas softer grades are virtually transparent, harder grades are more opaque.

When the materials are processed, the granules should be dry. The main problem, particularly with the softer grades, is a tendency to stick to the moulds. The use of silicone-based mould releases is not recommended as this may cause delamination of the mouldings and it is preferable to use a grade containing a proprietary mould-release agent or to mould at lower injection pressures.

Applications of the material include keyboard pads for computers and word processors, sports footwear, pump membranes, loudspeaker gaskets and, when filled with an inorganic filler, watch straps.

16.7 THERMOPLASTIC POLYOLEFIN RUBBERS (TPO)

These materials differ from those discussed earlier in this chapter since they consist essentially of physical blends of a thermoplastics material, usually polypropylene (PP), with an ethylene–propylene rubber (EPM or EPDM). However, as with the other materials, it is possible to produce TPOs with a range of hardness from that of polypropylene at one extreme to that of a raw EPM rubber at the other. Where polypropylene is the major component it forms the continuous phase and the material may be considered as a modified thermoplastic. Where the rubber is the major component it will not necessarily form the continuous phase because of the difference in melt viscosities of the two components.

TPOs are marketed by several companies, many of whom manufacture either polypropylene or ethylene–propylene rubbers, or both, as well as by a number of compounders.

16.7.1 Formulation and Structure

The TPO rubbers are prepared by intensive mixing of EPM or EPDM with polypropylene above the melting point of the latter. In some grades there is a measure of cross-linking, usually within rubber particles, in order to enhance the elasticity.

Either EPM or EPDM may be used as the choice of rubber; rather more crucial is the level of propylene units in the copolymer. An ethylene: propylene ratio of about 70:30 is usually preferred for rubbery grades since this allows some crystallinity in the rubber, giving improved green strength as well as probably allowing some tie points between the rubber and the polypropylene. Better green strength, high- and low-temperature resilience, extrusion performance, creep resistance and elastic recovery are to be obtained when higher-molecular-weight grades of elastomer (Mooney ML 1+4 (100°C) of 70–100) are used.

Whilst a wide range of polypropylenes may be blended with the rubber, a fairly low-molecular-weight polypropylene homopolymer is generally preferred as this helps to lower the melt viscosity of the blend as well as providing substantial levels of crystallinity to act as reinforcing sites and tie points with the rubber. For use as rubbers, as opposed to elastomer-modified thermoplastics, the polypropylene content should not exceed 30%.

A number of additives have been used in TPO rubbers. For example, some suppliers have offered grades containing some polyethylene as a partial replacement of the polypropylene whilst other market grades contain paraffinic oils as plasticizers. Fillers may also be used. The addition of such materials may, however, adversely alter the morphology of the rubbers and most TPO rubber suppliers discourage users from undertaking their own compounding.

16.7.2 General Properties

The TPO rubbers are non-polar, aliphatic, virtually saturated rubbers containing some crystallinity. Free of additives, they therefore exhibit the following desirable characteristics:

1. Resistance to most polar chemicals and solvents;
2. good outdoor ageing (when properly stabilized) and excellent ozone resistance;
3. excellent electrical insulation properties over a wide frequency range;
4. negligible water absorption and excellent resistance to hydrolysis;
5. flexibility at low temperatures (down to $-60°C$);
6. low density (*c*. 0·89 g cm^{-3} for unfilled grades).

They do, however, suffer a number of disadvantages, which include:

1. Poor elastic recovery, particularly above 60°C;
2. poor resistance to swelling by hydrocarbons;
3. lack of affinity to common room-temperature adhesives;
4. poor scratch and abrasion resistance;
5. lower strength and hardness than most other thermoplastic rubbers.

In recent years many attempts have been made to improve elasticity and creep recovery. In particular, this has led to techniques based on blending and cross-linking within the rubber in a single operation known as dynamic cure blending. In such a process an intractable scorched rubber network is to be avoided and small (as low as 1–3 μm diameter) dispersed cross-linked particles are to be encouraged. Suitably formulated grades may retain their elasticity up to 100°C and, as with other more conventional rubbers containing particles of cross-linked materials, improved processing characteristics.

16.7.3 Applications

Because of their restricted elastic recovery, TPOs tend to be used for applications where low short-term deformations are encountered. This nevertheless covers a wide range of uses, including

convoluted bellows and side bumper strips for auto applications;
extruded profiles for windows;
cable insulation;
footwear.

16.8 HALOGENATED POLYOLEFIN ALLOY THERMOPLASTIC RUBBERS

In 1985 Du Pont announced a new thermoplastic rubber, Alcryn, which is said to be based on a partially cross-linked halogenated polyolefin alloy with some trade literature specifically mentioning the presence of a chlorinated polyolefin. Compared with the thermoplastic polyolefin rubbers, these materials possess the extra dimension of oil resistance (equivalent to that of a medium-acrylonitrile NBR). As with the TPOs, the materials have good resistance to weathering, ozone and heat. Compression set is only moderate but claimed to be better than for many other thermoplastic elastomers.

Some typical properties of three grades of Alcryn are given in Table 16.8.

These rubbers may be processed on conventional thermoplastics processing equipment such as injection-moulding machines and extruders. However, because of the halogenated nature of the material, care has to be taken in a number of respects. For example, injection-moulding equipment should use similar corrosion-resistant materials to those used for PVC and processing melt temperatures should be kept below 180°C to avoid evolution of hydrochloric acid. It is also most important that contamination with polyacetal thermoplastics be avoided as hydrochloric acid catalyses the depolymerization of the latter into gaseous formaldehyde which, when trapped within the processing equipment, gives rise to the risk of an explosion.

TABLE 16.8
Typical properties of Alcryn rubbers
(data for compression-moulded samples using ASTM test methods)

Property	*1201 B-60*	*1201 B-70*	*1201 B-80*
Specific gravity	1·19	1·18	1·25
Hardness (D2240 Durometer A)	61	69	77
Tensile properties at 24°C			
100% modulus, MPa	3·6	5·5	7·6
Tensile strength, MPa	11·0	12·8	13·9
Elongation at break, %	320	230	210
Compression set (ASTM D 395, method B), %			
After 22 h at 24°C	18	18	18
After 22 h at 100°C	50	51	52
Brittleness temperature (ASTM D 746), °C	−53	−50	−44
Fluid resistance (D 471): volume change at 100°C, %			
7 days in ASTM Oil No. 1	−14	−12	−10
7 days in ASTM Oil No. 3	+12	+11	+11
7 days in water	+9	+8	+9

The halogenated polyolefin thermoplastic rubbers have found early acceptance for use in cable sheathing. Other areas of interest are seals and gaskets, weatherstripping, sheet goods, mechanical goods, tubing and hose.

16.9 THERMOPLASTIC NATURAL RUBBER BLENDS

Whilst blends of natural rubber with polyolefin thermoplastics have been known since the late 1970s (Campbell *et al.*, 1978; Elliott, 1982) they have (to date) not achieved much market penetration. In general they possess most of the advantages and disadvantages of thermoplastic polyolefin rubbers but with the additional problem that the unsaturation provides a point for oxidative and ozone attack. Recent developments (Tinker, 1987) suggest that the adoption of dynamic cross-linking and careful compounding can give products with good ozone and weathering resistance.

As with other thermoplastic rubbers, it is possible to recognize a range of products which at one end may be regarded as rubber-modified thermoplastics and at the other as thermoplastic rubbers. Initial commercial application appears to have been in India for interlocking industrial floor tiles and for motor-cycle baggage boxes.

BIBLIOGRAPHY

Whelan, A. & Lee, K. S. (eds) (1982). *Developments in Rubber Technology—3*. Applied Science, London.

REFERENCES

Arcozzi, A., Diani, E. & Vitali, R. (1975). *Proc. Int. Rubb. Conf.*, The Society of Rubber Industry, Japan, Tokyo, p. 135.
Brydson, J. A. (1982). In *Developments in Rubber Technology—3*, ed. A. Whelan & K. S. Lee. Applied Science, London, Chapter 1.
Campbell, D. S., Elliott, D. J. & Wheelans, M. A. (1978). *NR Technology*, **9**, 21.
Chung, C. I. & Gale, J. C. (1976). *J. Polym. Sci.—Phys.*, **14**, 1146.
Chung, C. I. & Lin, M. I. (1978). *J. Polym. Sci.—Phys.*, **16**, 545.
Elliott, D. J. (1982). In *Developments in Rubber Technology—3*, ed. A. Whelan & K. S. Lee. Applied Science, London, Chapter 7.
Gilby, G. W. (1982). In *Developments in Rubber Technology—3*, ed. A. Whelan & K. S. Lee. Applied Science, London, Chapter 4.
Harrop, D. J. (1982). In *Developments in Rubber Technology—3*, ed. A. Whelan & K. S. Lee. Applied Science, London, Chapter 5.
Holden, G., Bishop, E. T. & Legge, N. R. (1969). *J. Polym. Sci.*, **C26**, 37.
Lloyd, I. R. (1982). In *Developments in Rubber Technology—3*, ed. A. Whelan & K. S. Lee. Applied Science, London, Chapter 6.
Pedemonte, E., Turturro, A. & Dondero, G. (1974). *Br. Polym. J.*, **6**, 277.
Pedemonte, E., Alfonso, G., Dondero, G., De Candia, F. & Araimo, L. (1977). *Polymer*, **18**, 191.
Streib, H., Pump, W. & Riess, R. (1977). *Kunststoffe*, **67**, 118.
Tinker, A. J. (1987). *NR Technology*, **18**, 30.

Chapter 17

Rubber Additives: Basic Principles

17.1 INTRODUCTION

From the time of the introduction of the vulcanization process in the 1840s until the 1930s the rubber industry was based almost entirely on the use of one polymer, natural rubber. Furthermore, during this period flexible thermoplastics such as polyethylene and plasticized PVC did not exist and the natural polymer was used for an incredibly diverse number of applications. By appropriate use of additives, products which called for widely differing properties in the rubber were made. Examples include elastic bands, battery boxes, tyres, latex foam upholstery, hot-water bottles, bicycle and car pedals, rubber flooring, vee-belts, proofed goods such as rainwear, surgeons' gloves and swimming caps.

The need to modify properties and to reduce cost meant that over the years an enormous range of liquids, resins and powders were incorporated into rubber. Hoffer (1883) mentioned the use of chalk to reduce cost, and zinc white and iron oxide to change colour. Murphy (1870) had patented the use of phenolic antioxidants, by which time the use of factice had been long established. An indication of the bizarre nature of some additives proposed is seen in the following abstract of a French patent: 'The factice is composed of sulphite lye and dissolved cellulosic material, e.g. old paper dissolved by the aid of sodium silicate. Other materials such as blood, casein, latex, solid rubber, cork or wood flour may be added' (E. Rheinberger). Books with such titles as *The Black Art of Rubber Compounding* were also published.

Over most of this time practical usage preceded scientific understanding and it is probably fair to say that even today many materials are found to be effective without the reason for it being fully understood. Nevertheless much painstaking systematic experimentation has led to better use of materials and a sense of order in their use has been established.

In this chapter we will concern ourselves with the principles of the use of additives, considering in particular common problems such as compatibility, synergism, economics and toxicity, with a more detailed discussion of the types of additive in the following four chapters. The final chapter of the book, concerned with compound design, will then show how these diverse materials may be assembled into compound formulations.

17.2 CLASSIFICATION OF ADDITIVES

The first basic essential for all rubbers other than the thermoplastic elastomers is a cross-linking or vulcanizing system. In the case of the diene hydrocarbon rubbers, EPDM, IIR and NBR, which comprise about 95% of the market, this usually means a sulphur-based system involving a vulcanizing agent, an accelerator, an accelerator activator and a fatty acid. There are however a number of alternative systems which are not only necessary for saturated rubbers but which may also be used with benefit on occasion with unsaturated elastomers.

The second group of additives are those which improve ageing behaviour such as antioxidants, antiozonants, metal-ion chelators and UV absorbers. With some rubbers their use is almost always essential, with some they may further improve ageing behaviour in order to meet stringent product requirements, whilst for a select few speciality rubbers they are quite unnecessary.

There are a small number of rubber products made from compounds containing only polymer, vulcanizing system and antioxidant. Such compounds are known as *gum-stocks.*

However, most rubber compounds contain fillers, of which there are many types. In some cases their main function is simply to reduce cost, in others to improve mechanical properties such as tensile strength, tear resistance and abrasion resistance. In the first instance they are known as *inert fillers* and in the second case as *reinforcing fillers.*

Such fillers are usually powders, in which case they would be described as particulate inert or reinforcing fillers. Fillers also exist in fibre form; yarns and textiles, in particular, have a key role in many important rubber products, particularly as they make it possible to control stiffness in different directions.

Whereas fillers tend to stiffen rubber compounds, during both processing and service, softeners and plasticizers make the compounds more flexible. Somewhat intermediate between fillers and plasticizers are low-molecular-weight resins which help to lower viscosity during processing but which may in a number of instances stiffen the rubber (often without any significant increase in density).

The above classes of additives are not only used almost universally but additionally they will form the bulk, if not all, of a rubber compound. However, from time to time other additives have a special role and some 14 types are reviewed in Chapter 21.

From what has been said above, it can be seen that a rubber compound may contain the ingredients listed in Table 17.1.

TABLE 17.1
Additives that may be encountered in rubber compounds

Component	*Comments*
Polymer vulcanizing system	Most commonly sulphur-based and therefore comprising: Sulphur Accelerator Activator Fatty acid
Antiageing additives:	
Antioxidants	
Antiozonants	Only in unsaturated rubbers
Chelators	Usually antioxidants
UV absorbers	Usually carbon black
Fillers:	
Inert particulate	
Reinforcing particulate	
Fibres and fabrics	Not considered in this book
Softeners and plasticizers	
Resins and process aids	
Miscellaneous special additives:	Not used in the majority of applications
Abrasives	
Antistats	
Blowing agents	
Colourants and white pigments	
Cork	
Electroactive materials such as magnetic powders and radiation absorbers	
Factice	
Fire retardants	
Friction lubricants	
Peptizing agents	
Reclaim	
Reodorants	
Tackifiers	

17.3 RECURRING THEMES

In general it may be expected that additives should have the following features unless by their specific function such requirements are excluded.

1. They should be efficient in their function.
2. They should be stable under processing conditions.
3. They should not bleed or bloom.
4. They should not adversely interfere with the functioning of other additives and more generally not adversely affect the properties of the vulcanizate.
5. They should be cheap.
6. They should be non-toxic and not impart taste or odour where this is relevant.

In reality additives will often fall short in some respects and it will be necessary to make some compromise.

Most of the above requirements are self-explanatory but the terms *bleeding* and *blooming* require some explanation.

Bleeding is the phenomenon of migration of a polymer additive from the polymer into an adjacent material. This may be another polymer, even part of the same moulding, perhaps of different colour, or it may be a totally unrelated solid or liquid, perhaps a foodstuff. Bleeding occurs when the additive is wholly or partly soluble in both the polymer in which it is incorporated and the adjacent material. When these conditions allowing bleeding are fulfilled, the rates of bleeding will be determined by the laws of diffusion. Diffusion of matter through a mass of material requires that there are 'holes' in the material providing a pathway for the diffusing molecules. In addition the diffusing molecules should be small enough to pass down the pathway. Compared with many plastics, rubbers possess two characteristics that facilitate such diffusion. Firstly in their service temperature range rubbers are well above the T_g. Since the free volume, i.e. unoccupied space, in a polymer mass increases rapidly above T_g, diffusion is that much easier. Secondly the continual motion of polymer segments will tend to be continually opening up pathways in front of the diffusing molecules, which will be tending to move towards zones of lower concentration.

To avoid problems of bleeding it may be possible to use an additive with a different solubility parameter from that of the polymer and will thus be insoluble in it. Alternatively it may be possible to use a chemical retaining the relevant active groups but which has a higher molecular weight and therefore can only diffuse through the rubber with difficulty. This latter approach has been particularly successful with phenolic antioxidants.

A separate phenomenon which is also associated with solubility is *blooming*. It occurs when an additive has totally dissolved in the polymer at processing temperature but is only partially soluble at ambient temperature. As a result some of the additive is thrown out of solution on cooling and some of this may collect on the surface of the polymer mass. If the additive is only partially soluble at the processing temperature the residual material may form nuclei around which the additive molecules that are thrown out of solution can congregate. In this case much less of the additive will bloom onto the surface. It is also possible that another additive might form suitable nuclei in each of the two above instances. Blooming will not occur either if the additive is totally insoluble at processing temperature or totally soluble at ambient or operating temperatures. Solubility should be considered with respect to the polymer compound as a whole including, for example, plasticizers and not just the polymer itself.

Blooming is usually an undesirable phenomenon which sometimes gives rise to spectacular crystalline deposits. It can cause contamination, be unsightly and interfere with hand-building and other plying-up operations. On the other hand it is made use of when crystalline waxes are allowed to bloom out of diene rubbers in order to provide a protective layer against zone attack.

Table 17.2 summarizes the blooming expectancies of additives in polymer compounds but there may be exceptions to these very general rules.

17.3.1 Synergism

The term *synergism* is used in several branches of science. As an example a dictionary of psychology uses it to describe the acting together of

TABLE 17.2
Blooming in polymer compounds

Example	*Concentration of additive*	*Solubility at ambient temp.*	*Solubility at process temp.*	*Expected effect*
1	x	0	0	No blooming
2	x	$<x$	$<x$	No blooming
3	x	$<x$	$>x$	Blooming unless nucleated *in situ* by another additive
4	x	$>x$	$>x$	No blooming

processes or influences and it is still in this sense that it is used in polymer technology. One example is where two antioxidants may exhibit a co-operative effect and be more powerful than either antioxidant alone, even when the total antioxidant concentration is the same. Other well-known examples are the co-operative effect of two accelerators (e.g. MBT and DPG), cadmium–barium stabilizers for PVC, antimony oxide and tritolyl phosphate fire retardants, and antiozonant–wax combinations in diene rubbers.

There can be a variety of reasons for such synergism. For example, one component may operate by one mechanism, the second by another; one component might regenerate the other as it is being consumed when acting out its role; one component might be a catalyst for the second; or the components may combine *in situ* to form a more powerful third additive. Whatever the mechanism, synergism occurs often enough for synergistic combinations to be continually sought after.

Whilst there is some minor variation in detailed interpretation of the term synergism, Table 17.3 indicates most common usage along with some related terms.

Table 17.3 is somewhat formalistic and requires some additional points to be made.

1. In practice, doubling the concentration of a single additive does not always double the effect.
2. Maximum synergism does not necessarily occur with equal levels of additive.
3. Some additives that themselves function by more than one mechanism are probably autosynergistic.
4. An additive that functions synergistically with another additive may be antagonistic ('antisynergistic') with a third. A well-known example is carbon black, which can have a synergistic effect with phenolic sulphides as an antioxidant in polyolefins but is antagonistic with many phenols and amines.
5. Whilst a pair of additives may be strongly synergistic, they may not be a preferred combination if there is a large disparity in price and it is more efficient economically to use larger quantities of the cheaper additive.

17.3.2 Toxicity

A material may be regarded as toxic if it has a poisoning effect on the body and a consequent adverse effect on health. Often it is not difficult to prove

TABLE 17.3
Definitions of synergism and related terms

Description of effect	Additive concentration		Effect
	[X]	[Y]	
	1	—	x
	2	—	$2x$
	—	1	y
	—	2	$2y$
Additive effect	1	1	$x+y$
Synergistic effect	1	1	$>(x+y)$
Antagonistic effect	1	1	$<(x+y)$
Practical synergism	1	1	$>(x+y)$, and also $>2x$ and $>2y$

that a material is toxic but it is very difficult to prove that a material is non-toxic. Tobacco was smoked for many centuries before the dangerous effects of cigarette smoking were appreciated, whilst for many years vinyl chloride was believed to be free from serious toxic hazards and it was only in the 1970s that it was found to have carcinogenic effects. In general terms it is to be expected that materials that are reactive chemically may also be reactive biologically although the converse is not necessarily the case. Problems may arise with hard fillers such as silicas and asbestos, whilst impermeable films of quite inert polymers in certain parts of the body may give rise to problems.

Toxic chemicals may enter the body in various ways, in particular by skin absorption, inhalation and swallowing. Skin absorption may lead to dermatitis and this can be a most annoying complaint. Whereas some chemicals have an almost universal effect on human beings, others may attack only a few. A person who has worked with a particular chemical for many years may suddenly become sensitive to it and from then on be unable to withstand the slightest trace of the material in the atmosphere. Some chemicals may also attack the skin with little absorption; particular mention should be made here of the necessity of eye protection.

In many instances toxic materials may have an almost instant effect whilst in others it may take many years, as in the case of certain naphthylamines. Some toxic materials may be purged out of the body and, providing they do not go above a certain concentration, there may be no lasting ill-effect. On the other hand, other materials such as lead salts may accumulate in the body and eventually a lethal dose may be present.

The mention of any polymer or additive in this book without reference to its toxicity does not imply that it is free from toxic effects, and no material should be used without reference to health and safety literature provided by the raw-material supplier. Many state and trade organizations (such as the US Foods and Drugs Administration and the British Rubber Manufacturers' Association) also have available useful databanks. Potential users must also be aware of legislation which governs the use of materials and which may vary from country to country.

REFERENCES

Hoffer, R. (1883). *A Practical Treatise on Caoutchouc and Gutta Percha.* Sampson Low, Philadelphia, London (translated from the German by W. T. Brannt).
Murphy, J. (1870). US Patent 99 935.
Rheinberger, E. French Patent 825244.

Chapter 18

Additives I: Curing Systems

18.1 INTRODUCTION

For most applications (exceptions include solution and latex adhesives, putty-like products, chewing-gum bases and CSM rubber sheeting) it is necessary to convert rubbery molecules into a three-dimensional network in order that recovery after deformation is essentially complete. Such networks may be held together by small crystalline bundles (as in raw NR), by ionic bonds or by a two-phase system in the case of the thermoplastic rubbers discussed in Chapter 16. For most elastomers, however, networks are a result of covalently bonded cross-linking (vulcanization).

In principle covalent cross-linking may be achieved by a number of routes as illustrated in Fig. 18.1, although not all are of importance in rubber technology.

Most important of all are the chemicals that act as bridging agents and are typified by the accelerated sulphur vulcanizing systems used for unsaturated hydrocarbon rubbers, and the di-isocyanates used to produce polyurethanes from polyols. Cross-linking initiators are important with polyester laminating resins while catalytic systems are also more important with thermosetting plastics such as the urea–formaldehyde and epoxide resins. An example of active-site generators are the peroxides used to cross-link EPM and silicone rubbers.

In preceding chapters a number of curing systems highly specific to individual elastomers have been described. This chapter will concern itself with those vulcanization systems useful with a range of rubbery materials, namely:

1. Accelerated sulphur systems;
2. sulphur donor systems;
3. peroxides;
4. phenolic resin systems;
5. nitroso and urethane systems.

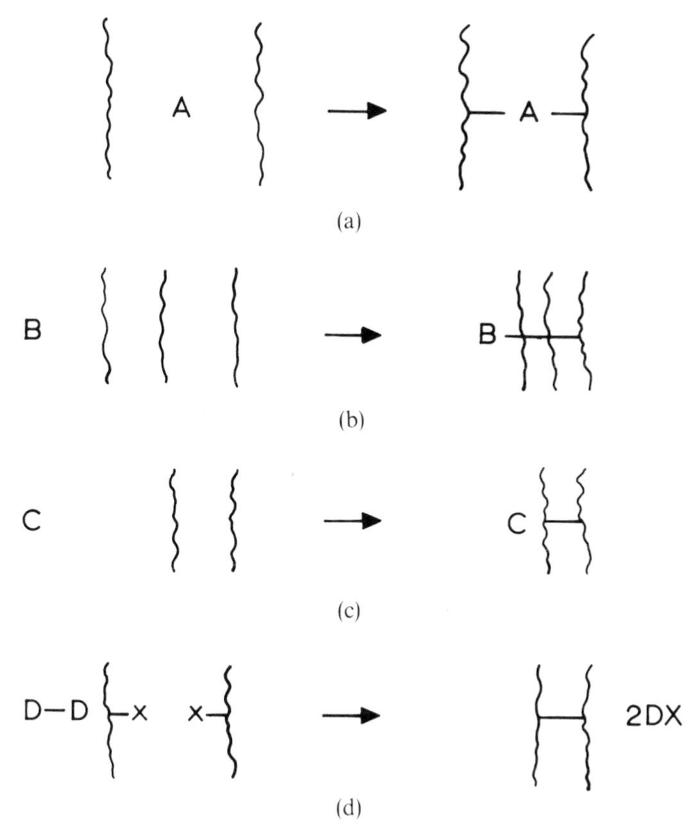

FIG. 18.1. Mechanisms of cross-linking: (a) bridging agents; (b) cross-linking initiators; (c) catalytic cross-linking agents; (d) active-site generators.

18.2 ACCELERATED SULPHUR SYSTEMS

The discovery that the properties of natural rubber changed on heating it with sulphur is generally credited to Charles Goodyear in 1839. Goodyear however failed to exploit the process, this being achieved by Thomas Hancock in London in 1843. It was a friend of Hancock, William Brockeden, who coined the name vulcanization (after Vulcan, the Roman god of fire) to describe the process.

On its own, sulphur is not a very good vulcanizing agent. It is slow and much of it is inefficiently used, for example in intramolecular cyclic reactions. In 1881 Rowley found that ammonia could speed up the curing reaction and subsequent developments by Oenslager, Wa and Wo Ostwald and others led to the availability of more practical accelerators by the beginning of World War I. It was also discovered that the use of accelerator activators such as zinc oxide in conjunction with a fatty acid such as stearic acid could further improve the reaction. Thus by the beginning of World War II a typical curing recipe for natural rubber would have been

		pphr
	Natural rubber	100
Vulcanizing agent	Sulphur	3
Accelerator	MBT	1·5
Accelerator activator	ZnO	5
Fatty acid	Stearic acid	1·5

Some 50 years later changes are largely a matter of degree, the main differences being:

1. A tendency to decrease sulphur levels but increase those of accelerators;
2. widespread use of delayed-action accelerators;
3. greater use of synergistic combinations of accelerators, particularly with the slower-curing synthetic rubbers (see Section 17.3.1);
4. the introduction of prevulcanization inhibitors.

Modern accelerated sulphur systems are today the principal approach for vulcanizing the following rubbers and thus clearly dominate rubber technology:

Natural rubber.
Synthetic polyisoprene rubber.
Polybutadiene rubber.
SBR.
Nitrile rubber.
Butyl rubber.
Halogenated butyl rubbers.
EPDM rubbers.

That these systems should attain such pre-eminence is a combination of a number of factors, including:

1. Economics are comparatively reasonable;
2. there is considerable flexibility in formulations, giving considerable control over the curing process and enabling scorching (premature vulcanization) to be avoided;

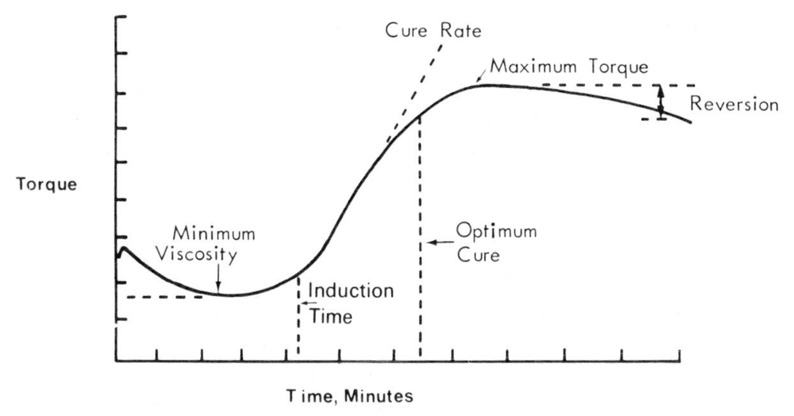

FIG. 18.2. Rheometer curve. (From Rodger, 1979)

3. better physical properties, such as tensile strength, can be obtained than with most other systems;
4. the systems can be handled using standard equipment;
5. toxic hazards are generally well-known and understood.

There are, today, several dozen accelerators available which may be used either alone or in combination, and each of these variations may affect processing characteristics. It is therefore advisable to consider these in turn before looking at individual systems.

18.2.1 Processing Characteristics

The vulcanization characteristics of a rubber are usually followed using a rheometer. In one version of such a device the sample of rubber is enclosed within a heated chamber. Embedded within the rubber is a small rotor which oscillates through a small angle of arc (1°, 3° or 5°) and its resistance to this oscillation, which is a measure of modulus, is recorded as a function of time to give a characteristic cure curve as shown in Fig. 18.2. This curve exhibits a number of features:

1. An initial period during which the rubber softens due simply to a warming-up effect.
2. A delay or induction time before the torque or resistance value begins to rise. (Because the onset of this rise is difficult to determine precisely, it is normal to note the point at which the torque rises to a prescribed level above the minimum. Suitably chosen, this provides a measure of the scorch time at the curing temperature. Also of

importance is the scorch time at lower shaping temperatures. This may be obtained by using this type of equipment at lower temperatures or, as is common practice, by using a Mooney viscometer, described in Appendix B).

3. A rise in the value of torque with time, the slope of the curve giving a measure of cure rate.
4. A reduction in the cure rate as it approaches its maximum value. Since this final approach may be lengthy and the point at which the maxima is achieved is not very precise, it is common to specify a technical or optimum cure time to obtain a specified percentage of the maximum change in torque, usually 90%.
5. With some rubbers the torque does not level out but continues to rise and this effect is sometimes known as 'marching modulus'. Also observed with some curing systems is a reduction in modulus at long cure times, which is generally known as *reversion* although this does not necessarily imply that the curing reactions have gone into reverse.

From the processing viewpoint, the following features of an accelerator are usually desirable.

1. Good processing safety, i.e. long scorch times to reduce the chance of premature vulcanization during mixing and shaping.
2. Rapid curing at vulcanization temperatures.
3. High cross-linking efficiency.
4. Solubility in the rubber so that there is good dispersion and freedom from blooming.
5. Freedom from reversion effects.
6. Good storage stability.
7. No adverse reactions with other additives.
8. Freedom from health and handling problems.

Occasionally there are exceptions to these requirements. For example, in the rubber lining of pipes and large vessels for chemical plant, it is sometimes possible to vulcanize by allowing cure to take place over several weeks at room temperature. Another example is the case of thick mouldings where the use of strong delayed-action accelerators may unduly delay the time it takes to cure the centre of the moulding since cure will not commence until the centre has reached a sufficiently high temperature.

Some typical cure curves using a variety of accelerators are given in Fig. 18.3. The scorch curves for these systems using the Mooney viscometer at the more relevant temperature of 121° are given in Fig. 18.4. Of particular

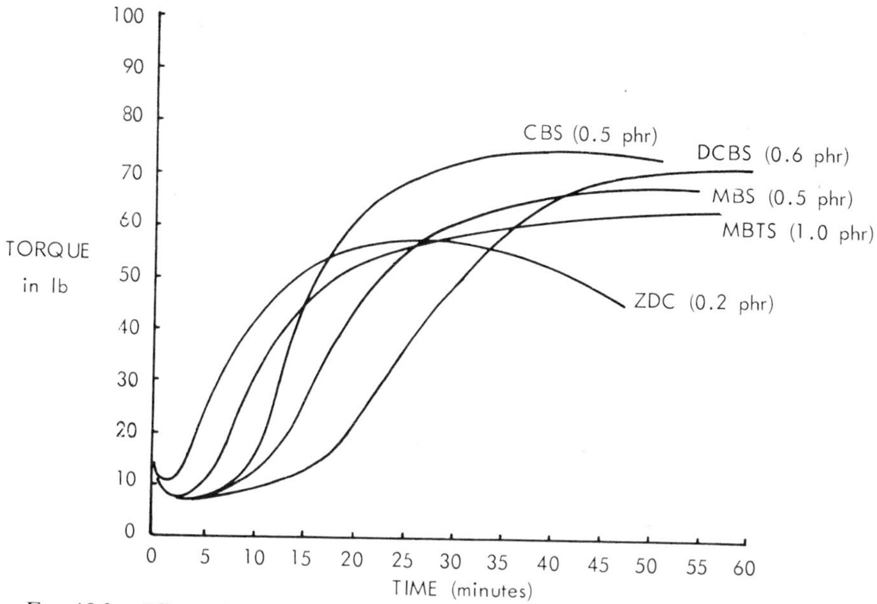

FIG. 18.3. Effect of cure system on cure characteristics. Rheometer at 140°C. Formulation: NR, 100; N330, 50; processing oil, 8; zinc oxide, 5; stearic acid, 2; 6PPD, 2; sulphur, 2·5. (From Rodger, 1979.)

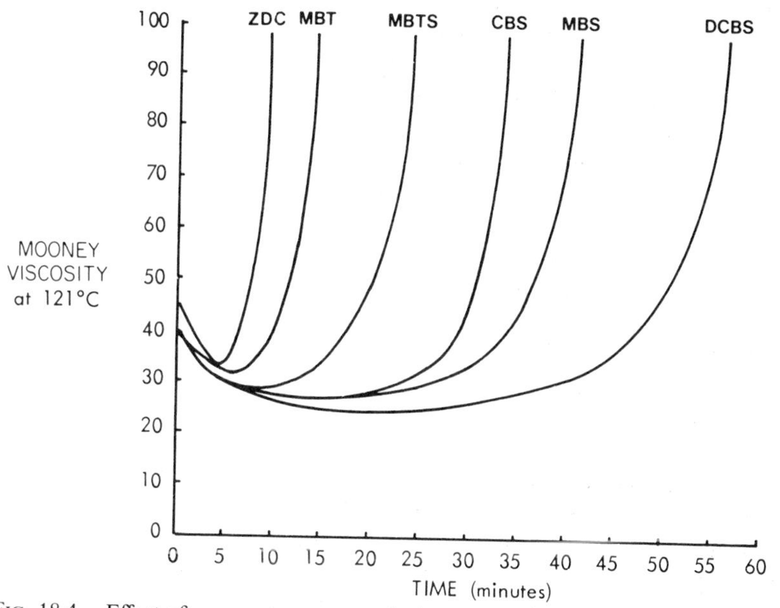

FIG. 18.4. Effect of cure system on scorch time. Formulation: NR, 100; N330, 50; processing oil, 8; zinc oxide, 5; stearic acid, 2; 6PPD, 2; sulphur, 2·5. (From Rodger, 1979.)

note in these curves is the delayed-action but rapid cure of the system containing the accelerator CBS, one of the most widely used accelerators on the market.

18.2.2 Vulcanizate Characteristics

It is important to appreciate that not all properties reach their 'best' value at the same cure time. This is illustrated schematically in Fig. 18.5. Other examples were given in Chapter 5. The inter-relation between formulation, curing schedule and vulcanizate properties is discussed further in Section 18.2.7.

18.2.3 Review of Commercial Accelerators

Most commercial accelerators fall into one of the following classes.

1. Guanidines.
2. Thiazoles.
3. Sulphenamides.
4. Thiurams.
5. Dithiocarbamates.
6. Xanthates.
7. Thiophosphates.
8. Triazines.

Of these the guanidines are seldom used alone but can be very effective in synergistic combinations with other accelerators. They were initially replaced by the thiazoles, which are fast-acting and confer good ageing properties but tend to be rather scorchy. This became a particular problem when the furnace blacks began to replace the channel blacks soon after the end of World War II and led to the development of the sulphenamides. Today this is the most important class of accelerator on account of its fast cure, good processing safety and high solubility in the rubber. The thiurams, dithiocarbamates and xanthates are generally very powerful and tend to be too fast for use in dry natural rubber technology as primary accelerators. They are however used in latex technology and as part of a synergistic combination with the slower-curing synthetic diene rubbers, EPDM and IIR. The thiophosphates and triazines have so far only enjoyed limited acceptance. The principal commercial accelerators are listed in Table 18.1.

18.2.4 Retarders and Prevulcanization Inhibitors

For many years a number of formulations included a retarder, such as salicylic acid, benzoic acid or *N*-nitrosodiphenylamine, at a level of 0·5–2·0 pphr. They were used to try and reduce scorch problems but at the same time they did tend to cause a slowing down in cure rate and modification of modulus. The introduction of *prevulcanization inhibitors* in

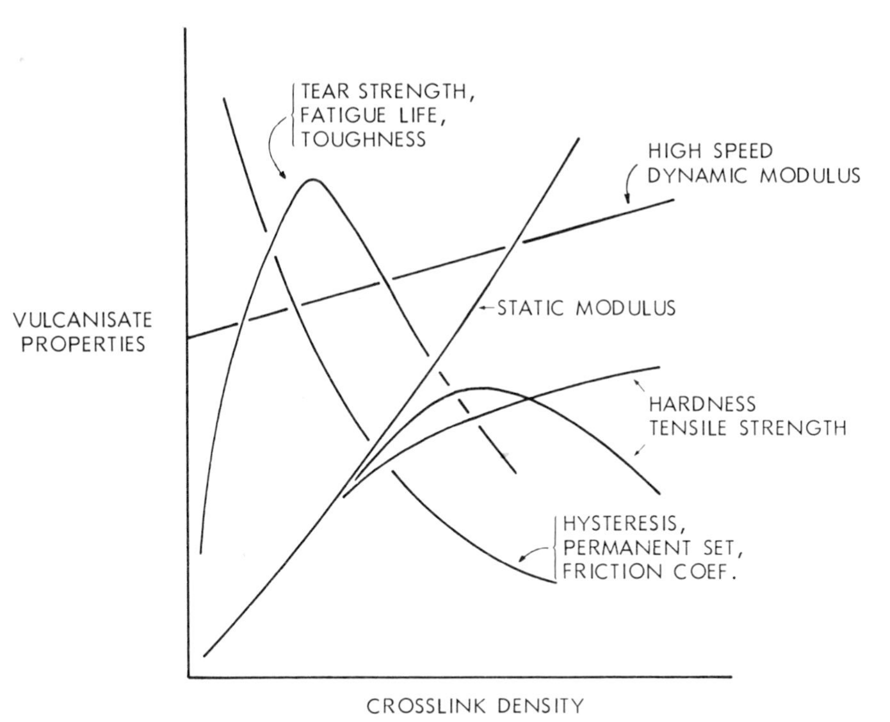

FIG. 18.5. The dependence of various vulcanizate properties on cross-link density. (From Rodger, 1979.)

the early 1970s provided an important tool to reduce scorching but without having a serious effect on cure rate. Anand *et al.* (1977) systematically studied the effect of the most important prevulcanization inhibitor, *N*-(cyclohexylthio)phthalimide (**I**) on the induction and cross-link insertion reactions and on reversion.

O
N—S—
O

(I)

They found that an increase in level of use of this additive greatly lengthened the induction period at typical processing temperatures (*c.* 120°C) but by a much smaller amount at typical vulcanization temperatures (*c.* 150°C). The effect on the cross-link insertion reaction

TABLE 18.1
Principal accelerators used for accelerated sulphur vulcanization of diene rubbers

Class and abbreviation	*Chemical name and formula*	*Notes on usage*
Guanidines		
DPG	Diphenylguanidine	Slow–medium speed accelerator. Synergistic with MBT, MBTS and CBS. Rather scorchy with MBT.
DOTG	Di-*ortho*-tolylguanidine	Slightly faster than DPG.
TPG	Triphenylguanidine	Slow accelerator.
Thiazoles		
MBT	2-Mercaptobenzothiazole	Powerful but apt to be scorchy, particularly with furnace blacks.
MBTS	Dibenzothiazole disulphide	Powerful but some delayed action effect. Often used synergistically with other accelerators.
ZMBT	Zinc mercaptobenzothiazole	Slightly less scorchy than MBT but not as safe as MBTS.
SMBT	Sodium salt of MBT	Water-soluble and used in latex technology.

Sulphenamides		
CBS	*N*-Cyclohexylbenzothiazole-2-sulphenamide 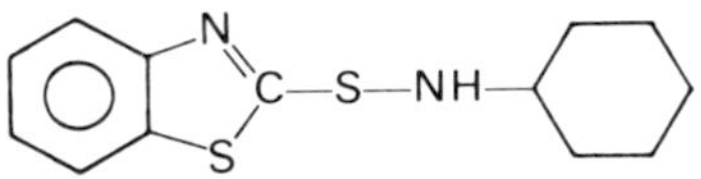	Very widely used delayed-action fast accelerator. Virtually the standard by which other accelerators are judged.
NOBS	*N*-Oxydiethylbenzothiazole-2-sulphenamide 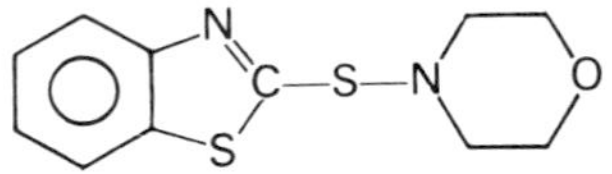	Similar to CBS.
TBBS	*N*-*t*-Butylbenzothiazole-2-sulphenamide	
MBS	*N*-Morpholinothiobenzothiazole-2-sulphenamide	
OTOS	*N*-Oxydiethylenethiocarbamyl-*N*-oxydiethylenesulphenamide	Good processing safety and fast cure but possibly inferior ageing, cf. CBS and TBBS.
DCBS	*N*,*N*-Dicyclohexylbenzothiazole-2-sulphenamide	
Thiurams		
TMT (or TMTD)	Tetramethylthiuram disulphide 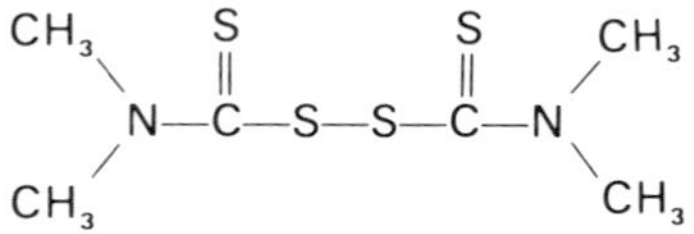	Very fast accelerator—tends to be scorchy in NR. Often used in synergistic combinations with slower-curing rubbers. Confers good heat resistance. Can act as vulcanizing agent.
TETD	Tetraethylthiuram disulphide	Slightly less scorchy than TMT.
TMTM	Tetramethylthiuram monosulphide	Safer processing than TMT.
DPTT	Dipentamethylenethiuram tetrasulphide	Scorchy. Can act as a vulcanizing agent.

(continued)

TABLE 18.1—*contd.*

Class and abbreviation	*Chemical name and formula*	*Notes on usage*
Dithiocarbamates		
ZDC (or ZEDC)	Zinc diethyldithiocarbamate $\left[(C_2H_5)_2N{-}C({=}S){-}S^- \right]_2 Zn^{2+}$	Very fast—widely used in latex technology.
ZDMC	Zinc dimethyldithiocarbamate	Similar to ZDC.
ZDBC	Zinc dibutyldithiocarbamate	Similar to ZDC.
SDC	Sodium diethyldithiocarbamate	Water-soluble.
PPD	Piperidine pentamethylene dithiocarbamate	Water-soluble.
Xanthates		
ZIX	Zinc isopropyl xanthate $\left[(CH_3)_2CH{-}O{-}C({=}S){-}S^- \right]_2 Zn^{2+}$	Ultra-fast.
ZBX	Zinc butyl xanthate	Ultra-fast
SIX	Sodium isopropyl xanthate	Water-soluble. Used in latex technology.

(which dominates the cure rate) was very small but an increase in level did lead to some increase in reversion, which indicated that the material might not be wholly satisfactory for thick sections and other products for which 'peaky' cures are to be avoided. The chemistry and technology of prevulcanization inhibitors have been reviewed (Trivette *et al.*, 1977).

18.2.5 Structure of Accelerated Sulphur Vulcanizates

The very wide range of structures that may be present as a result of accelerated sulphur vulcanization of natural rubber were shown schematically in Fig. 4.3. It is now known that the relative extents of the various reactions will depend not only on the curing formulation but also on the time of cure.

As a general rule, it is desirable that the sulphur be used only in the formation of cross-links. These may be monosulphidic, disulphidic or polysulphidic (i.e. more than two sulphur atoms in the cross-link). One way of quantifying the cross-linking efficiency is to use the efficiency parameter E introduced by Moore & Trego (1961) and given by

$$E = \frac{\text{gram-atoms of total combined sulphur per gram rubber}}{\text{gram-molecules of chemical cross-links per gram rubber}}$$

Some typical figures for this parameter collected by Saville & Watson (1967) are given in Table 18.2. These results clearly show the low efficiency of non-accelerated sulphur systems with $E = 40$–55 with the moderate efficiency of a typical sulphur–MBT-accelerated system and the high efficiency of systems using higher accelerator/sulphur ratios together with lauric acid.

If the vulcanizate is treated with triphenylphosphine the disulphide and polysulphide links are converted to monosulphide. It is now possible to calculate a new value for the efficiency E'. The value $(E' - 1)$ will now be a measure of 'the combined sulphur not in cross-links', i.e. in cyclic or pendant sulphides. While it is tempting to use $[E - (E' - 1)]$ as a measure of the number of sulphur atoms in cross-links, there is a possible error in that the triphenylphosphine may remove sulphur from cyclic and pendant structures.

Figures 18.6 and 18.7 illustrate how the ratio of the various types of cross-link changes with time of cure, for both a conventional system and a so-called 'efficient vulcanization' system using a low-sulphur/high-accelerator ratio of 0·4:6·0. ($M_{c,chem}$ is the molecular-weight average between chemical cross-links.)

It is worth noting that in a real network there are a variety of

TABLE 18.2

Selected values for the efficiency parameter (E) determined by MRPRA workers and collected together by Saville & Watson (1967)

Vulcanizing system[a]	*Cure temp.* (°C)	*Range of* E values[b]
NR, 100; sulphur, 6–10	140	40–55
NR, 100; sulphur, 1·5; MBT, 1·5; zinc oxide, 5·0; lauric acid, 1·0	140 100	15–21 11–14
As above, with lauric acid, 10 phr	100	5·7–1·6 8·2–1·9
NR, 100; sulphur, 2·5; CBS, 0·6; zinc oxide, 5·0; lauric acid, 0·7	140	12–22
NR, 100; sulphur, 2; zinc oxide, 2; ZDMC, 2	100	7–18
cis-PI, 100; TMTD, 4; zinc oxide, 4	140	13·5–3·2
cis-PI, 100; TMTD, 4; zinc oxide, 4; lauric acid, 1·5	140	7·9–2·1
NR, 100; Sulfasan R, 0·6; TMTD, 1·45; MBTS, 1·32; zinc oxide, 4; lauric acid, 3	140	10·4–3·7

[a] Parts by weight of ingredients.
[b] Latter figures refer to an advanced state of cure.

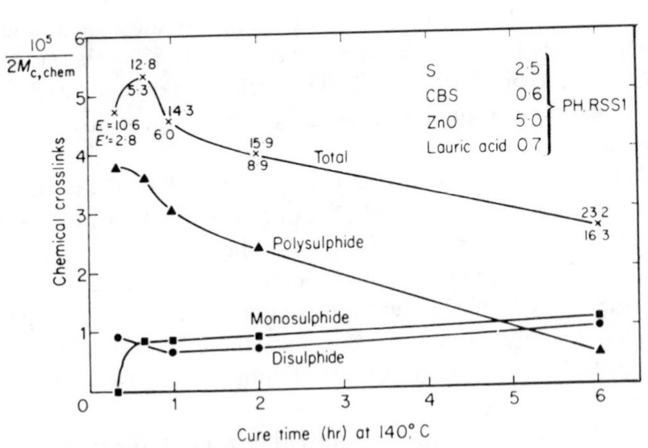

FIG. 18.6. Course of the CBS-accelerated sulphur vulcanization of natural rubber at 140°C with a high sulphur/accelerator ratio (2·5: 0·6). (×) Total cross-links; (▲) polysulphide; (■) monosulphide; (●) disulphide. The upper and lower figures on the curve relating to total cross-links are values of E and E', respectively. (From Moore (1965), reproduced by permission of the Malaysian Rubber Producers' Research Association.)

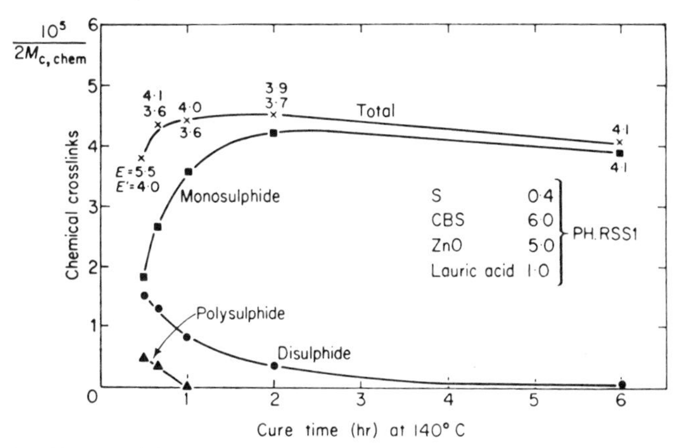

FIG. 18.7. Course of the CBS-accelerated sulphur vulcanization of natural rubber at 140°C with a low sulphur: accelerator ratio (0·4:6·0). (×) Total cross-links; (▲) polysulphide; (■) monosulphide; (●) disulphide. The upper and lower figures on the curve relating to total cross-links are values of E and E', respectively. (From Moore (1965), reproduced by permission of the Malaysian Rubber Producers' Research Association.)

macrostructural features as indicated in Fig. 18.8. The free chain ends will not be stress-bearing and they become important where the rubber before cross-linking is of low molecular weight, such as with highly masticated NR. An entanglement may act like a cross-link but if it slips to become adjacent to a true chemical cross-link then it will be less effective, as will the individual cross-links in a cluster.

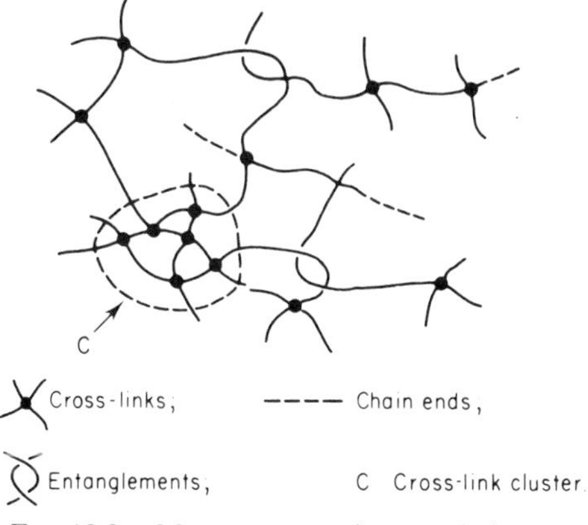

FIG. 18.8. Macrostructural network features.

18.2.6 Sulphur Donor Systems

The high-accelerator/low-sulphur systems (the efficient vulcanization systems) were initially of interest because of the improved ageing behaviour compared with conventional networks. In order to maintain modulus, a reduction in sulphur had to be linked with an increase in accelerator level. However, once the sulphur level is reduced below 0·5 pphr the modulus can no longer be maintained by further increase of the accelerator level.

It has been found that certain sulphur compounds can act as sulphur donors during cure, and these are sometimes used in formulations where no elemental sulphur is included in the recipe. It appears that the improved ageing is linked with the absence of free sulphur in the compound, so that sulphur donor systems, which can almost be considered as a variant of the efficient vulcanization systems, can confer very good ageing properties.

Sulphur donors are of two types.

1. Chemicals which act both as sulphur donors and accelerators. The most important of these is tetramethylthiuram disulphide (TMTD) but others include TETD, DPTT, BDTM, and ZBDP.
2. Chemicals that are not also accelerators. The most important of these is 4,4′-dithiodimorpholine (DTDM); others include ETPT (bis(diethylthiophosphoryl) trisulphide) and DTBC (*N*,*N*′-dithiobishexahydro-2,4-azepinone).

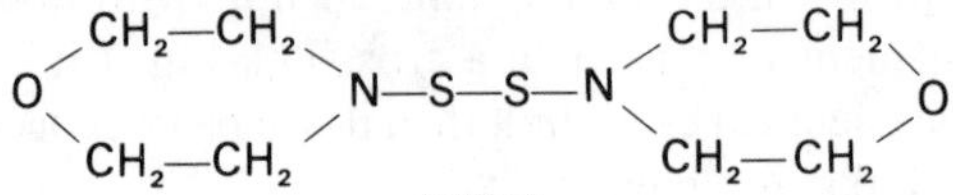

$$O\langle\begin{matrix}CH_2-CH_2\\CH_2-CH_2\end{matrix}\rangle N-S-S-N\langle\begin{matrix}CH_2-CH_2\\CH_2-CH_2\end{matrix}\rangle O$$

DTDM

Sulphur donors are primarily used with the aim of improving thermal and oxidative ageing but they may also be used to reduce the possibility of sulphur bloom. Partial replacement of sulphur by a sulphur donor can also sometimes increase the scorch time without extending cure.

18.2.7 Formulations for Accelerated Sulphur Vulcanization

Different rubbers have different levels of reactivity so that what may be a fast-curing system for one rubber may be too slow for practical use in another. Some measure of the relative reactivities of five important sulphur-cured rubbers may be obtained by considering the basic 'starter' formulations given in Table 18.3.

Variations in such formulations may be made in order to alter cure time, scorch time, modulus, ageing, fatigue life, reversion resistance, crystallization tendencies and low-temperature properties, or in order to cure the

TABLE 18.3
Typical vulcanization systems for black-loaded rubbers
(After Rodger, 1979)

Component	*Formulation (pphr)*				
	NR	*SBR*	*NBR*	*IIR*	*EPDM*
Sulphur	2·5	2·0	1·5	2·0	1·5
Zinc oxide	5	5	5	3	5
Stearic acid	2	2	1	2	1
TBBS	0·6	1	—	—	—
MBTS	—	—	1	0·5	—
MBT	—	—	—	—	1·5
TMTD	—	—	0·1	1	0·5

rubber at different temperatures. In respect of the last-mentioned for example, chemical plant linings may be cured over several days at room temperature whilst continuous vulcanization of extrudates or injection-moulded products may be cured in the range 180–240°C. The factors above are interrelated and will also depend on other additives such as reinforcing fillers and plasticizers, but the following guidelines may be useful.

1. *To reduce the cure time (with increase in modulus)*: Use a faster accelerator or a synergistic mixture, or simply increase the accelerator level. Where a delayed-action accelerator such as MBS is used there should be very little increase in scorchiness.
2. *To reduce the cure time without increasing modulus*: Increase the accelerator level but reduce the sulphur. Figure 18.9 indicates the sulphur/accelerator levels for equal modulus in NR. This may reduce scorch time and it may be necessary to use a prevulcanization inhibitor.
3. *To improve ageing resistance*: A high accelerator/sulphur ratio (i.e. an efficient vulcanization or EV system) which gives a high proportion of monosulphidic links confers better ageing in NR. It is interesting to note that more conventional accelerator/sulphur ratios in SBR give quite substantial levels of monosulphidic linkages, these being increased further by the use of an EV system. This is seen in Table 18.4.
4. *Effect of sulphur/accelerator ratio on fatigue life*: In NR the EV systems tend to give a poor fatigue life to the vulcanizates but in SBR they often confer significantly better fatigue resistance than

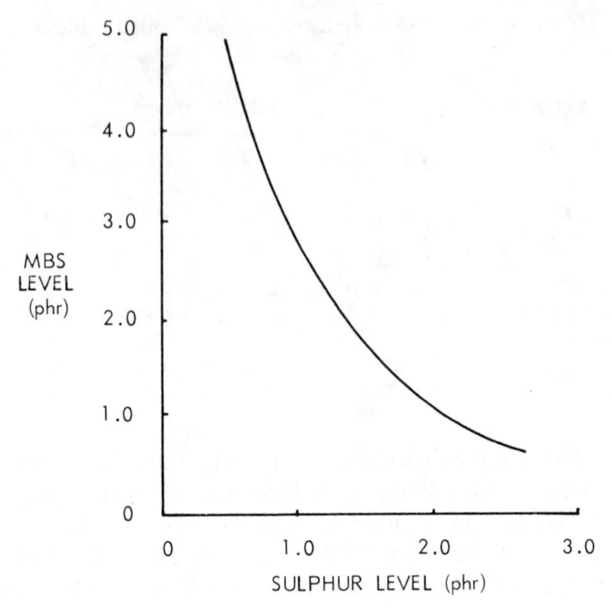

FIG. 18.9. Sulphur/accelerator levels for equal modulus in 100 NR. (From Rodger, 1979.)

TABLE 18.4
Cross-link distributions in NR and SBR
(After Rodger, 1979)

Vulcanization system	*Cross-link type (%)*			
	NR		*SBR*	
	S_1	$S_2 + S_x$	S_1	$S_2 + S_x$
Conventional[a]	0	100	38	62
EV[b]	46	54	86	14

[a] Conventional for NR: sulphur, 2·5; MBS, 0·6. Conventional for SBR: sulphur, 2·0; CBS, 1·0 pphr.
[b] EV for NR: CBS, 1·5; DTDM, 1·5; TMTD, 1·0. EV for SBR: CBS, 1·5; DTDM, 2·0; TMTD, 0·5 pphr.

conventional systems. The adverse effect on fatigue life has limited the use of EV systems in NR for applications where good dynamic properties are important.

5. *Effect on reversion*: Since many conventional systems tend to exhibit reversion in NR the compromise of a semi-EV system can be very beneficial since these systems have a less severe adverse effect on fatigue life. Overcure can be a problem on the surface layers of thick-section mouldings and in the press curing of long belts and other parts where 'overlap' pressing may lead to a double cure.
6. *Effect on low-temperature properties*: The more regular monosulphidic cross-linked rubbers have a greater tendency to crystallization with NR and thus have higher brittle points than those obtained with conventional systems. EV systems therefore tend to be avoided when good low-temperature properties are required, e.g. in bridge bearings.
7. *Dependence of formulation on curing temperature*: As the vulcanization temperature is increased the time taken to reach optimum cure

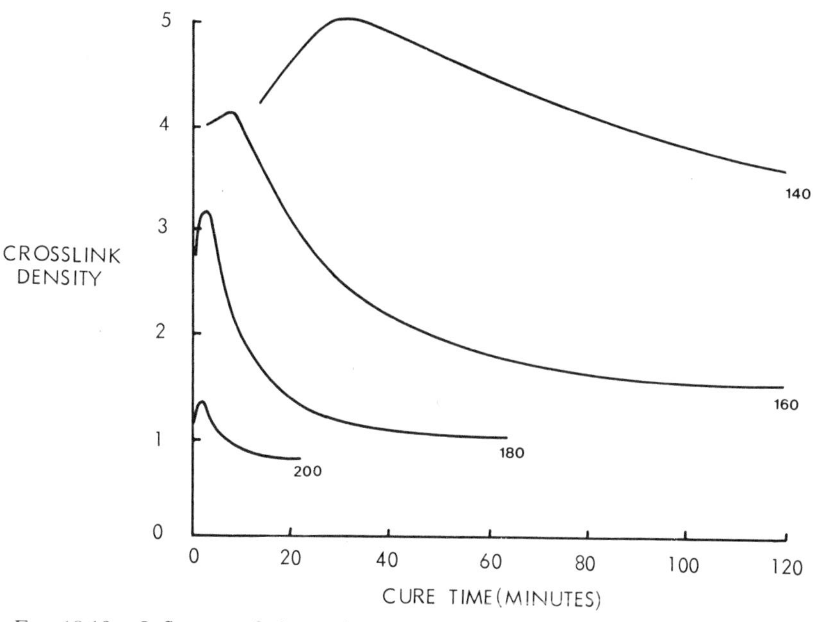

FIG. 18.10. Influence of the curing temperature on cross-link density in NR. Sulphur, 2·5; CBS, 0·6 pphr. (From Rodger (1979), reproduced by permission of Monsanto Co.)

is reduced. However, in the case of NR the optimum cross-link density decreases with cure temperature (see Fig. 18.10). Because of the reduction in cross-link density tensile strength, tear strength, modulus, hardness and resilience all deteriorate.

It has been found that although the total amount of combined sulphur remains approximately constant, not only is the cross-link density reduced but that the cross-links become primarily monosulphidic. Hence a substantial proportion of the sulphur must become involved in the formation of cyclic and pendant sulphide groups.

Such an adverse effect is most pronounced in NR but can also occur to a lesser extent in SBR and SBR/BR blends. In order to achieve high cross-link densities at high temperatures it is usually necessary to increase the level of accelerator without reducing the sulphur content. However, this will not necessarily lead to an increase in tensile strength since this goes through a maximum with cross-link density (Greensmith *et al.*, 1963) (Fig. 18.11).

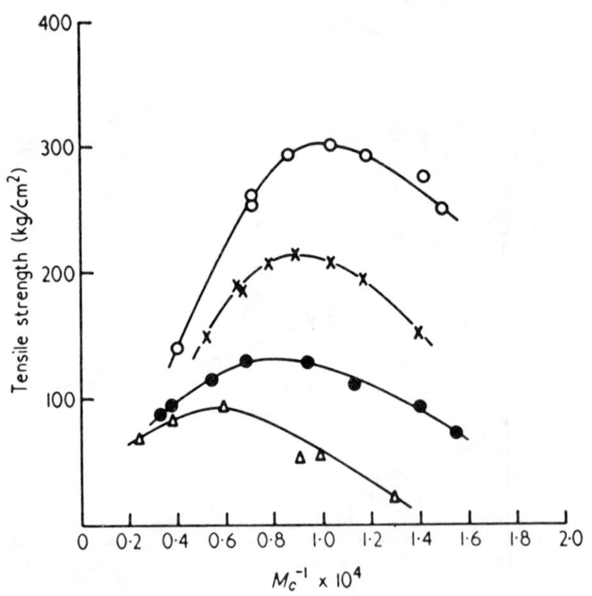

FIG. 18.11. Tensile strength of pure gum natural rubber vulcanizates plotted against $1/M_c$ for various vulcanizing systems. ○, Accelerated sulphur; ×, TMT sulphurless; ●, peroxide; △, high-energy radiation. (From Greensmith *et al.* (1963), reproduced by permission of the Malaysian Rubber Producers' Research Association.)

It is now recognized that the tensile strength depends on the type of cross-link, decreasing in the order

polysulphidic > monosulphidic > carbon–carbon

i.e. in inverse order to the weakest bond. This has led to the hypothesis which states that the weakest bond gives the strongest rubber because it is labile and able to break under stress, thus providing a yield mechanism for the dissipation of stress at the critical point of failure.

18.3 PEROXIDES

Peroxides were first introduced as cross-linking agents for diene rubbers as long ago as 1915 by Ostrosmislenskii, benzoyl peroxide being used in conjunction with natural rubber. They remained laboratory curiosities until the introduction of dicumyl peroxide (Braden *et al*, 1954; Braden & Fletcher, 1955). Natural rubber cross-linked by this peroxide showed low compression set and creep, could be compounded for high-temperature use and did not exhibit the staining problems, e.g. in contact with copper, encountered with sulphur-cured rubbers. However, the system had some serious disadvantages, which included:

1. Low tensile and tear strengths of the vulcanizates;
2. absence of any delayed action before cure, which led to problems of scorching;
3. a somewhat low rate of cure as compared with a typical accelerated sulphur system.

For these reasons the use of peroxides as curing agents for diene rubbers remained very limited.

For a number of rubbers, however, the technologist had little choice but to use peroxides, EPM (see Chapter 7) being the best-known example. Whilst this restriction did lead to the development of the sulphur-curable EPDM rubbers it also led to improved peroxide-based systems involving the use of coagents. More recently work has been reported which allowed some delayed action to the curing process.

As with sulphur vulcanization, curing with peroxides is chemically very complex. The chemistry of peroxide vulcanization has been reviewed by Loan (1967), by Baldwin & VerStrate (1972) and, more briefly, by the author (Brydson, 1978). Because of the variety of the reactions, widely different behaviour occurs with different rubbers and different peroxides whilst the presence of coagents can have considerable influence. It is thus useful, albeit briefly here, to consider the relevant chemistry in order to have a better understanding of the process.

The first stage of the process is the breakdown of the peroxide into free radicals,

$$\text{R—R} \longrightarrow 2\text{R}\cdot$$

e.g. with dicumyl peroxide:

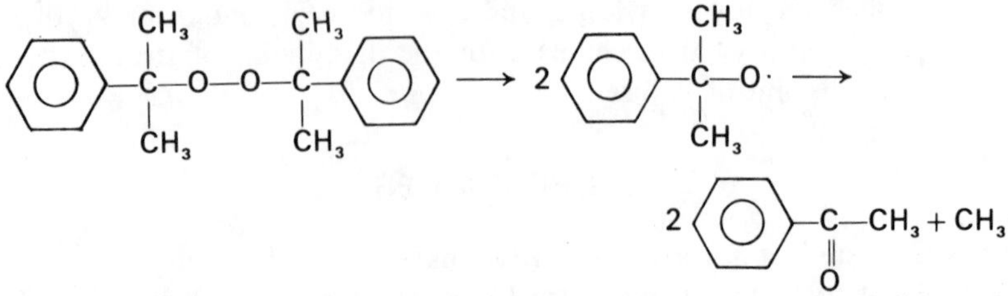

$$\text{C}_6\text{H}_5\text{—C(CH}_3)_2\text{—O—O—C(CH}_3)_2\text{—C}_6\text{H}_5 \longrightarrow 2\,\text{C}_6\text{H}_5\text{—C(CH}_3)_2\text{—O}\cdot \longrightarrow$$

$$2\,\text{C}_6\text{H}_5\text{—C(=O)—CH}_3 + \text{CH}_3\cdot$$

The following reactions may then be envisaged:

1. Abstraction of hydrogen from a polymer chain PH:

$$\text{R}\cdot + \text{PH} \longrightarrow \text{RH} + \text{P}\cdot$$

In the case of a diene rubber this is likely to occur allylic to the double bond. (It is also possible that the peroxide fragment may attach itself to a double bond as in a polymerization initiation process but this does not appear to occur with dicumyl peroxide.)

2. Cross-link formation by radical combination:

$$\text{P}\cdot + \text{P}\cdot \longrightarrow \text{P—P}$$

This is clearly most desirable for efficient vulcanization.

3. Chain scission by unimolecular reaction without radical destruction, e.g.

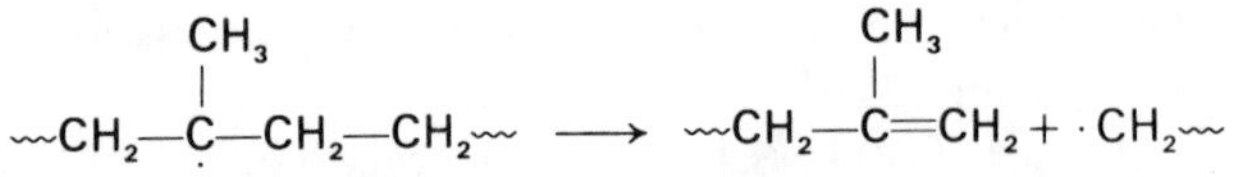

$$\sim\text{CH}_2\text{—}\dot{\text{C}}\text{(CH}_3)\text{—CH}_2\text{—CH}_2\sim \longrightarrow \sim\text{CH}_2\text{—C(CH}_3)\text{=CH}_2 + \cdot\text{CH}_2\sim$$

This is clearly very undesirable, but it is quite common, and with IIR dominant.

4. Radical destruction by bimolecular disproportionation:

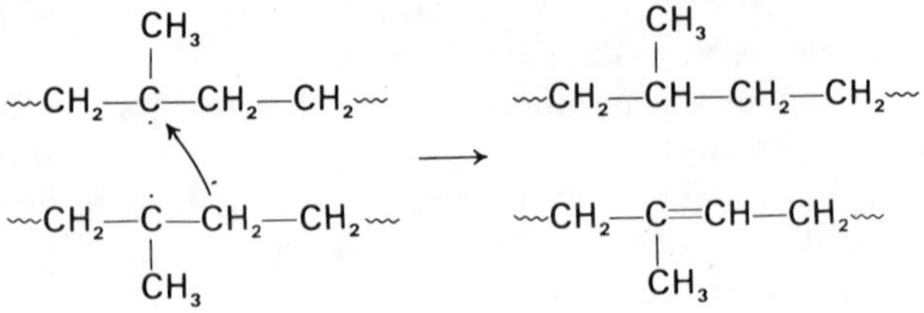

$$\begin{array}{l}\sim\text{CH}_2\text{—}\dot{\text{C}}\text{(CH}_3)\text{—CH}_2\text{—CH}_2\sim \\ \sim\text{CH}_2\text{—}\dot{\text{C}}\text{(CH}_3)\text{—CH}_2\text{—CH}_2\sim\end{array} \longrightarrow \begin{array}{l}\sim\text{CH}_2\text{—CH(CH}_3)\text{—CH}_2\text{—CH}_2\sim \\ \sim\text{CH}_2\text{—C(CH}_3)\text{=CH—CH}_2\sim\end{array}$$

5. Radical destruction by combination with peroxide fragments:

$$P\cdot + R\cdot \longrightarrow P\text{—}R$$

6. Intra-chain cyclization, e.g.

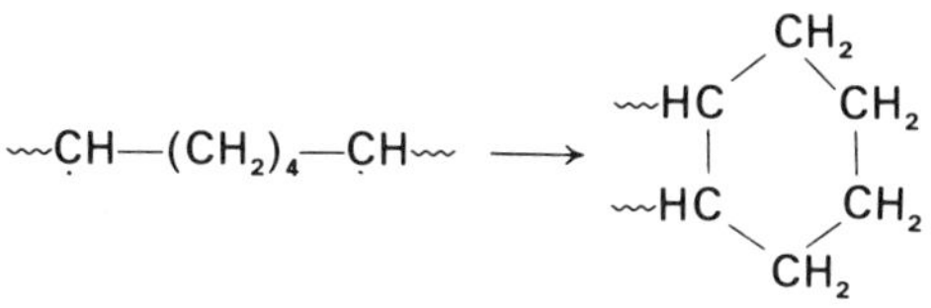

$$\sim\dot{C}H\text{—}(CH_2)_4\text{—}\dot{C}H\sim \longrightarrow \begin{array}{c} \sim HC \quad CH_2 \quad CH_2 \\ | \qquad\qquad | \\ \sim HC \quad CH_2 \quad CH_2 \end{array}$$

Also possible are a number of radical-transfer reactions which, with EPM rubbers, could lead to radicals which may be methyl (a), methylene (b) or methine (c) in character. It is believed that steric factors may hinder coupling, i.e. cross-linking, of methylene and, particularly, methine radicals so that if possible these should be avoided.

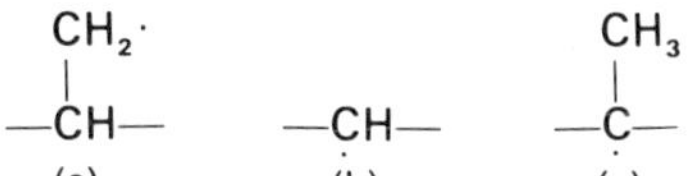

$$\underset{(a)}{\text{—}\overset{\overset{\displaystyle CH_2\cdot}{|}}{C}H\text{—}} \qquad \underset{(b)}{\text{—}\dot{C}H\text{—}} \qquad \underset{(c)}{\text{—}\overset{\overset{\displaystyle CH_3}{|}}{\dot{C}}\text{—}}$$

In general it is of interest to have information on:

1. The yield of cross-links per peroxide molecule which has undergone rupture into radicals (a measure of initiator efficiency);
2. the ratio of cross-link to scission reactions (a measure of cross-link efficiency).

Agents which suppress non-coupling reactions such as chain scission are known as *coagents* or sometimes as promoters or adjuvants. One way in which they function is by involving radical transfer reactions to points on the coagent where coupling reactions are not hindered. The earliest coagent was the ubiquitous sulphur but this has now been largely replaced by such materials as triallyl cyanurate, bismaleimides and acrylic esters. In the case of bismaleimides such as *N,N'*-(*m*-phenylene) dimaleimide, the peroxide appears to initiate a curing reaction by the maleimide.

The scorchiness of compounds containing peroxides has been reduced by the use of radical scavengers such as *N*-nitrosodiphenylamine (Chow & Knight, 1977). These seem to be particularly useful with a peroxide–maleimide system. This is because scorchiness is largely a function of the peroxide initiation stage and can be controlled by controlling the peroxide concentration and by the use of scavengers, while the cure rate is largely controlled by the maleimide concentration. In this way scorch behaviour and cure rate can, to an extent, be made independent of each other.

The technology of peroxide vulcanization has been reviewed by Elliott & Tidd (1974) and updated by Elliott (1979).

18.4 PHENOLIC RESINS

The use of phenolic resins as curing agents for diene rubbers was first investigated over 50 years ago (Hönel, 1936). Systematic study of the phenolic resin vulcanization of natural rubber was carried out at the Dutch research institute, Rubber-Stitching, during World War II (Van der Meer, 1944, 1945) and more recently at the National College of Rubber Technology (Fitch, 1978; Roca, 1983).

The best-known use of phenolic resins as rubber vulcanizing agents is probably that to give butyl rubber vulcanizates of excellent heat resistance used for tyre curing bladders (see Chapter 8). Less well known is their use in adhesive formulations based on diene rubbers.

One virtue of phenolic resins is that suitable grades may be blended in any proportion with diene rubbers to give products of widely varying hardness. Additionally, unlike other rubber reinforcing resins such as cyclized rubber and the so-called high-styrene resins, they are quite fluid at processing temperatures and at this stage may act as processing aids. They are thus useful in producing easy-processing vulcanizates of high hardness but low density. In such applications they are not used primarily as vulcanizing agents, accelerated sulphur systems still being employed, but at elevated temperatures, usually above 120°C, reactions occur leading to scorch-type phenomena. For use in hydrocarbon rubbers the benzene rings in the phenolic resins may be alkyl-substituted to improve compatibility, but this may not be necessary where the resin is to be used in conjunction with a polar rubber.

18.5 NITROSO AND URETHANE SYSTEMS

Nitroso compounds react with diene polymers to produce a variety of cross-links. One such probable reaction is summarized in the equation:

$$2\ \sim CH_2{-}C(CH_3){=}CH{-}CH_2\sim \;+\; ON{-}C_6H_4{-}NO \;\longrightarrow\; \sim CH{=}C(CH_3){-}C({=}NO{-}C_6H_4{-}NO{=})C{-}\ldots \;+\; HON{=}C_6H_4{=}NOH$$

Whilst the nitroso compounds may be added directly to the rubber it is preferable, in order to reduce toxic hazards, to form the nitroso compounds *in situ* either by reduction of a nitro compound or oxidation of an oxime. Whilst not of importance with the major diene rubbers, there has been some use for *p*-quione dioxime as the vulcanizing agent for butyl rubber (see Chapter 8).

Related to the nitroso systems are the urethane cross-link systems developed by the Malaysian Rubber Producers' Research Association and subsequently marketed under the trade name Novor. The principle is to attach a nitrosophenol onto a diene rubber (usually NR) in order to produce a pendant aminophenol group on the rubber. The phenol groups are then linked by means of an isocyanate to give urea cross-links. In practice, for toxicity considerations, it is usual to add an adduct of an isocyanate and a nitroso compound to the rubber. This then reacts *in situ* and dissociation of the adduct is followed by tautomerization and pendant group formation followed by cross-linking with the liberated isocyanate. The process is shown for a di-adduct (p. 372). For better scorch safety the di-adduct may be replaced by a mono-adduct.

It is usual to use the adduct in conjunction with a drying agent based on calcium oxide to prevent hydrolysis of the isocyanate by traces of moisture in the rubber mix, and some ZDMC as a catalyst for the reaction.

Even in the presence of ZDMC, urethane cures are very slow. However it has been found that mixed urethane/sulphur curing systems are synergistic, giving fast cures and higher cross-link densities than are obtained with either system alone and yet retaining most of the advantages of the urethane cross-link systems. ZDMC is too scorchy a catalyst with the mixed systems and may be replaced by TMT.

Thermal dissociation

Tautomerization

Pendant group formation

Cross-linking

The principal advantages of the urethane systems are:

1. A marked increase in reversion resistance and an ability to withstand vulcanization temperatures up to 200°C with some formulations. These characteristics may be of use in curing thick sections, in curing belting where double curing of overlaps may be a problem, in high-temperature molten salt bath vulcanization and injection moulding.
2. Very good fatigue resistance on ageing both at moderate temperature (40–70°C) and at higher temperatures up to 125°C.

A typical 50/50 mixed system would be:

	pphr
NR	100
Adduct (Novor 924)	2·7
TMT	1·0
Sulphur	1·0
TBBS	0·2

BIBLIOGRAPHY

Alliger, G. & Sjothun, I. J. (eds) (1964). *Vulcanization of Elastomers*. Reinhold, New York.

Bateman, L., Moore, C. G., Porter, M. & Saville, B. (1963). Chemistry of vulcanization. In *The Chemistry and Physics of Rubber-like Substances*, ed. L. Bateman. Applied Science, London, Chapter 15.

Brydson, J. A. (1978). *Rubber Chemistry*. Applied Science, London.

Elliott, D. J. & Tidd, B. K. (1974). Developments in curing systems for natural rubber. In *Annual Report of the Progress of Rubber Technology 1973–74*. Institution of the Rubber Industry, London.

Hoffman, W. (1967). *Vulcanization and Vulcanizing Agents* (English translation of publication issued by Farbenfabriken Bayer in 1965). Applied Science, London.

Porter, M. (1968). The chemistry of sulphur vulcanization of natural rubber. In *The Chemistry of Sulphides*, ed. A. V. Tobolsky. Interscience, New York.

Rodger, E. R. (1979). In *Developments in Rubber Technology—1*, ed. A. Whelan & K. S. Lee. Applied Science, London, Chapter 3.

REFERENCES

Anand, R., Blackley, D. C. & Lee, K. S. (1971). Paper presented to the International Rubber Conference, Brighton, UK, May 1977 (preprints produced by the Plastics and Rubber Institute, London).

Baldwin, F. P. & VerStrate, G. (1972). Polyolefin elastomers based on ethylene and propylene. Rubber Reviews for 1972. *Rubber Chem. Technol.*, **45**, 709.

Braden, M. & Fletcher, W. P. (1955). *Trans. Inst. Rubber Ind.*, **31**, 155.
Braden, M., Fletcher, W. P. & McSweeney, G. P. (1954). *Trans. Inst. Rubber Ind.*, **30**, 44.
Brydson, J. A. (1978). *Rubber Chemistry*. Applied Science, London.
Chow, Y. W. & Knight, G. T. (1977). Paper presented to the International Rubber Conf., Brighton, UK, May 1977 (preprints produced by the Plastics and Rubber Institute, London).
Elliott, D. J. (1979). In *Developments in Rubber Technology—1*, ed. A. Whelan & K. S. Lee. Applied Science, London, Chapter 1.
Elliott, D. J. & Tidd, B. K. (1974). In *Annual Report of the Progress of Rubber Technology 1973–74*. Institution of the Rubber Industry, London.
Fitch, A. (1983). Ph.D thesis, National College of Rubber Technology.
Greensmith, H. W., Mullins, L. & Thomas, A. G. (1963). In *The Chemistry and Physics of Rubber-like Substances*, ed. L. Bateman. Applied Science, London, Chapter 10.
Hönel, H. (1936). French Patent 804 552.
Loan, L. D. (1967). *Rubber Chem. Technol.*, **40**, 149.
Moore, C. G. (1965). In *Proceedings of the NRPRA Jubilee Conference, Cambridge, 1964*, ed. L. Mullins. Maclaren, London.
Moore, C. G. & Trego, B. R. (1961). *J. Appl. Polymer Sci.*, **5**, 299.
Roca, R. (1983). PhD thesis, National College of Rubber Technology, London.
Rodger, E. R. (1979). In *Developments in Rubber Technology—1*, ed. A. Whelan & K. S. Lee. Applied Science, London, Chapter 3.
Saville, B. & Watson, A. A. (1967). *Rubber Chem. Technol.*, **40**, 100.
Trivette, C. D., Morita, E. & Maender, O. W. (1977). *Rubber Chem. Technol.*, **50**, 570.
Van der Meer, S. (1944). *Rubber Stichting Communications*, Nos 47, 48.
Van der Meer, S. (1945). *Rubber Chem. Technol.*, **18**, 853.

Chapter 19

Additives II: Antidegradants

19.1 INTRODUCTION

Early users of vulcanized natural rubber soon became aware of the liability of the material to deteriorate under a variety of conditions and in a number of ways. These included the change to a sticky mass on general ageing (often referred to at one time as 'perishing'), the formation of deep cracks in a direction perpendicular to the application of a stress (now associated with ozone attack), deterioration in contact with copper wire (serious when natural rubber was widely used as an electrical insulator) and the surface hardening that could be observed after exposure to light.

That deterioration of rubber could be associated with oxygen absorption was demonstrated over 100 years ago (Hoffman, 1861) whilst phenol appears to have been proposed as an antioxidant in the 1870s. A German patent (Ostwald & Ostwald, 1908) reported that oxidation was inhibited by aromatic amines, but commercial exploitation only became extensive after World War I.

Today it is recognized that most of the degradation encountered with natural rubber and the synthetic diene rubbers is due either to oxygen or to ozone, and this has led to extensive development of both *antioxidants* and *antiozonants*. The non-diene rubbers with few, if any, double bonds are much more resistant to ozone and, generally, to oxygen. It is, however, to be noted that double bonds are by no means necessary for oxidation, as can be demonstrated very clearly by unprotected polypropylene. Whilst some synthetics may be degraded in other ways, for example by hydrolysis, this chapter will concentrate on the two types of additive used to combat the more common causes of degradation, i.e. oxidation and ozone attack.

19.2 AUTOXIDATION OF HYDROCARBON POLYMERS

In sulphur-cured diene vulcanizates containing antioxidants, the oxidation process is extremely complex. It is therefore unwise to rely unduly on incomplete theory when developing new antioxidant systems but preferable to undertake substantial systematic trials. It is nevertheless instructive to consider the oxidation of a pure hydrocarbon compound as this does give some insight into the mode of operation of an antioxidant.

Hydrocarbon oxidation is a chain reaction (autoxidation) with initiation, propagation and termination stages. These may be summarized in the following scheme:

Initiation

$$\underset{\text{hydrocarbon}}{RH} \longrightarrow \underset{\text{radical}}{R\cdot} \tag{19.1}$$

Propagation

$$R\cdot + O_2 \longrightarrow \underset{\text{peroxide radical}}{RO_2\cdot} \tag{19.2a}$$

$$RO_2\cdot + RH \longrightarrow \underset{\text{hydroperoxide}}{ROOH} + R\cdot \tag{19.2b}$$

$$ROOH \longrightarrow RO\cdot + \cdot OH \tag{19.2c}$$

$$2ROOH \longrightarrow RO\cdot + RO_2\cdot + H_2O \tag{19.2d}$$

Termination

$$2RO_2\cdot \longrightarrow \text{non-radical products} \tag{19.3a}$$

$$2R\cdot \longrightarrow R{-}R \tag{19.3b}$$

$$RO_2\cdot + R\cdot \longrightarrow ROOR \tag{19.3c}$$

The initiation reaction (19.1) may be brought about as a result of other reactions arising in a number of ways, for example by mechanical shear, UV radiation and thermal decomposition of weak bonds. Most importantly, they are brought about by reactions (19.2c) and (19.2d), in which the hydroperoxide molecule formed in reaction (19.2b) generates radicals which then in turn react in (19.2b). Thus *each cycle* of the propagation stage also produces hydroperoxides which act as a source of free radicals.

Of the termination reactions, (19.3b) and (19.3c) clearly lead to cross-linking. Variations of reaction (19.3a) could lead to chain scission.

There are three approaches to minimize damage due to oxidation:

1. To use a polymer and/or curing systems that reduce the number of initiation reactions (outside the scope of this chapter);
2. to employ additives which prevent the initiation reaction (19.1) from taking place (preventive antioxidants);
3. to employ additives that interrupt the propagation reactions (chain-breaking antioxidants).

(To these three could also be added the removal of oxygen from the environment, for example by surrounding the rubber in nitrogen gas.)

Preventive antioxidants may be of a number of types and include:

1. Sequestering agents which deactivate harmful metal ions such as those of copper and manganese;
2. UV absorbers;
3. peroxide decomposition agents, such as trisnonylphenyl phosphite, which convert the hydroperoxide into a harmless inert product;
4. radical traps.

Whilst such preventive antioxidants, and the peroxide decomposition agents in particular, are widely used in thermoplastics materials, they are seldom used deliberately in rubbers. Carbon blacks may, however, have such a role in addition to their prime function as reinforcing fillers. The presence of carbon black is not necessarily beneficial as it may also be involved in a number of reactions which reduce the oxidation resistance of the compound.

The chain-breaking antioxidants may be involved in the following reactions:

$$AH + O_2 \longrightarrow A\cdot + \cdot OOH \tag{19.4}$$

$$RO_2\cdot + AH \longrightarrow ROOH + A\cdot \tag{19.5a}$$

$$A\cdot + RH \xrightarrow{O_2} AOOH + RO_2\cdot \tag{19.5b}$$

$$RO_2\cdot + A\cdot \longrightarrow ROOA \tag{19.6a}$$

$$A\cdot + A\cdot \longrightarrow A{-}A \tag{19.6b}$$

In reaction (19.4) the material AH actually acts as a *pro-oxidant* rather than as an antioxidant. If reaction (19.5a) is more likely than reaction (19.2b) and also the rate of reaction (19.5b) is low, AH can function as a retarder of oxidation. By means of the terminating reactions (19.6a) and (19.6b) the material acts as a chain-breaking antioxidant.

A single antioxidant may act in all these ways more or less simultaneously so that whilst in some circumstances the material may appear primarily as an antioxidant, in other circumstances it may show pro-oxidant tendencies. It has been found (Shelton & Cox, 1954*a*,*b*) that if used in natural rubber above a critical concentration, phenyl-β-naphthylamine causes the oxidation rate to increase (see Fig. 19.1).

19.3 CHAIN-BREAKING ANTIOXIDANTS

The requirements of an antioxidant will vary from compound to compound. In terms of antioxidant efficiency alone the requirements may

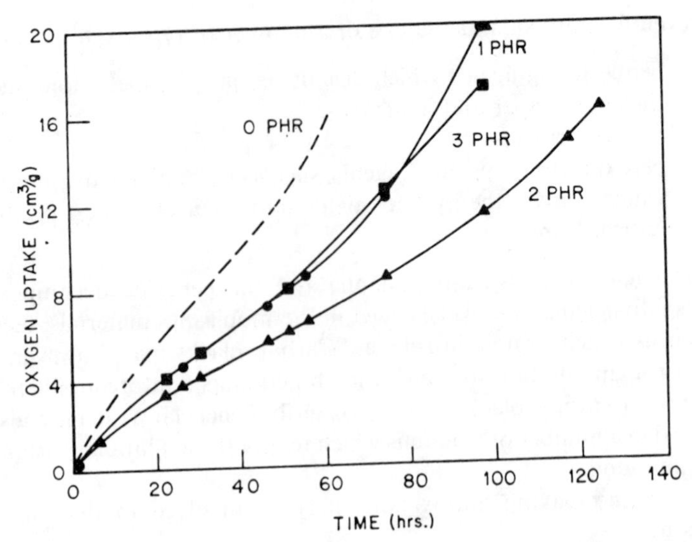

FIG. 19.1. Oxygen absorption of carbon-black-loaded natural rubber compounds containing phenyl-β-naphthylamine (90°C, 760 mm O_2). (Reproduced by permission from *Rubber Chem. Technol.*, **27** (1954) 672. Copyright American Chemical Society.)

vary and an antioxidant which confers good heat resistance will not necessarily confer good flex cracking resistance or metal inhibition (resistance to copper, manganese, etc.). In addition, the importance of the following factors must be considered when selecting an antioxidant system:

1. Tendency to stain or discolour the rubber on ageing;
2. tendency to bloom or bleed;
3. volatility and leaching resistance;
4. toxicity and acceptance for good and pharmaceutical use;
5. physical form;
6. price.

It must also be appreciated that the polymer type and the vulcanization system used will influence the efficiency of the antioxidant. To quote Ashworth & Hill (1979): 'Unless an enormous amount of experimental work is to be carried out, the compounder must rely in the first instance on information provided by the (antioxidant) supplier, although he should always satisfy himself on very specific aspects of his service requirements such as migration, staining or contamination of contact fluids.'

There are two basic classes of antioxidant:

1. Amines—generally more powerful but staining;
2. phenols—non-staining.

19.3.1 Amine Antioxidants

The staining antioxidants, which may all be considered to be derivatives of aniline, dominate the market. This is because for the bulk of applications of rubber, often black, their power as an antioxidant outweighs staining considerations. Toxicity considerations have, however, had an adverse effect on the use of some types.

The amine antioxidants may be subdivided into the following groups:

1. Phenylnaphthylamines;
2. ketone–amine condensates;
3. substituted diphenylamines;
4. substituted *p*-phenylene diamines.

At one time the *phenylnaphthylamines*, particularly phenyl-β-naphthylamine (PBN), were amongst the most important of commercial antioxidants. However, modern analytical techniques revealed that these materials contained small quantities of β-naphthylamine, the carcinogen known to cause papilloma of the bladder. Although grades have been made available with β-naphthylamine levels as low as 1 ppm and the author is not aware of any results of epidemiological studies which link PBN to any higher-than-normal incidence of bladder cancer, the use of these materials has declined. In a similar way aldehyde–amine condensates, particularly aldol–naphthylamines, are no longer of importance.

The *ketone–amines*, on the other hand, remain one of the two major classes of commercial antioxidant conferring very good heat resistance to diene rubbers, in some cases very good flex cracking resistance but little improvement in metal inhibition or ozone resistance. As a result they are widely used in tyres, cables and heavy-duty mechanical goods where heat resistance is an important requirement. Rather similar are the *polymerized quinolines*, which show less staining than other common amine antioxidants.

Also showing low levels of staining are the *alkylated diphenylamines* such as octylated diphenylamine. As antioxidants in natural rubber they are not as powerful as the two previous classes, but they have been found to be particularly effective in polychloroprene compounds.

The disubstituted p-*phenylene diamines* are very important commercially. Diaryl compounds such as di-β-naphthyl-*p*-phenylene diamine confer very good heat resistance and metal inhibition but the specific antioxidant mentioned has gone out of favour because of problems of trace amounts of β-naphthylamine present. The diphenyl compound is not very compatible in NR but finds use in polychloroprenes. Alkyl–aryl derivatives are good antioxidants and, in larger quantities, good antiozonants (Section 19.4). The dialkyls are mainly used as antiozonants.

Because of the variety of roles in which antioxidants are called to perform, they are often used in combination. For example, the dihydroquinoline derivatives (which are in effect acetone–aniline condensates), with their particularly good long-term heat resistance, are widely used in conjunction with the *p*-phenylene diamines in tyre compounding.

Combinations of chain-breaking and preventive antioxidants may also be synergistic although the latter are not widely used in rubbers. Important examples, used in plastics, are trisnonylphenyl phosphite (rather subject to hydrolysis) and dilauryl thiodipropionate, but interest to date in rubber technology appears to lie mainly in SBS thermoplastic rubbers. Genuine synergism also appears to occur when a conventional amine antioxidant is blended with 2-mercaptobenzimidazole (MBI) and is particularly useful against metal-catalysed oxidation as well as improving resistance to heat ageing and flex cracking.

Some antioxidants may be considered as *autosynergistic*. For example, *p*-phenylene diamines not only act as chain stoppers but are also peroxide decomposers, metal deactivators and antiflex cracking agents whilst some are also good antiozonants.

The properties of the main classes of amine antioxidant are summarized in Table 19.1.

19.3.2 Phenolic Antioxidants

For a minority of rubber applications it is important that additives should not discolour or cause staining through migration. It is this requirement which has led to the use of phenolic antioxidants, already widely used in polyolefin thermoplastics, in rubber goods. They also find additional use as stabilizers for raw polymer to improve stability before compounding and in formulations where toxicity is a consideration.

TABLE 19.1
Comparison of amine antioxidants

Class	*Heat resistance*	*Flex-cracking resistance*	*Metal inhibition*	*Ozone activity*	*Staining and discolouration*	*Volatility*	*Solubility*	*Chemical stability*
Aryl naphthylamines								
e.g. PBN (naphthyl—NH—phenyl)	Good	Good	Good	None	Considerable	Low	No bloom	Oxidizes
Ketone–amine condensates								
1. Acetone–diphenylamine condensates	Very good	Good–very good	None	None	Considerable	Moderate	No bloom	Oxidizes
2. Polymerized 1,2-dihydro-2,2,4-trimethylquinoline	Very good	Moderate	None	None	Slight	Low	No bloom	Slight oxidation
Substituted diphenylamines e.g.								
C_8H_{17}—(phenyl)—NH—(phenyl)—C_8H_{17}	Good	Moderate	None	None	Slight	Very low–low	No bloom	Slight oxidation
Substituted *p*-phenylene diamines								
R—NH—(phenyl)—NH—R′								
1. R = R′ = aryl, e.g. β-naphthyl	Very good	Weak	Very good	Some very good in CR	Considerable	Very low–low	Possible bloom	Oxidizes
2. R = R′ = alkyl	Good	Moderate	Very good	Good–very good	Severe	Very low–low	No bloom	Oxidizes
3. R = alkyl; R′ = aryl	Very good	Very good	Very good	Very good	Severe	Very low–moderate	No bloom	Oxidizes

There are two main classes:

1. Simple hindered phenols (also known as mononuclear phenols or as monohydric phenols) of general type

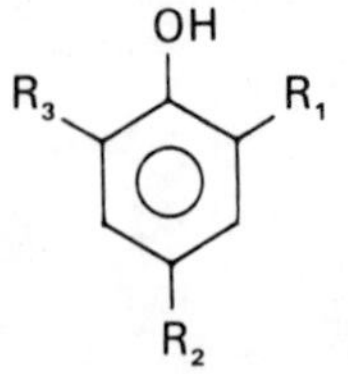

   (the dihydric phenols such as 2,5-di-*t*-amylhydroquinone which discolour and are used mainly in stereorubbers as stabilizers may be considered to be a variant).
2. Bridged hindered phenols (or polynuclear phenols). In turn these may be subdivided into:

   (a) Bisphenols of the phenyl alkane type:

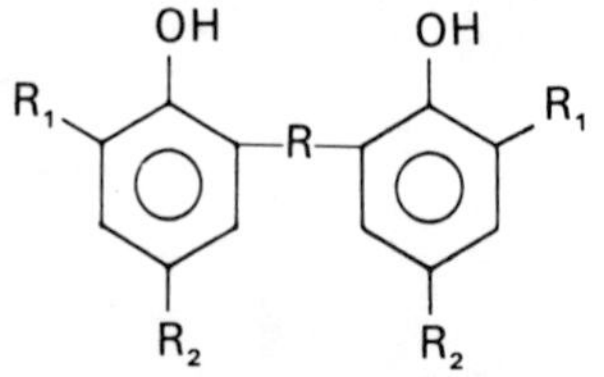

   (b) Bisphenols of the phenolic sulphide type:

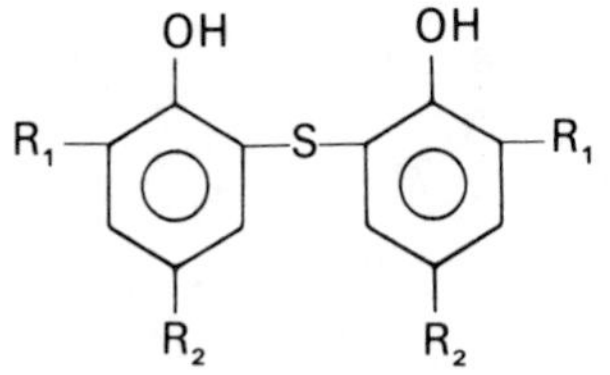

   (c) Polyphenols, e.g.

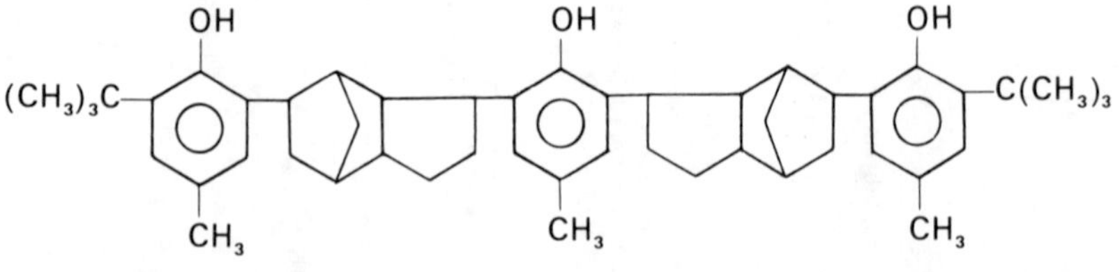

The simple hindered phenols do not show particularly good heat resistance and, for equivalent molecular weight, are more volatile than the amines. They also tend to be easy to extract and this can cause staining of adjacent materials. Some simple phenols can show very good protection against light and flex cracking. The substituent groups are of the electron-releasing type in order to increase antioxidant efficiency. For example, 2,6-di-*t*-butyl-4-methylphenol is 100 times more effective as an antioxidant in petrol (used as a model system) than phenol itself.

These monohydric phenols are widely used for polymer stabilization, for light-coloured goods and in some instances to meet FDA requirements. Most commercial materials are complex mixtures.

By linking phenolic structures, larger molecules are created which are less volatile. At the same time antioxidant activity depends widely on the nature and position of the bridge. In rubber vulcanizates, but not necessarily in other polymers, the best results are usually obtained with an

TABLE 19.2
Comparison of phenolic antioxidants

Class	*Heat resistance*	*Flex-cracking resistance*	*Ozone activity*	*Staining and discoloration*
Simple hindered monohydric phenols e.g. 2,6-di-*t*-butyl-4-methylphenol	Moderate	Poor–moderate	None	Slight
Phenol alkane bisphenols e.g. 2,2-methylene-bis-(4-methyl-6-*t*-butylphenol)	Moderate–very good	Poor–moderate	Negligible	Moderate
Phenol sulphide bisphenols e.g. 4,4′-thiobis-(6-*t*-butyl-3-methylphenol)	Moderate–good	Poor–moderate	Negligible	Moderate
Polyhydric phenols	Moderate–very good	Poor–moderate	Negligible	Slight

Class	*Volatility*	*Solubility*	*Chemical stability*
Simple hindered monohydric phenols e.g. 2,6-di-*t*-butyl-4-methylphenol	Low–moderate	No bloom	Stable
Phenol alkane bisphenols e.g. 2,2-methylene-bis-(4-methyl-6-*t*-butylphenol)	Very low–low	No bloom	Slight oxidation
Phenol sulphide bisphenols e.g. 4,4′-thiobis-(6-*t*-butyl-3-methylphenol)	Very low–low	No bloom	Slight oxidation
Polyhydric phenols	Very low	No bloom	Slight oxidation

ortho bridge. In the case of phenol alkanes, highest activity is obtained with a methane link. Higher linkages such as those of ethane, propane and so on will reduce the tendency to impart pinking discolouration to the compound but also reduce activity. The hindering groups (R_1) may be methyl but for peak activity *t*-butyl and α-methylcyclohexyl are widely used.

The phenolic sulphides are not particularly powerful and do cause some discolouration. In plastics technology they have been found to improve in efficiency in the presence of small amounts of carbon black. In rubbers their main use appears to be in latex compounding.

The higher-molecular-weight polynuclear phenols have similar antioxidant activity to the bisphenols but generally lower volatility.

The properties of phenolic antioxidants are summarized in Table 19.2.

19.4 ANTIOZONANTS

As mentioned earlier in this chapter, deterioration of rubbery materials upon oxidative degradation can take many forms. In particular, photooxidation can cause a number of surface effects variously described as sun checking, sun crazing, light-oxidized cracking, alligatoring, crazing and mud cracking. Whilst as long ago as 1885 Thompson had reported that ozone generated in the laboratory could crack stretched vulcanized rubber it was only as a result of much later work, largely by Newton (1945), that it was appreciated that a combination of ozone and mechanical stress could induce deterioration which had some of the characteristics of photooxidative degradation.

Ozone is produced in the upper atmosphere some 12–22 miles above the earth's surface by photolysis of oxygen. At this level it exists in concentrations of the order of 500 parts per 100 million (pphm). Just above the earth's surface, where organic materials tend to react with and destroy the ozone, normal levels are of the order of 0–10 pphm. In polluted areas, for example in Los Angeles where photochemical smogs occur, the level may be of the order of 40–100 pphm. Concentrations above the normal terrestrial level may also occur near electrically-operated machinery and over the sea where there is an absence of organic material to react with the ozone.

The most common characteristic of atmospheric ozone attack is the formation of cracks perpendicular to the direction of stress in a strained piece of rubber. However, where the sample is stressed simultaneously in two directions, as can happen in a tyre sidewall, small square crack patterns may occur which could be confused with photo-oxidative degradation.

The use of waxes to protect against ozone attack appears to be over a century old (Kreusler & Budde, 1881) although it was not probably realized at the time that the protection afforded was against ozone. It was, however, eventually realized that waxes, on their own, failed to protect against ozone attack under dynamic service conditions, and this led to the search for materials which could function more effectively. By 1939 Tuley found that certain amines had some influence but 15 more years passed before the effectiveness of certain substituted *p*-phenylene diamines was established (Shaw *et al.*, 1954).

Extensive studies on substituted *p*-phenylene diamines as antiozonants led to the following general observations (Ambelang *et al.*, 1963).

1. Ring-substituted *p*-phenylene diamines (ppds) were less effective than unsubstituted ppds.
2. An increase in the number of rings between —NH— groups reduced the effectiveness.
3. The effectiveness as an antiozonant decreases in the sequence *para* > *ortho* > *meta*.
4. In the case of alkyl-substituted compounds, the secondary alkyls appear the most effective.

Whilst many materials of the *p*-phenylene diamine type were found to exhibit antiozonant behaviour, problems of toxicity, excessive volatility and serious interference with cure have limited the choice in practice. As it is, all the important commercial antiozonants exhibit staining and discolouration.

The substituted *p*-phenylene diamines fall into three classes:

1. Diaryl ppds;
2. dialkyl or cycloalkyl ppds;
3. aryl–alkyl ppds.

The diaryl ppds show either poor compatibility or poor antiozonant activity, or both, in diene hydrocarbon rubbers. Some have, however, found use in polychloroprenes such as *N,N'*-ditolyl-*p*-phenylene diamine (DTPD). This particular antiozonant has also been claimed to confer good long-term antiozonant properties although its initial efficiency is modest.

The dialkyl ppds confer the best ozone resistance to diene hydrocarbon rubbers *under static conditions* and are typified by DCPD and B(DMP)PD.

Under dynamic conditions the alkyl–aryl compounds such as IPPD and

6PPD are clearly superior. Furthermore, their behaviour under static conditions may be measurably improved by their use in conjunction with waxes. Additionally these compounds are often easier to handle than the dialkyl materials so that the alkyl–aryl ppds tend to dominate the market.

Brief mention should also be made of 6-ethoxy-2,2,4-trimethyl-1,2-dihydroquinoline, which is the only commercial material effective in both static and dynamic conditions that is not a substituted *p*-phenylene diamine.

Data on some commercial antiozonants is given in Table 19.3 (after Ashworth & Hill, 1979).

19.4.1 Prevention of Ozone Attack

The severity of ozone attack on a rubber compound does not simply depend on the type and level of antiozonant used. Obviously the choice of rubber will be very significant, but also important are the conditions under which the product will be used.

As has already been stated, exposure of a diene rubber vulcanizate to atmospheric ozone will cause cracks to occur in a direction perpendicular to the applied stress. It has been observed that the number of cracks increases as the extension increases. If many cracks are formed they tend to interfere with each other and reduce stress concentrations. On the other hand, if there are few cracks catastrophic failure can occur. Thus the problem is often greatest at low strains. It has also been noted that under a strain regime that would cause many cracks in NR compounds only a few cracks will occur in SBR vulcanizates.

Braden & Gent (1960*a*,*b*, 1962) considered ozone cracking as having two stages:

1. Crack initiation;
2. crack growth.

They noted that there was a threshold stress below which cracking did not occur and that this varied with the square root of the modulus of the elastomer. The rate of crack initiation, in NR, reaches a maximum at about 70% strain.

It was also found that there was a critical stress at which crack growth reached a maximum, equivalent to a strain of the order of 3–5%, and that this was inversely proportional to the cross-link density. Rate of crack growth also tends to rise with temperature and plasticizer content.

From the above comments it will be recognized that such factors as polymer type, cross-link density, vulcanizate modulus and levels of stress and strain in service are all important to the successful use of a rubber in the presence of atmospheric ozone.

The mechanism by which an antiozonant operates is not fully understood. Because ozone effects are surface-based, the antiozonant should clearly function at the surface. Waxes may be assumed to do this simply by blooming to the surface and producing a barrier between ozone and rubber. Not surprisingly, surface flexing will cause cracks to occur in the barrier and thus under dynamic conditions waxes are not very effective alone.

According to the *scavenger theory*, antiozonants are effective because they diffuse to the surface and scavenge ozone owing to their greater reactivity to ozone than the rubber. Analysis of kinetic studies does not, however, support this theory, nor does the observation that a number of chemicals very reactive to ozone, such as dilauryl selenide, are not effective antiozonants. More in accord with experimental observations is the *relaxation theory* which states that the antiozonant reacts with an already ozonized rubber surface in order to produce a relaxed and strain-free surface. One consequence is that the agent will take some time to work, and great care should be taken in assessing short-term accelerated tests using high ozone concentrations.

19.4.2 Use of Waxes and Saturated Polymers for Ozone Protection

As mentioned earlier, waxes have been used for over a century, much of the time unwittingly, to protect diene rubbers from ozone. This is brought about by the diffusion of the crystalline wax, which has limited compatibility with the polymer, onto the surface to form a protective layer not susceptible to ozone attack and effectively impervious to ozone.

Pure paraffin wax is not satisfactory as it blooms too rapidly and flakes off the rubber surface. On the other hand, many microcrystalline waxes bloom too slowly and take unduly long to become effective, so that today combinations of waxes are frequently used.

There is some dispute as to what constitutes the best wax. Van Pul (1958) suggests the wax should have a melting point of 65–72°C and 30–50% side chain branching. On the other hand, Winkelman (1952) has argued that a linear chain wax is preferable to a branched one and that a blend of a low-melting microcrystalline wax with a high-melting paraffin wax is desirable. What is clear is that accelerated testing at temperatures above the wax melting point, or even more generally at elevated temperatures, can give misleading results.

TABLE 19.3
Some commercially available antiozonant structures
(From Ashworth & Hill, 1979)

	Melting point (°C)	Mooney scorch at 120°C (min)[a]	NR protection		SBR protection		De Mattia flexing (kc)[b]
			Static	*Dynamic*	*Static*	*Dynamic*	
Alkyl/aryl-substituted *p*-phenylenediamines							
N-Isopropyl-*N'*-phenyl-*p*-phenylenediamine (IPPD)[c]							
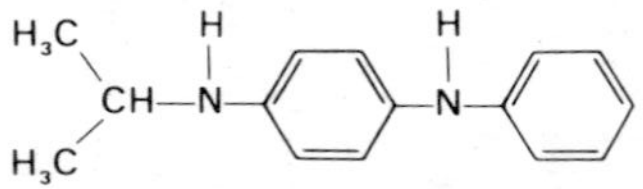	75	23	100	100	100	100	430
N-1,3-Dimethylbutyl-*N'*-phenyl-*p*-phenylenediamine (6PPD)							
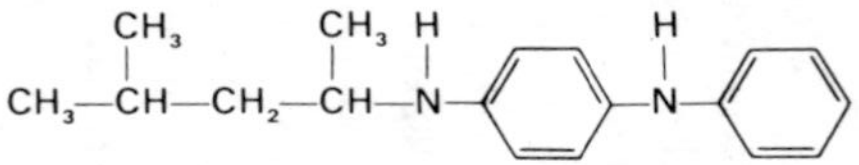	45	26	100	90	85	80	300
N-Cyclohexyl-*N'*-phenyl-*p*-phenylenediamine (CPPD)							
	118	24	130	80	85	80	300
Dialkyl-substituted *p*-phenylenediamines							
N,N'-Bis(1,4-dimethylpentyl)-*p*-phenylenediamine [B(DMP)PD]							
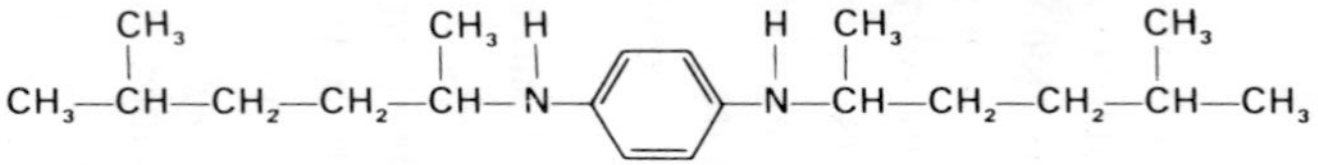	Liquid	19	160	65	130	75	200

Diaryl *p*-phenylenediamines							
N,N'-Dicyclohexyl-*p*-phenylenediamine (DCPD)							
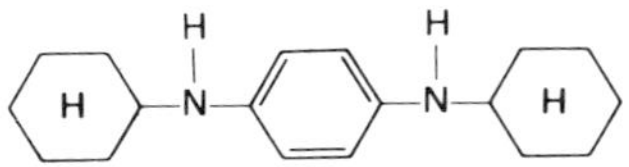	104	15	150	65	125	70	225
N,N'-Ditolyl-*p*-phenylenediamine (DTPD)							
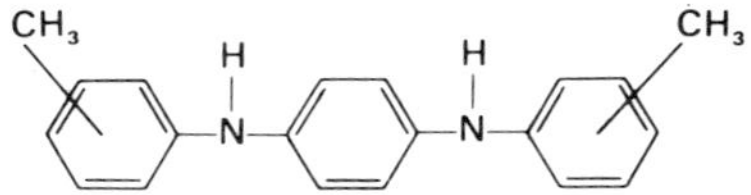	95	25	50	70	50	40	280
Others							
6-Ethoxy-2,2,4-trimethyl-1,2-dihydroquinoline (ETQ)							
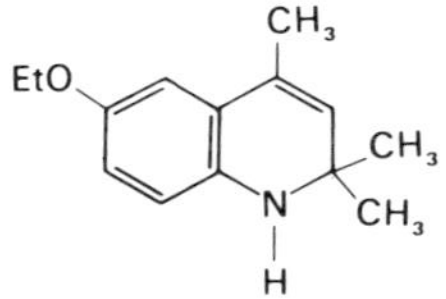	Liquid	21	45	50	50	60	230

[a] Mooney scorch at 120°C—the times quoted are those observed for a rise of 10 Mooney units above the minimum value. The result was obtained in an NR tyre treadstock with 2% of the antiozonant. The value obtained with the blank was 27 min.
[b] De Mattia flexing—the results quoted were obtained in an NR tyre treadstock with 2% of the antiozonant. A 3-in free sample length was used, and the results quoted are the number of kilocycles required to produce deep cracks. The result obtained with the blank was 87 kc.
[c] IPPD was taken as standard and given a rating of 100.

The level of wax used is critical. If insufficient it may allow a few cracks to develop and this could be more serious than the large number of cracks without protection. On the other hand, an excess will lead to flaking.

Waxes may be used on their own, for example when antiozonants cannot be used for reasons of staining or toxicity. They may also be used beneficially in conjunction with an antiozonant, typical levels being in the range 1–5%.

Saturated polymers are also sometimes used to protect a diene rubber by physical means. Examples include PVC in nitrile rubber and EPDM in IIR. Substantially higher quantities are however required.

BIBLIOGRAPHY

Ashworth, B. T. & Hill, P. (1979). In *Developments in Rubber Technology—1*, ed. A. Whelan & K. S. Lee. Applied Science, London, Chapter 7.
Hawkins, W. L. (ed.) (1972). *Polymer Stabilization*. John Wiley, New York.
Leyland, B. N. (1972). *Prog. Rubber Technol.*, **36**, 19.
Parks, C. R. & Spacht, R. B. (1977). *Elastomers*, **1977**(5), 25–34.
Scott, G. (1965). *Atmospheric Oxidation and Antioxidants*. Elsevier, Amsterdam.

REFERENCES

Ambelang, J. C., Kline, R. H., Lorenz, O., Parks, C. R. & Wadelin, C. (1963). *Rubber Chem. Technol.*, **36**, 1497.
Ashworth, B. T. & Hill, P. (1979). In *Developments in Rubber Technology—1*, ed. A. Whelan & K. S. Lee. Applied Science, London, Chapter 7.
Braden, M. & Gent, A. N. (1960*a*). *J. Appl. Polym. Sci.*, **3**, 90.
Braden, M. & Gent, A. N. (1960*b*). *J. Appl. Polym. Sci.*, **3**, 100.
Braden, M. & Gent, A. N. (1962). *J. Appl. Polym. Sci.*, **6**, 449.
Hoffman, A. W. (1861). *J. Chem. Soc.*, **13**, 87.
Kreusler & Budde (1881). German Patent 18 740.
Newton, R. G. (1945). *J. Rubber Res. Inst. Malaya*, **14**, 27, 41.
Ostwald, W. & Ostwald, W. (1908). German Patent 221 310.
Shaw, R. F., Ossefort, Z. T. & Touhey, W. J. (1954). *Rubber World*, **130**, 636.
Shelton, J. R. & Cox, W. L. (1954*a*). *Ind. Eng. Chem.*, **46**, 816.
Shelton, J. R. & Cox, W. L. (1954*b*). *Rubber Chem. Technol.*, **27**, 671.
Thompson, W. (1885). *J. Soc. Chem. Ind.*, **4**, 710.
Tuley, W. F. (1939). *Ind. Eng. Chem.*, **31**, 714.
Van Pul, B. I. C. F. (1958). *Rubber Chem. Technol.*, **31**, 866.
Winkelman, H. A. (1952). *Ind. Eng. Chem.*, **44**, 841.

Chapter 20

Additives III: Plasticizers and Fillers

20.1 INTRODUCTION

In some respects the roles of plasticizers and fillers are contrary, in others they are similar. Thus the softening effect of plasticizers in vulcanizates contrasts with the stiffening effect of fillers. On the other hand, both plasticizers and fillers may reduce cost and can aid processing. Intermediate between fillers and plasticizers are some resins which are sufficiently fluid at processing temperature to reduce viscosity and thus aid flow, but then either cross-link at the same time as the rubber is vulcanized or simply harden on cooling to give a stiffer vulcanizate than would have been obtained in their absence.

20.2 PLASTICIZERS

In this chapter the term *plasticizer* is being used in an all-embracing way to cover a range of products. They have the common feature of pushing apart the rubber molecules and hence leading to a reduction in the glass transition temperature, melt viscosity of the uncured compound and vulcanizate hardness. In order to function successfully there may be some interaction between plasticizer and polymer, as discussed in the next section.

The rubber industry also frequently uses terms such as softener, processing aid and extender, and it is appropriate to say a few words about each. The term *softener* has been widely used for hydrocarbon oils employed in hydrocarbon rubbers, whilst at one time the term plasticizer was more confined to polar materials in polar polymers (a usage previously established with thermoplastics such as PVC). If a distinction was required between these two terms as then used, it could be argued that there was little interaction chemically between softeners and rubbers whilst with plasticizers and polar polymers the interaction was significant. Such a distinction is rather artificial and in recent publications there has been a tendency to drop the term softener: this approach is adopted in this chapter. Notwithstanding this convention it will be appreciated that plasticizers have an important softening effect in vulcanizates.

A *processing aid* is, by definition, an additive that aids processing. It may do this by reducing the compound viscosity or by making the raw compound less elastic or nervy and hence easier to extrude, mould or calender. Since fillers as well as a variety of other additives may also be effective in the latter role, the term 'processing aid' should be used with care. A further complication is that some authorities use the term to embrace tackifiers and colourant dispersion aids.

The term *extender* has many meanings in polymer technology and also must be used carefully. In PVC technology it is used for a material which is insufficiently compatible with the polymer to act as a true plasticizer but which is capable of diluting or extending plasticizer already present without making the whole plasticizer–extender blend incompatible with the polymer. In adhesives technology it is often used in lieu of the term filler for such additives as blood albumen and wood and nut flours. In rubber technology the term is most commonly used to describe compatible hydrocarbon oils employed in substantial quantities as a cost-reducing 'liquid filler', as for example in oil-extended SBR.

20.2.1 Plasticizer Action

It is not proposed to review here the various theories of plasticizer action. For practical purposes the author has found it convenient to consider plasticizers as non-volatile solvents for a polymer with the possibility, particularly with polar polymers and plasticizers, of some chemical interaction between the two. In practice a plasticizer needs to meet the following requirements.

1. The solubility parameter should be within 2 $MPa^{1/2}$ of that of the polymer for adequate compatibility. (Solubility parameter data on polymers, solvents and plasticizers are given in Chapter 3.)
2. The molecular weight of the plasticizer should exceed 300 to reduce volatility.
3. There should be some polymer–plasticizer interaction, particularly if either polymer or plasticizer have any tendency to crystallize.

Whilst some writers have described the interaction between polar polymers and plasticizers as due to dipole attraction, Small (1953) produced strong evidence, in the case of PVC at least, that the interaction was a form of weak hydrogen bonding. Small argued that PVC acted as a weak proton donor, using the hydrogen atom opposite the chlorine. On the other hand, esters and ethers acted as proton acceptors. Phosphates were stronger acceptors than carboxylic esters which in turn were more powerful than ethers, whilst ring structures were more effective than aliphatic plasticizers. This difference in bonding interaction largely accounted for the difference between aromatic phosphates, phthalates and aliphatic esters in PVC and NBR.

20.2.2 Petroleum Oil Plasticizers

There are few published data on the solubility parameters of petroleum oils, the data in Chapter 3 being estimated by the author using Small's method (Small, 1953). In general, more sensitive methods are required to discriminate between the materials.

The petroleum oil plasticizers are complex mixtures with individual molecules containing various ratios of unsaturated aromatic rings, saturated rings (known as naphthenic rings) and paraffinic side chains. The oils are then classified as being aromatic, naphthenic or paraffinic according to the ratios of the three types of structure present, schematically illustrated in Fig. 20.1. It should, however, be made clear that an oil

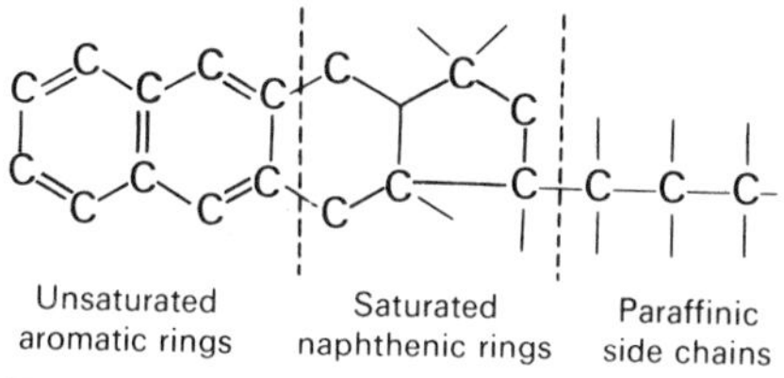

Unsaturated aromatic rings | Saturated naphthenic rings | Paraffinic side chains

FIG. 20.1. Typical molecular structures present in petroleum oils.

designated as aromatic will not be totally aromatic, with similar comments for the so-called paraffinic and naphthenic oils. This can be seen clearly from the data in Table 20.1.

It has been found possible to carry out a carbon-type analysis using a correlation relating the carbon-type distribution to the viscosity–gravity constant and the refractivity intercept. The viscosity–gravity constant (VGC) is a measure of aromaticity calculated from the equation

$$\mathrm{VGC} = \frac{10G - 1{\cdot}0752\log(V - 38)}{10 - \log(V - 38)}$$

where G is the specific gravity at 15°C and V is the Saybolt viscosity at 38°C.

TABLE 20.1
Typical properties of petroleum oils for rubber
(After Morris, 1979)

	Paraffinic	*Naphthenic*	*Aromatic*
Viscosity, cSt			
At 40°C	19·7	110·2	763·5
At 100°C	4·0	8·0	17·0
Specific gravity at 15°C	0·861	0·932	1·018
VGC	0·809	0·885	0·980
Refractive index	1·475 1	1·516 7	1·580 4
Refractivity intercept	1·045 7	1·050 3	1·072 1
Carbon-type analysis, %			
C_A	3·5	21	45
C_N	31·0	37	18
C_P	65·5	42	37
Molecular-type analysis, %			
Asphaltenes	0	0	0
Polar compounds	0·4	2·8	7·8
Aromatics	12·1	42·8	80·0
Saturates	87·5	54·4	12·2
Aniline point, °C	96·0	75·0	38·2

The refractivity intercept is given by

$$\text{Refractivity intercept} = N_D^{20} - 0{\cdot}5d^{20}$$

where N_D^{20} is the refractive index at 20°C for the sodium D line and d^{20} is the specific gravity at 20°C.

The relationship between these two experimentally determinable quantities and the carbon-type ratios is shown in the triangular graph (Fig. 20.2).

Commercial oils also contain non-hydrocarbon materials with molecules containing nitrogen, sulphur or oxygen. These may affect polymer degradation and vulcanization rates and tend to be highest in the aromatic oils.

When selecting an oil there are five main factors to take into account:

1. The viscosity–gravity constant—a measure of aromaticity;
2. the oil viscosity—a function of both structure and molecular weight;
3. presence of heterocyclics and other materials containing nitrogen, oxygen or sulphur;
4. oil loading;
5. price.

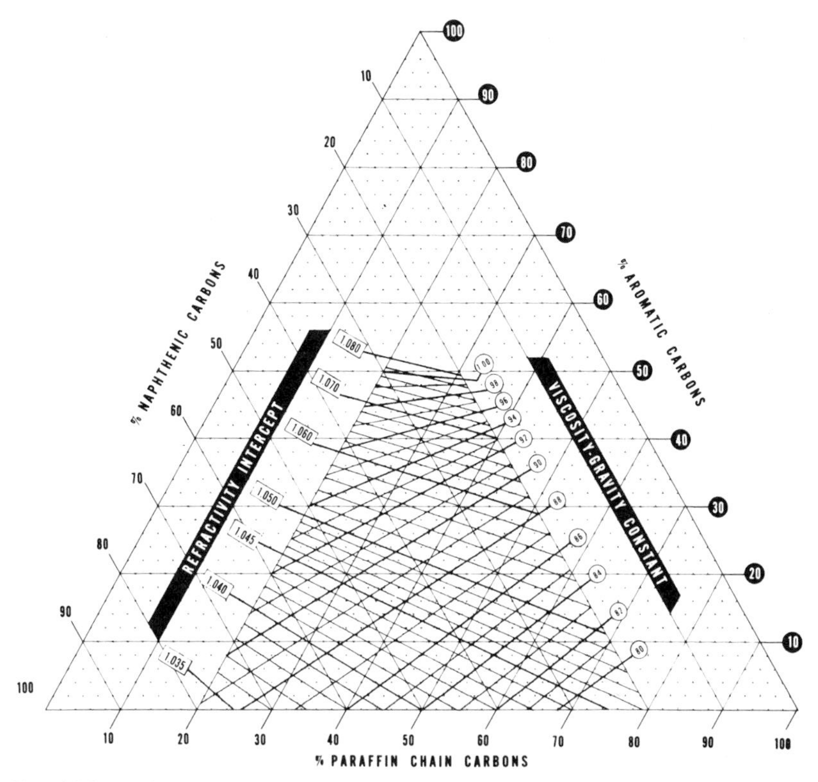

FIG. 20.2. Triangular graph used for relating VGC and refractivity intercept to carbon type in petroleum oils. (From Morris, 1979.)

The first four of these factors will affect both processing and vulcanizate properties.

As might be expected, the more paraffinic oils with the low VGC values tend to be more compatible with the more 'paraffinic' rubbers such as EPDM and IIR. On the other hand, the more polar rubbers such as CR and NBR are more compatible with the aromatic oils. Natural rubber and SBR, holding an intermediate position, are compatible with a broad range of oils.

The effect of oils on processability is complex since this will not only involve a reduction in compound viscosity but also extrudate smoothness, power consumption during compounding, and shaping and filler dispersion. An increase in the plasticizer level will reduce stock viscosity and power consumption.

A high level of shearing stresses is required to disperse fillers, particularly the finer carbon blacks. Thus the addition of oil before the black will lead to a soft compound in which filler dispersion will be difficult. On the other hand, addition of the black in total before the oil may cause other problems such as undue power consumption and heat build-up. Addition of portions of oil and black alternately is usually the best approach. Aromatic oils give better dispersion of black fillers, this probably a combination of their higher viscosity and their greater interaction and compatibility with carbon black. Aromatic oils should, however, be used with care because heterocyclic impurities inevitably present may affect the rate of cure.

The effects of viscosity–gravity constant on vulcanizate properties have been described by Morris (1979) with specific reference to SBR. As a rule, the higher the VGC of the oil,

1. the higher the tensile strength of the vulcanizate;
2. the higher the tear strength;
3. the higher the heat build-up and the lower the resilience in dynamic applications;
4. the greater the tendency to crack growth.

Hardness, modulus and elongation at break are only slightly affected.

The oil viscosity has its greatest effect on low-temperature properties. At a given temperature the vulcanizate properties will depend on the viscosity of the plasticizing oil at that temperature and this will depend on both the type of oil and its molecular weight. The paraffinic oils not only have lower viscosities than corresponding aromatics at normal ambient temperatures but the viscosities are also less temperature-sensitive. For these reasons low-molecular-weight paraffinic oils are preferred for low-temperature applications.

At the other end of the temperature scale the aromatics also have a low flash point and hence a high volatility. Thus for high-temperature work paraffinic oils are also usually preferred, but in this case oils of higher molecular weight to reduce volatility.

The presence of heterocyclic materials, more prevalent in the high-aromatic grades, adversely affects oxidation stability. For this reason paraffinic, and to some extent naphthenic, oils are preferred when good heat and/or light stability are required. As has already been mentioned, the presence of heterocyclics can accelerate cure in diene rubbers and can also cause problems in the peroxide curing of EPM.

Although most grades of OESBR used 37·5 pphr oil, further oil may be added by the compounder, often alongside higher loadings of carbon black fillers. In the case of EPDM some grades are able to accept up to 200 pphr oil, usually of the high-molecular-weight paraffinic type. As pointed out in Chapter 5, oil extension was largely introduced to reduce the price, a mixture of oil and a higher-molecular-weight rubber giving a cheaper material than a conventional unextended elastomer.

20.2.3 Ester and Ether Plasticizers

The ester plasticizers were developed primarily for PVC but have also proved to be useful in a number of rubbers with similar solubility parameters, NBR in particular. Five main classes may be recognized.

1. Phosphates.
2. Phthalates.
3. Aliphatic esters.
4. Polyesters.
5. Polymerizable esters.

As a class the *phosphates* have the greatest interaction with NBR. They thus do not lower the cold flex temperature as much as phthalate and aliphatic esters. Today their main interest is in their fire-retarding properties. At one time tritolyl phosphate and trixylyl phosphate predominated, but the petroleum-based tri-isopropylphenol phosphates are today more widely used.

A wide range of *phthalates* has been developed over the years, being typified by branched alkyl phthalates with some 7–9 carbon atoms in the side chain such as di-2-ethylhexyl phthalate and di-iso-octyl phthalate and, importantly, esters from mixed alcohols with either 7–9 carbon atoms or 8–10 carbon atoms. The phthalates show good all-round properties such as a better reduction in cold flex temperature than with a phosphate.

For optimum low-temperature properties, in effect to have the greatest depressing effect on the T_g, *aliphatic esters* are used. At one time they were typified by materials such as dioctyl sebacate and dioctyl adipate; it is now more common to use the cheaper esters made by reacting unseparated mixtures of acids and alcohols of appropriate average length.

Where low volatility and extractability are important, low-molecular-weight *polyesters* such as polypropylene adipates may be used.

A number of ester plasticizers are available that are also polymerizable. These include ethylene glycol dimethacrylate, 1,3-butylene glycol dimethacrylate and trimethylolpropane trimethacrylate. In the presence of peroxides these polyfunctional materials are capable of cross-linking. They are best known as coagents during peroxide cures where they become involved in the vulcanizing reaction. They can act as processing aids during processing but after cure contribute to an increase in hardness and stiffness.

Thioether and thioester plasticizers also find use in CR and NBR materials; these include ether thioethers of undisclosed composition (e.g. Vulkanol OT—Bayer), dibutyl methylene bisthioglycollate and di-2-ethylhexyl ester thiodiglycollate.

Epoxidized oils such as epoxidized soya bean oil may also be used in NBR materials on account of their low volatility.

20.3 PARTICULATE FILLERS

Fillers in the form of small particles have been used in rubber compounds since the early days of the rubber industry. Originally their primary function was to reduce cost but it was soon appreciated that the fillers used such as chalk and china clay were also useful processing aids in that they reduced die swell and gave smooth extruded and calendered products. In 1904 Mote, Mathews and others working in Silvertown, London, used carbon black in rubber and this led to the discovery that certain fine-particle-size fillers had a *reinforcing effect* in that properties such as modulus, abrasion resistance, tear resistance and tensile strength were enhanced. Today only a few rubber compounds, the so-called gum stocks, are prepared without substantial quantities of filler being present.

Particulate fillers are often subdivided into two groups—the inert fillers and the reinforcing fillers—but this is an oversimplification. The *inert fillers*, typified by untreated chalk (whiting), clays and barytes, do not have a marked reinforcing effect although the finer-particle-size grades may give some marginal reinforcement, particularly at low loadings. They do, however, increase hardness and act as processing aids, as mentioned above. Additionally they may help to control density, and influence electrical properties and radiation absorption.

The *reinforcing fillers*, dominated by carbon black, are of major importance to the rubber industry. White reinforcing fillers such as silicas and silicates are also available for more specialized uses.

Surface treatment of inert fillers has led to products with intermediate levels of reinforcement. Particularly good results can sometimes be achieved by the use of *coupling agents* that chemically link the rubber molecule to the filler particle. Another approach is to treat the filler particle with a selected polymer.

The principal factors that affect filler behaviour in a rubber are:

1. The surface nature of the filler particle which may be due either to the bulk nature of the particle or to a coating such as a coupling agent applied to the particle;
2. the particle size and size distribution;
3. the particle shape;
4. the structure of agglomerates of particles.

The major classes of filler will be discussed in the following sections.

20.3.1 Carbon Black

Carbon black may be formed by the incomplete combustion of many organic substances, be they solid, liquid or gaseous, but today it is produced mainly from petroleum oil or natural gas. The original use of carbon black, as a pigment, appears lost in the mists of time, one source (Donnet & Voet, 1976) mentioning its use by the Chinese and the Hindus in the third century AD whilst more recently Medalia *et al.* (1979) suggest that it was used by the Chinese over 5000 years ago. Of the more modern processes, that for channel black dates from 1872 and for furnace black from 1922.

World capacity for carbon black production is of the order of four million tpa; that is, in broad terms, about half of annual world rubber production. Since about 95% of the carbon black produced is used in the rubber industry (about 75% in tyre products alone) and taking into account the difference between output and capacity, this indicates that on average about 35–40 parts of carbon black are used for every 100 parts of rubber.

A variety of processes are available to produce carbon blacks. The most important types are

1. furnace blacks;
2. thermal blacks;
3. channel blacks;
4. acetylene blacks;
5. lamp black.

The *furnace blacks* comprise around 95% of the market. In the older gas-furnace process, a diffusion flame is created by burning part of the gas with air in a firebrick-lined furnace. Because of the insufficiency of the air the rest of the gas thermally decomposes in the flame at temperatures in the range 1260–1420°C. The quality and yield is controlled by the temperature, which is in turn regulated by the gas/air ratio (typically 1:5) and by turbulent conditions in the gas mixture. The gas-furnace process is used to produce coarser grades of black with particle diameters greater than 50 nm (1 nm = 10^{-9} m).

In the oil-furnace process, used for the finer grades, liquid fuel is injected into the flame, which itself may be produced either from the liquid fuel or from natural gas. The black formed in each process is swept out of the furnace by the gas at a high predetermined velocity and quenched by water spray before being collected by a series of electrostatic precipitators and cyclone collectors, prior to pelletizing and bagging.

The *thermal blacks*, which are coarser than the furnace blacks, take about 4% of the market and are produced by a cyclic process. In this process a stoichiometric mixture of gas and air is burnt in a furnace, raising the temperature to about 1300°C. The air supply is then switched off and the incoming gas decomposes into carbon and hydrogen. After about 4 min the furnace temperature has dropped to a point where the air is switched on again and the reheating, and indeed the whole cycle, are repeated.

Channel blacks were at one time very important, being obtained by impinging burnt natural gas on water-cooled iron channels. Although very fine particles were obtained by this process, yields were as low as 1–5% and because of environmental pollution problems the process is virtually obsolete.

Acetylene black is a form of thermal black produced using acetylene as a feedstock. Heat is used to start the process but since acetylene decomposition is exothermic the process proceeds at about 800–1000°C without the need for intermittent cycling.

Lamp black, used primarily as a pigment, is made by burning aromatic oils in shallow open pans in a limited air supply.

20.3.1.1 Classification of Carbon Blacks

Over the years the various types of black were subdivided largely according to particle size and given such unsystematic names as intermediate super abrasion furnace black (ISAF), fast extrusion furnace (FEF), easy processing channel (EPC), medium thermal (MT) and even intermediate intermediate super abrasion furnace (IISAF).

This led to the emergence of a variety of more logical, yet still imperfect, classification systems (reviewed anonymously in the July/August 1968 edition of *The Rubber Journal*) with the ASTM system (ASTM D-1765) becoming widely accepted although the older non-systematic names continue to be used.

The ASTM system is based on four digits, a letter followed by three numbers. The letter indicates cure rate, either S for slow-curing acidic

TABLE 20.2
Second-digit classification in ASTM system[a]

Second digit	*Particle diameter (nm)*	*Old code*	*Type of black*
0	1–10	—	—
1	11–19	SAF	Super abrasion furnace
2	20–25	ISAF	Intermediate super abrasion furnace
3	26–30	HAF	High abrasion furnace
		EPC, MPC	Easy and medium processing channel
4	31–39	FF	Fine furnace
5	40–48	FEF	Fast extrusion furnace
6	49–60	GPF	General-purpose furnace
		HMF	High modulus furnace
7	61–100	SRF	Semi-reinforcing furnace
8	101–200	FT	Fine thermal
9	201–500	MT	Medium thermal

[a] The last two digits are arbitrarily chosen and assigned by an ASTM committee. The selection may be tentative or definitive.

channel blacks or N for normal-curing neutral or basic blacks. The second digit (i.e. the first number) is indicative of the particle size, as indicated in Table 20.2. The third and fourth digits are arbitrarily assigned although for 'standard' structure levels the third digit is the same as the second, as in N330 but not N347. With higher-structure blacks becoming more common this begins to lose some meaning. Table 20.3 provides a comparison of ASTM and traditional nomenclature.

20.3.1.2 The Nature of Carbon Black

Carbon blacks are essentially elementary carbon in particles which are partly graphitic in nature. Bound oxygen, hydrogen and sulphur will, however, be present. The graphitic elements appear to assemble in concentric shells with a higher degree of order in the outer layers, as shown in Fig. 20.3. The individual particles are basically spheroidal. The particle sizes may be determined by electron microscopy, by nitrogen adsorption using the method of Brunauer, Emmett and Teller (Brunauer *et al.*, 1938) (the BET method) or more simply by iodine adsorption. Typical diameters for the different blacks are given in Table 20.2.

The individual particles may combine to form chains or loose three-dimensional aggregates in the flame during manufacture, the phenomenon

TABLE 20.3
ASTM designation and surface area limits for commercial blacks

ASTM designation	*Older type*	*SA limits* ($m^2 g^{-1}$)	*ASTM designation*	*Older type*	*SA limits* ($m^2 g^{-1}$)
N110	SAF	125–155	S301	MPC	105–125
N219	ISAF-LS[a]	105–135	N440	FF	43–69
N220	ISAF	110–140	N550	FEF	36–52
N242	ISAF-HS[b]	110–140	N601	HMF	26–42
N285	IISAF-HS[b,c]	100–130	N660	GPF	26–42
N326	HAF-LS[a]	75–105	N770	SRF	17–33
N330	HAF	70–90	N774	SRF-NS[e]	17–33
N347	HAF-HS[b]	80–100	N880	FT	13–17
S300	EPC	95–115	N990	MT	6–9
N339	HAF-HS (NT)[b,c]	90–105			

[a] LS, Low structure.
[b] HS, High structure.
[c] IISAF, 'Intermediate' intermediate super abrasion black.
[d] NT, New technology.
[e] NS, Non-staining.

being referred to as *structure*. (There is an additional association of these aggregates known as *secondary structure* which largely breaks down on pelletization during manufacture and then completely during mixing into the rubber.) Whilst structure is low in thermal blacks, it can be high in furnace blacks. This has some influence on vulcanizate properties and a more significant influence on processing characteristics of the uncured compound. Structure levels may be determined by measuring the void volume between the particles by filling it with an inert liquid such as dibutyl

FIG. 20.3. Schematic of Heckman & Harling's 1966 model of carbon black microstructure. (Reproduced by permission from *Rubber Chem. Technol.*, **39** (1966) 1. Copyright American Chemical Society.)

phthalate (Eaton & Middleton, 1965; Horn, 1969). It is, however, to be recognized that the structure has two characteristics, the number of particles in the aggregate and the degree of 'openness' of the aggregate. An *open* aggregate is one in which the aggregate has a high bulk whilst a *clustered* aggregate is much more compact for a given number of particles.

20.3.1.3 Effect of Carbon Black in Rubber

Einstein (1906, 1911) explained how incorporation of spherical particles in a liquid could lead to an increase in viscosity. This explanation was adopted by Guth & Gold (1938), who proposed the following relationship for low particle concentrations:

$$\eta = \eta_0(1 + 2{\cdot}5c + 14{\cdot}1c^2) \tag{20.1}$$

where η and η_0 are the dispersion and vehicle viscosities, and c the volume concentration of the dispersed phase. Kraus (1965) found that this relation held for MT blacks in rubber up to concentration levels of 0·2. However, it does not hold for the structure blacks which are not only non-spherical but may also contain occluded rubber in the voids. Qualitatively, however, an increase in black loading leads to a substantial increase in viscosity, as seen in Fig. 20.4 (Smit, 1969).

Die swell decreases with an increase in black loading and is also dependent on structure.

The effects of particle size and structure on processing characteristics have been summarized by Horn (1969) (Table 20.4).

TABLE 20.4
Effects of changes in particle size and structure of carbon black on processing behaviour
(After Horn, 1969)

Processing properties	*Decreasing particle size*	*Increasing structure*
Loading capacity	Decreases	Decreases
Incorporation time	Increases	Increases
Oil extension potential	Little	Increases
Dispersability	Decreases	Increases
Mill bagging	Increases	Increases
Viscosity	Increases	Increases
Scorch time	Decreases	Decreases
Extrusion shrinkage	Decreases	Decreases
Extrusion smoothness	Increases	Increases
Extrusion rate	Decreases	Little

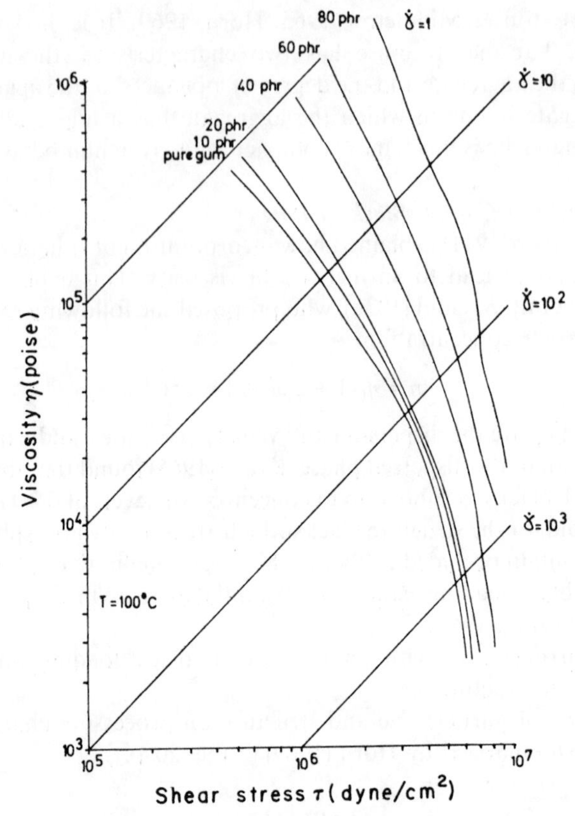

FIG. 20.4. Apparent viscosity of SBR filled with varying amounts of HAF at 100°C. (From Smit (1969), reproduced by permission of Steinkopff-Verlag, Darmstadt.)

The properties of the filled vulcanizate will depend on:

1. The particle size of the black;
2. structure;
3. black loading;
4. the type of rubber used.

The general effects of particle size and structure have also been summarized by Horn (1969) (Table 20.5). As is well known, decreasing the particle size (i.e. increasing the specific surface, the surface-to-volume ratio) increases tensile strength, tear strength, abrasion and fatigue resistance. At the same

TABLE 20.5
Effects of changes in particle size and structure of carbon black on vulcanizate properties
(After Horn, 1969)

Vulcanizate properties	*Decreasing particle size*	*Increasing structure*
Rate of cure	Decreases	Little
Tensile strength	Increases	Decreases
Modulus	Increases to maximum then decreases	Increases
Hardness	Increases	Increases
Elongation	Decreases to minimum then increases	Decreases
Abrasion resistance	Increases	Increases
Tear resistance	Increases	Little
Cut growth resistance	Increases	Decreases
Flex resistance	Increases	Decreases
Resilience	Decreases	Little
Heat build-up	Increases	Increases slightly
Compression set	Little	Little
Electrical conductivity	Increases	Little

time there is an increase in hysteresis during compounding and in dynamic applications of the vulcanizate.

The newer, more open, structured blacks exhibit higher hysteresis and much improved abrasion resistance. Vulcanizates may also exhibit some increase in modulus and hardness which may lead in some dynamic applications to inferior cracking and fatigue resistance.

A modification of the Guth–Gold relationship has been found to relate the vulcanizate modulus to the filler loading in the case of structure-free MT blacks. This has the form

$$E = E_0(1 + 2{\cdot}5c + 14{\cdot}1c^2) \qquad 20.2)$$

where E and E_0 are the moduli for the filled and unfilled compounds, and c the concentration of filler. As with stock viscosity, the simple relationship does not hold for the much more important structured blacks.

The relationship between black loading and tensile strength is interesting in that with most rubbers there is an optimum loading for peak tensile strength (Fig. 20.5) (Boonstra, 1971). Quite clearly the value and position of the peak depends on both filler and elastomer type. It may also be noted that the peak may be moved to lower loadings at higher strain rates, e.g. under dynamic service conditions.

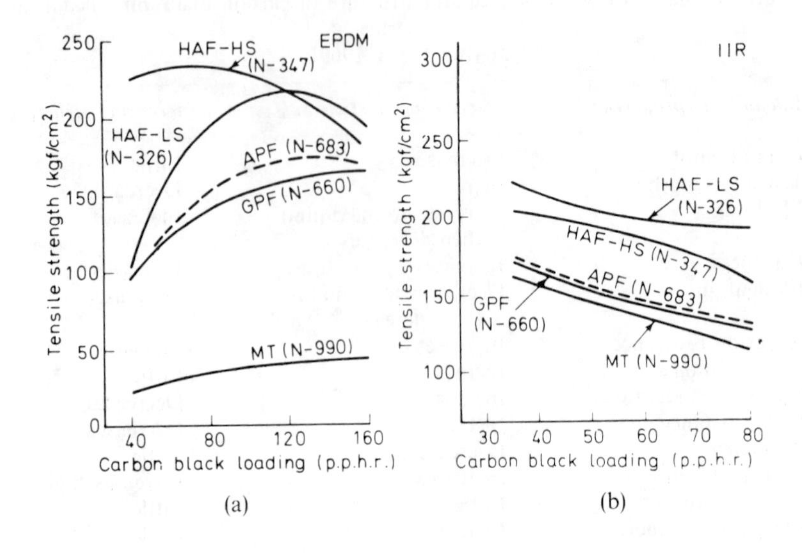

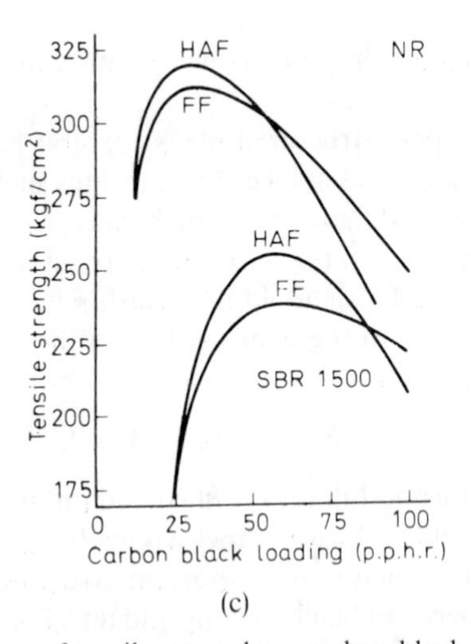

(c)

FIG. 20.5. Dependence of tensile strength on carbon black loading in (a) EPDM, (b) IIR, and (c) NR and SBR 1500. (From Boonstra (1971), reproduced by permission of the Plastics and Rubber Institute.)

TABLE 20.6
Effect of particle size of carbon black at 50 pphr loading on mechanical properties of SBR
(After Boonstra, 1971)

Type	*ASTM designation*	*Average particle size (nm)*	*Tensile strength (MPa)*[a]	*Relative laboratory abrasion*[b]	*Relative roadwear*[b]
SAF	N110	20–25	24·51	1·35	1·20
ISAF	N220	24–33	22·55	1·25	1·15
HAF	N330	28–36	22·05	1·00	1·00
EPC	S300	30–35	21·57	0·80	0·90
FEF	N550	39–55	18·13	0·64	0·72
HMF	N683	49–73	15·68	0·56	0·66
SRF	N770	70–96	14·70	0·48	0·60
FT	N880	180–200	12·25	0·22	—
MT	N990	250–350	9·80	0·18	—

[a] The original data for tensile strength were in kgf cm^{-2}. The conversion into MPa has not been rounded more than to the second decimal place.
[b] Relative to the HAF vulcanizate.

Some comparative data on the effect of different carbon blacks on some mechanical properties of SBR vulcanizates are given in Table 20.6.

20.3.2 Calcium Carbonate Fillers

Calcium carbonate is in many ways a direct contrast to carbon black. In its natural forms of ground limestone and ground chalk (whiting), the particles are coarse (of the order of 0·5–30 μm for limestone and 0·2–10 μm for chalk), are off-white (limestone) or white (chalk) and confer very little, if any, reinforcing effect. These carbonates do give some increase in hardness and provide a useful base for light-coloured goods. Because of their low cost they are frequently used in non-black compounds of only modest strength specifications.

Somewhat superior materials may be obtained by calcining the raw chalk to give calcium oxide, slaking it in water, filtering off impurities and bubbling in carbon dioxide to re-form calcium carbonate, which is precipitated. Such *precipitated calcium carbonates* are of finer particle size (as low as 0·1 μm) and give compounds of improved properties compared with those based on ground whiting.

The precipitated calcium carbonates may be treated with up to 3% of stearic acid or a stearate to aid dispersion. Such *activated calcium carbonates* are sometimes used in higher-grade coloured compounds.

It is now recognized that the presence of many small molecules, including stearates, on the particle surface may interfere with the interaction between filler particle and rubber molecule and thus prevent calcium carbonate from reaching its true potential as a reinforcing filler. In contrast to the silicas and silicates, little success has been achieved by the use of coupling agents of the silane and titanate types. Recently, however, it has been found (Hutchinson & Birchall, 1980) that coating the calcium carbonate filler with a carboxylated unsaturated polymer produces a filler with a reinforcing effect between that of an MT and an SRF black. A filler using such a coating is now commercially available (Rothon, 1984).

20.3.3 China Clays

The china clays are hydrated aluminium silicates resulting from the weathering of feldspars. Like the calcium carbonates they are often considered as low-cost inert fillers but grades are available that provide a fair measure of reinforcement at low cost.

Compared with calcium carbonates the basic grades of china clay show the following vulcanizate characteristics:

1. Low pigmenting power;
2. slightly reinforcing at low levels;
3. lower resilience;
4. the plate-like particles can lead to anisotropy in products and also to poor tear resistance;
5. good electrical insulation properties;
6. good resistance to mineral acids.

The china clays are frequently classified into the following groups.

A. *Soft clays* of particle size greater than 2 μm. They have a marginal reinforcing effect and are used mainly in mechanical goods.
B. *Hard clays* of finer particle size (less than 2 μm) and which confer a significant reinforcing effect to the rubber. As with other fine-particle-size silicates they may have a retarding effect on cure, particularly at high loadings.
C. *Calcined clays.* These are hard clays calcined to remove water and which confer higher tensile strength, hardness and electrical resistivity as well as a good white colour. They may be used where colour and good electrical insulation are particularly important. They may have a less retarding effect on cure.

D. *Treated clays*, which may be of many types. At one time such diverse materials as amines and polyglycols were used but are no longer of much importance. Much more significant are the silane coupling agents, particularly the mercapto-terminated silanes which are particularly effective with the calcined clays. They function by providing a link between filler and rubber molecule.

20.3.4 Other Natural Mineral Particulate Fillers

Several other naturally occurring minerals have also been used over the years in rubber compounds but today are of lesser importance.

Barytes is a naturally occurring barium sulphate of specific gravity about 4·4. Dry ground materials have a maximum particle size of about 60 μm whilst wet ground materials are finer (20 μm). Purer and often finer materials may be obtained by converting barium ores to the soluble chloride, to which is added sodium sulphate to precipitate the barium sulphate (often referred to as blanc fixe, which is also obtained as the by-product in the manufacture of hydrogen peroxide). It is useful where a high-density material is required such as in flooring compounds, for X-ray opaque compounds and for chemical-resistant applications.

The natural *magnesium carbonates*, magnesite and dolomite ($MgCO_3 \cdot CaCO_3$), are too coarse to be used directly as fillers and commercial magnesium carbonates are usually made by precipitating a magnesium salt, usually the sulphate, with sodium carbonate. The more important light magnesium carbonate is made by precipitation from dilute solutions at low temperature. It confers some measure of reinforcement to rubber, and moreover stiffens uncured extruded products and reduces collapse in open steam cures. Because it has a similar refractive index to natural rubber it is possible to make translucent filled compounds.

Talc, or French chalk, is a magnesium aluminium silicate used largely to prevent uncured stocks from sticking to each other. As a filler it is sometimes used in heat-resistant compounds, whilst laminar forms help to reduce gas permeability.

Sand and most other mineral silicas are too coarse as fillers for rubber but *kieselguhr* (diatomaceous earth) is used in heat-resisting compounds. It is abrasive, however, and can damage mould surfaces.

20.3.5 Synthetic Silicas and Silicates

Fine silicas, of finer particle size than most carbon blacks, may be prepared synthetically either by precipitation or by pyrogenic processes.

In the precipitation process a mineral acid is added to an alkali silicate solution, usually of sodium silicate, to cause the silica to precipitate. After filtration, washing and drying, the product is composed of 86–88% SiO_2 with 10–12% water, about half of which is chemically bound. It also contains some salt residues and traces of other metal oxides. Particle sizes are usually in the range 10–40 nm.

The thermal, or pyrogenic, silicas can be obtained by the electrical-arc vaporization of high purity silica sand or quartz at 3000–3500°C, coke being mixed with the sand to effect a reduction. The condensed silicon vapour is combined with oxygen at a lower temperature to produce a high-purity silica. Alternatively, silicon tetrachloride vapour is mixed with air and natural gas or hydrogen and burnt at about 1000°C. The product is sometimes known as fume silica.

Compared with carbon blacks the following features of silicas should be noted.

1. The surface of the silica particle is hydrophilic and much adsorbed water is present. This causes difficulties in achieving rapid wetting and dispersion. The longer mixing times may cause excessive heat generation (particularly since the compound viscosity is higher; see 3 below) and, in the case of IR and NR, polymer breakdown.
2. The particle surface is acidic and the active —OH groups on the surface tend to bond to and deactivate accelerator molecules. As a result of these effects cure is retarded, and in order to maintain a given cure rate it is necessary to use a higher level of accelerator than in corresponding black formulations. At one time it was common practice to add a substance which was preferentially adsorbed onto the surface of the filler particle at an early stage in the mixing process and before the addition of the accelerator. Diethylene glycol and polyethylene glycols were widely used for this purpose. These materials have now been largely replaced by silane coupling agents (see below).
3. Silicas generally have a greater surface area than a carbon black of the same particle size, thus indicating a greater porosity. This can lead to a greater stiffness for uncured stocks at a given filler loading. Additional plasticizer is therefore required to achieve parity of stock viscosity with the corresponding black compound.

High levels of reinforcement may be achieved with silica fillers, particularly where silane coupling agents are used. Most studies reported concur that the tensile strength of a silica-filled compound is comparable with a reinforcing black such as N339 and that the tear resistance is at least as good if not superior. Some workers (e.g. Stewart, 1977) state that replacement of carbon black with a fine silica will adversely affect abrasion resistance and lower the 300% modulus. On the other hand, Wagner (1977) states that these properties are comparable with the two types of filler. He has suggested that the difference is due to the choice, level and point of addition of the coupling agent. For example, Stewart used bis(triethoxylsilylpropyl) tetrasulphide whilst Wagner used mercaptopropyltriethoxysilane. Best results are obtained if the coupling agent is added to the filler before it is compounded into the rubber. Coupling agents may be added during the rubber compounding operation but it has been estimated that their efficiency is reduced by about 30%. If the coupling agent is added during the rubber compounding stage this should be done before the addition of the zinc oxide, which is best added in a separate later operation.

The major use of synthetic silicas is in shoe sole manufacture. There has been some increased use in tyre applications where tear resistance was more important than abrasion resistance, such as in off-the-road earthmover and truck tyres. Whilst they are more expensive than carbon black, they have also found use in a variety of applications where tough white or coloured articles are required.

Hydrated calcium and aluminium silicates may also be produced by modifications to the precipitation process. They are semi-reinforcing fillers which may have a slight retarding effect on cure. In recent times these materials have tended to be replaced by the less expensive mixtures of silica and china clay, particularly since the reinforcing power of the latter may be upgraded by the use of coupling agents.

20.4 RESINS AND PITCHES

A variety of resins and pitches has been used in rubber compounds for a variety of purposes. In most cases they have a lower viscosity than the rubber at processing temperatures and act as viscosity-depressant processing aids. However, at normal ambient temperatures they are harder than the rubber and act as stiffening agents. They may be natural or synthetic, the most important groups being:

1. Phenolic resins;
2. coumarone–indene resins;
3. high-styrene resins;
4. cyclized rubber;
5. coal tar pitch;
6. petroleum pitches and petroleum resins;
7. rosin-related materials.

The *phenolic resins* are reaction products of a phenol with an aldehyde, usually formaldehyde. Supplied as low-molecular-weight materials they are fluid at rubber processing temperatures and facilitate operations such as moulding and extrusion. If, however, a cross-linking agent for the phenolic resin, such as hexamethylene tetramine, is added, then the phenolic resin can be cross-linked at the same time as the rubber is vulcanized. The resulting product may be hard and leather-like, the hardness clearly depending on the level of resin used. For use in hydrocarbon rubbers the phenols should have long alkyl substituents on the benzene ring, but simpler phenols may be used where the resins are to be incorporated into NBR.

Coumarone–indene resins are prepared by cationic polymerization of a coal tar fraction boiling in the range 168–172°C and which consists primarily of coumarone and indene, two monomers bearing a strong formal resemblance to styrene. Sulphuric acid is commonly used as the catalyst to give a low-molecular-weight polymer. By varying the monomer ratio and the polymerization conditions a range of products are obtainable, varying from hard and brittle to soft and sticky, and in colour from pale straw to dark brown.

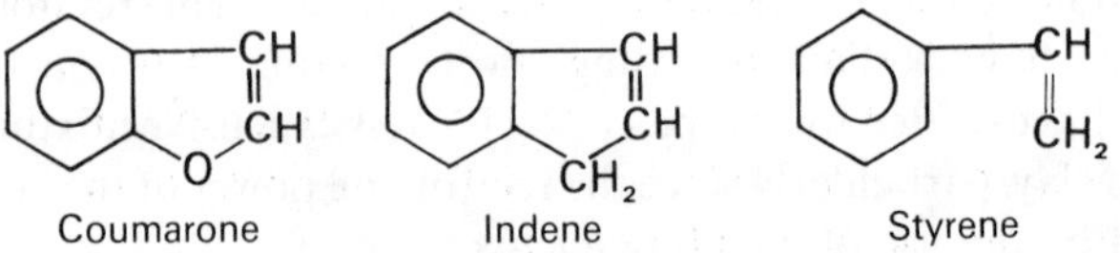

Coumarone Indene Styrene

The *high-styrene resins* are used primarily to increase vulcanizate hardness, which they can achieve with very little increase in density. They are copolymers of butadiene and styrene, like SBR, but with higher styrene contents in the range 50–75%. The T_g of the copolymer is roughly linear with the proportion of styrene. Since the T_g of polybutadiene is about -100°C and that of polystyrene about $+90$°C, a 50:50 copolymer will have a T_g of about -5°C and a 75:25 copolymer of about 40°C. For good dispersion it is important to blend the resin into the rubber well above the T_g of the resin. Used with silica fillers, the high-styrene resins have found use with shoe soling compounds. *Cyclized natural rubber* has many similar properties and may be used for similar purposes.

A number of industrial distillation residues have also been used in the rubber industry. *Coal tar pitch* can function as both a processing aid and a stiffener but has the disadvantage of being acidic and today it is not of importance in rubber technology. On the other hand, *petroleum pitches* continue to be of interest (Deviney *et al.*, 1974). They are less volatile than aromatic petroleum oils and also increase hardness and stiffness. They also confer lower resilience and higher heat build-up during service. Also of interest are a number of petroleum resins produced by the polymerization of olefins (alkenes) present in steam-cracked heavy hydrocarbon petroleum fractions.

Besides being the source of turpentine, the pine tree yields a number of products of interest to the rubber industry. Perhaps best known is *pine tar*, prepared by the destructive distillation of pine wood. It is a useful softener, tackifier and dispersing aid, usually employed at levels of 3–7 pphr. Also obtainable by a variety of processes such as the distillation of crude gum turpentine, extraction of aged dead tree stumps or as a by-product of paper manufacture is *rosin*, a resinous product consisting of abietic acid (35%), related materials (10%) and other products (10%). As it absorbs up to 10% oxygen it has a detrimental effect on the ageing of unsaturated rubbers. However, products of use in rubber compounds may be obtained as a result of modifying the rosin by such processes as disproportionation, hydrogenation, esterification (for example with glycerol to produce ester gum), dimerization or polymerization. These materials give good tack to a rubber formulation and are used in adhesive applications.

BIBLIOGRAPHY

Bernardo, J. J. & Burrell, H. (1972). Plasticization. In *Polymer Science*, ed. A. D. Jenkins. North-Holland, Amsterdam, Chapter 8.
Bruins, P. F. (1965). *Plasticizer Technology*. Reinhold, New York.
Donnet, J.-B. & Voet, A. (1976). *Carbon Black*. Marcel Dekker, New York.
Kraus, G. (ed.) (1965). *Reinforcement of Elastomers*. Interscience, New York.
Medalia, A. I., Juengel, R. R. & Collins, J. M. (1979). In *Developments in Rubber Technology—1*, ed. A. Whelan & K. S. Lee. Applied Science, London, Chapter 4.
Morris, G. (1979). In *Developments in Rubber Technology—1*, ed. A. Whelan & K. S. Lee. Applied Science, London, Chapter 6.
Smith, D. A. (1968). Compounding for flexibility. *Rubber J.*, **150**(Jan.), 21; (Feb.), 8.

REFERENCES

Boonstra, B. B. (1971). In *Rubber Technology and Manufacture*, ed. C. M. Blow. Butterworth, London, Chapter 7.
Brunauer, S., Emmett, P. H. & Teller, E. (1938). *J. Am. Chem. Soc.*, **59**, 1553.
Deviney, M. L., Weaver, E. J., Wade, W. H. & Gardner, J. E. (1974). *Rubber Chem. Technol.*, **47**, 837.

Donnet, J.-B. & Voet, A. (1976). *Carbon Black*. Marcel Dekker, New York.
Eaton, E. R. & Middleton, J. S. (1965). *Rubber World (N.Y.)*, **152**(3), 94.
Einstein, A. (1906). *Ann. Physik*, **19**, 289.
Einstein, A. (1911). *Ann. Physik*, **34**, 591.
Guth, E. & Gold, O. (1938). *Phys. Rev.*, **53**, 322.
Heckman, F. A. & Harling, D. E. (1966). *Rubber Chem. Technol.*, **39**, 1.
Horn, J. B. (1969). *Rubber Plast. Age.*, **50**, 457.
Hutchinson, J. & Birchall, J. D. (1980). *Elastomerics*, **112**(7), 17.
Kraus, G. (ed.) (1965). *Reinforcement of Elastomers*. Interscience, New York.
Medalia, A. I., Juengel, R. R. & Collins, J. M. (1979). In *Developments in Rubber Technology—1*, ed. A. Whelan & K. S. Lee. Applied Science, London, Chapter 4.
Morris, G. (1979). In *Developments in Rubber Technology—1*, ed. A. Whelan & K. S. Lee. Applied Science, London, Chapter 6.
Rothon, R. (1984). *Europ. Rubber J.*, **166**(10), 37.
Small, P. A. (1953). *J. Appl. Chem.*, **3**, 71.
Smit, P. P. A. (1969). *Rheol. Acta*, **8**, 277.
Stewart, E. J. (1977). *Rubber Chem. Technol.*, **50**, 444.
Wagner, M. P. (1977). *Rubber Chem. Technol.*, **50**, 356.

Chapter 21

Additives IV: Miscellaneous

21.1 INTRODUCTION

Whilst the most common additives in rubber compounds are the cure system components, fillers, plasticizers and antidegradants described in the three previous chapters, many other types of additive may be encountered. Some of these are used primarily as aids to processing, others to modify vulcanizate properties, yet others having a role in both areas. In some instances the additive may be specific for one or two polymers and where this is the case it will have been discussed in the appropriate chapter. In this chapter only those additives that may be used over a wide range of polymers will be discussed. For convenience of future reference they will also be considered in alphabetical order under the following headings.

21.2 Abrasives.
21.3 Antistats.
21.4 Blowing agents.
21.5 Colourants and white pigments.
21.6 Cork.
21.7 Electroactive materials (magnetic powders, radiation absorbers).
21.8 Factice.
21.9 Fire retardants.
21.10 Friction lubricants.
21.11 Glue.
21.12 Peptizing agents.
21.13 Reclaimed rubber.
21.14 Reodorants.
21.15 Tackifiers.

21.2 ABRASIVES

Abrasives may be encountered in two types of compound: erasers and compounds used in polishing and grinding equipment.

For erasers pumice and glass powder are most commonly used, although in the case of high-quality artists' erasers, which contain high loadings of factice (Section 21.8), no abrasive is present.

For polishing and grinding purposes carborundum powder is favoured.

21.3 ANTISTATS

There are a number of applications of rubber products where the build-up of static charge is undesirable because it attracts dust, distends moving webs of film or fibre by electrostatic attraction or provides an ignition hazard where inflammable materials such as hospital anaesthetics may be present. The dissipation of charge may involve bulk conduction through the body of the article, or surface conduction, or both. In many cases the use of a conductive carbon black will overcome the problem but, where this is not acceptable, proprietary antistatic agents, usually based on quaternary ammonium compounds, can be considered.

21.4 BLOWING AGENTS

The introduction of bubbles of gas into a mass of rubber can modify the properties of the rubber in a number of ways, including the following.

1. Whereas solid rubber is virtually incompressible (however soft), a cellular rubber has a low bulk compressibility. This can be very useful in some sealing applications.
2. The products have a low stiffness and high flexibility.
3. It is possible to obtain products of very low density.
4. The product can acquire very good shock absorption characteristics and show good cushioning behaviour.
5. Thermal insulation properties may be improved.
6. If there is an open cell structure, i.e. the cells are interconnected, the material can act as a sponge.

Bubbles may be introduced into a polymer by several methods, including:

1. The use of chemical blowing agents, i.e. chemicals that decompose on heating with the evolution of gas *in situ* in the polymer.
2. The use of volatile chemicals such as solid carbon dioxide or fluorocarbon gases which volatilize during the processing of the

polymer. This technique is widely used with polyurethane foams and plasticized PVC.

3. Beating of air into polymer when in a liquid state. This process was the basis of latex foam technology, the foamed latex then being gelled in the cellular state prior to vulcanization.
4. Impregnation of polymer with a gas, usually nitrogen, under very high pressures, and subsequent heating of the gas-saturated polymer at high temperatures, which softened the polymer and allowed the product to expand. One of the earliest applications for this approach was in the manufacture of expanded ebonite but today the process is largely restricted to plastics materials.
5. Incorporation of an additive such as a salt or starch which could be leached out by water was at one time (but not now) believed to be important with separators for battery boxes.
6. Use of a cross-linking system that evolved a gas during the cross-linking process. This was the original basis for the process of manufacture of polyurethane foams although today the foaming also involves volatiles (see 2 above).

Past practice has led to various terms being used somewhat haphazardly, but the following terminology is probably the most widely used and will be employed here:

1. *Cellular* refers to any product laden with bubbles.

2. *Expanded* refers to a closed cell product (i.e. a product where the bubbles or cells are not interconnected) made from a solid base.

3. *Foam* refers to an open cell product made from a liquid base such as rubber latex or a polyol–polyisocyanate mixture for making a polyurethane foam.

4. *Sponge* refers to an open cell product made from a solid base. An example would be where a chemical blowing agent was incorporated into a solid rubber compound which was then heated under pressure both to decompose the blowing agent and to vulcanize the compound, the resultant closed cell vulcanizate being *cracked* by passing between rollers. The term is probably, however, best avoided, particularly as in some countries its use in this context may contravene regulations which only allow the use of the word to refer to the natural marine product.

This section is concerned with chemical blowing agents, which today are only used for a small part of the cellular rubber market. Such agents have a number of requirements, of which the following are of particular importance.

1. The need for gases to be evolved within a narrow but clearly defined temperature range and in a controlled and reproducible manner.
2. The decomposition range should be suitable for the polymer; for example, it should not be above the maximum possible processing temperature that can be used if significant polymer degradation is not to occur, neither should premature blowing occur, for example during mixing.
3. Gases evolved should not corrode processing equipment.
4. Any toxicity of blowing agents and their decomposition products must be catered for.

Many hundreds of materials have been investigated as blowing agents for polymers but the number used, particularly for rubbers, is very small.

Sodium carbonate, which gives off carbon dioxide, is still widely used for open cell products, being low in cost but rather erratic in decomposition. It decomposes in the range 100–130°C yielding 125–130 $cm^3 g^{-1}$. Two other inorganic agents are ammonium carbonate and bicarbonate. Additional stearic acid may be incorporated into the compound to aid the decomposition.

Amongst the organic blowing agents, dinitrosopentamethylene tetramine, benzene sulphonyl hydrazide and azodicarbonamide are frequently quoted in the literature, with the first-named in particular being favoured for microcellular products such as shoe soling. Decomposing in the range 160–200°C, it is more powerful than sodium bicarbonate, yielding about 210 $cm^3 g^{-1}$. Once again slightly acidic conditions are necessary for it to function at maximum efficiency and the rubber compound should contain at least 4 pphr of stearic acid.

21.5 COLOURANTS AND WHITE PIGMENTS

In tonnage terms most rubber products are black, and a number of other products are left with their natural colour. Examples of rubber products that are coloured include wire insulation and seals (for identification), playballs, swimming hats, flooring, non-industrial rubber boots and other footwear parts. For a number of applications, such as shoe soles, colouring (including white and black) is not desirable since it may mark objects with which it comes into contact.

Whilst the market is clearly small, compared for example with the plastics industry, care should be exercised in the selection of colourants, not only to ensure the correct colour but also because some pigments may adversely affect ageing, others may affect cure whilst some may lead to a deterioration in electrical insulation properties. Problems such as bleeding, blooming and toxicity will also be important in some applications.

In principle there are four methods of colouring polymers. These are surface coating (for example by painting), surface dyeing, the introduction of colour-forming groups (chromophores) into the polymer molecules, and mass colouration. Painting with a flexible lacquer is sometimes used, for example with footwear and toys where a high glossy finish is required, the lacquers usually being based on either CSM or polyurethane polymers. Surface dyeing is used mainly with fibres but has been used occasionally with polar plastics such as casein, cellulose acetate and the nylons but the writer is not aware of any use in rubber technology. The deliberate introduction of chromophores may be considered only as an academic curiosity. It therefore follows that mass colouration is the usual approach. Colourants may, in principle, be divided into two classes, pigments which are insoluble and dyestuffs which are soluble in the polymer. In practice many so-called pigments are slightly soluble and this may lead to blooming and bleeding (see Chapter 17). The distinction is, however, rather pedantic and since true dyestuffs are rarely used in rubbers the term colourant and pigment tend to be employed interchangeably.

For a chemical to be a suitable pigment it must meet the requirements listed in Chapter 17 for additives. Thus it is desirable that the pigment does not bloom or bleed. (In some instances this can be minimized by using pigments of higher-than-usual molecular weight.) Stability to processing covers not only the obvious aspect of heat resistance but also shear, since some pigments break down under intensive shear with a change in colour. As already mentioned, suitability in service may require consideration of factors such as effect on stability, electrical properties and toxicity.

Coloured pigments are often divided into two types, inorganic and organic. It is sometimes said that the inorganics are more stable and the organics are brighter, but this is a simplification. Amongst inorganic pigments are iron oxides (red, orange and yellow), chromium oxide (dull green), nickel titanate (highly stable yellows), ultramarine blues, antimony sulphides, mercuric sulphide and cadmium sulphides and sulphoselenides. The use of some of these materials, particularly the above sulphides, may be restricted on grounds of toxicity.

A good range of organic and organometallic pigments are marketed for use in rubber goods. Over the years, materials have been developed with improved stability and general performance. These may vary according to the type of polymer and it is always advisable here to seek the advice of the colour manufacturers or suppliers.

White pigments may be used both for making white goods and as a base for pastel colours. Titanium dioxide is widely used in quality goods but lithopone, a mixture of zinc sulphide and barium sulphate, may be employed in cheaper compounds. Another alternative is zinc carbonate. Zinc oxide is a white pigment as well as being a reinforcing agent and accelerator activator. In large quantities, however, it does make uncured stocks sticky and difficult to handle.

21.6 CORK

Commercial cork is obtained from the outer layer of an evergreen oak, *Quercus suber*, and a sub-species, *Quercus suber occidentalis*. About 90% of commercial cork comes from the Iberian Peninsula and the North African states of Algeria, Morocco and Tunisia. The bark of most trees contains some cork cells but only in the cork oaks are the layers sufficiently thick to be of importance. Among alternative sources which have found some use are the Pao Santo tree of Brazil, the Japanese Abemaki tree and the birch.

A one-inch cube of cork has been estimated to contain about 200 000 000 air-filled cells with about 50% of the volume accounted for by the captive air cells, this resulting in a specific gravity of the order of 0·25. The cellular structure accounts for the buoyancy, compressibility, elasticity and low thermal conductivity of the product.

The incorporation of powdered cork gives rubber compounds a high degree of resilience and compressibility and finds some use in flooring, gaskets and tiles.

21.7 ELECTROACTIVE MATERIALS

The above heading is a rather unsatisfactory one used here to cover two classes of material:

1. Additives that give the material magnetic properties;
2. additives that increase impermeability to high-energy radiation.

In the first group there is one particular material to note—barium ferrite.

When used at high loadings, magnetic vulcanizates may be obtained. These are particularly useful for doors (such as for refrigerators) which can be opened from the inside, for example by a child who has crawled inside the equipment.

In the second class mention may be made of barium sulphate, which has proved useful in medical applications because of its impermeability to X-rays, and metallic lead and litharge, which reduce permeability to radiation.

21.8 FACTICE

The word *factice* is French and means substitute or artificial. There is some division of opinion as to whether it was first used by the French as a translation of the English expression *rubber substitute* or whether it was the other way round. Both English and French expressions indicate the original main purpose of the material which is today more commonly called 'factice' than 'rubber substitute'.

Factices are cross-linked or vulcanized vegetable or marine oils with rape-seed oil being the preferred material. This results from its high triglyceride content and medium unsaturation. Too low a level of unsaturation gives insufficient cross-linking; too high a level makes reaction difficult to control. Alone they are very friable materials with a negligible strength and are only of value when incorporated into rubber.

There are many classes of factice but these may be considered as variants of two basic types, brown or dark factice, and white factice. In a typical process for brown factice the rape-seed oil is heated with sulphur at 130–140°C for 5–8 h. Lighter golden factices may be obtained by reacting in the presence of an accelerator at a lower temperature. White factice is the reaction product of the oil with sulphur monochloride. This latter process was patented by Parkes (1855) whilst the method of preparation of dark factice is much older and its discovery is lost in the mists of time.

Dark factice is mainly used as a processing aid. Some 5–10 pphr will reduce power consumption during mixing and may alleviate bagging on a two-roll mill. In extrusion it is an aid to controlling die swell and reducing distortion and collapse during pan curing. It has advantages related to those of calendered products. It is interesting to note that this is yet another example of the beneficial effects of adding a cross-linked material as a processing aid, a phenomenon that has been noted with several materials (e.g. NR, NBR, CR and SBR). Larger loadings may be used in printers' rollers.

White factice has found its greatest use in 'cold-cured' rubber proofings, the factice not only aiding dissolution of the rubber compound but also giving the product improved drape and ageing properties. White factice is also the dominant component of artists' erasers.

21.9 FIRE RETARDANTS

The cost of fire damage each year is enormous. More seriously still, many people are killed or badly injured. It is therefore important to minimize fire risks. In doing so it is necessary to analyse fire hazards, and although this will not be done in detail here there are some aspects that are pertinent to the rubber compounder.

For example, in Great Britain approximately 1000 people are killed by fires each year, but of these only about 15% die through being burnt to death, the rest dying through inhalation of smoke and/or toxic products. Thus it is not sufficient to incorporate a fire retardant if in its operation it produces copious amounts of such products. It is little more than a statement of the obvious that fire retardancy in rubber products is particularly important where there are problems of escape from the fire, such as in space rockets, aircraft, underground mines and ships, particularly submarines.

Flame retardants appear to function by one or more of the following mechanisms:

1. By chemical interference with the mechanism of flame propagation;
2. by production of large volumes of incombustible gases that dilute the supply of air, specifically oxygen, to the fire;
3. by endothermic reaction, decomposition or state change which absorbs heat;
4. by formation of an impervious coating preventing access of oxygen to the polymer.

Where fire retardancy is important it is usual practice to use a rubber that itself has fire-retarding properties. However, many of these polymers do emit toxic gases at elevated temperatures, particularly halogens and halogen derivatives. Where this causes problems the addition of a fire retardant to a combustible polymer may be the solution. An example of this is the use of aluminium trihydrate to ethylene–acrylic rubbers for cable applications.

The selection of a fire retardant will also depend on its compatibility with the polymer; bearing this point in mind the following are the main types of fire retardant used in the rubber industry.

1. Phosphates such as tritolyl phosphate which may also have a plasticizing function. They are toxic and use is largely restricted to polar rubbers such as CR.
2. Halogen-containing compounds such as chlorinated paraffins which are useful in polar rubbers and which may also have a plasticizing role, at least as plasticizer extenders (i.e. not totally compatible alone but able to be used in conjunction with true plasticizers). Bromine compounds tend to be more powerful than chlorine compounds but the writer is not aware of any use in rubbers.
3. Antimony oxide is on its own a rather weak fire retardant although it appears to function by each of the four mechanisms listed above. It is, however, synergistic with both phosphorus and halogen compounds and is widely used. Alternatives or partial replacements are the oxides of titanium, zinc and molybdenum. Zinc borate is also sometimes used on account of its ability to form a crusty char on the surface, thus isolating the burning polymer from oxygen.
4. Hydrated alumina is useful where the processing temperatures are sufficiently low so that it does not decompose during processing. It may be incorporated into most rubbers in large amounts, although this will generally adversely affect physical properties, but in a fire it gives off copious quantities of water vapour, which is of course non-toxic.

21.10 FRICTION LUBRICANTS

The term *friction lubricants* is used here to describe materials that reduce the surface friction of rubber compounds, as opposed to *external lubricants*, which are incorporated into plastics materials to prevent sticking to processing machinery, and *internal lubricants*, which are sometimes used to aid the flow of rigid thermoplastics such as polystyrene and unplasticized PVC.

Just two materials are encountered: graphite and the more expensive but marginally superior molybdenum disulphide.

21.11 GLUE

Perhaps somewhat improbably, animal or fish glue may be added in hydrated or powder form to improve the hydrocarbon oil and fuel resistance of diene rubbers used for such purposes as hose linings and gaskets where only very moderate fuel resistance is required and at low cost. With the wide range of synthetic rubbers and flexible thermoplastics now available the author does not feel that this usage can continue to be recommended.

21.12 PEPTIZING AGENTS

The term *peptization* was originally introduced to describe the deflocculation of a suspensoid to produce a colloidal sol, on account of the superficial resemblance of the process to the conversion of insoluble protein into soluble peptone. Since much of the early work in rubber chemistry was undertaken in colloid science laboratories, it is not surprising that the term was then used to describe the softening of rubber by certain chemicals which, in effect, reduce the molecular weight. The chemicals are now known as peptizing agents and function as oxidation catalysts.

Peptizers will only function in elastomers that are amenable to such degradation and in practice their use is restricted to natural rubber and sulphur-modified polychloroprenes.

In the case of natural rubber the peptizing agents are useful in speeding up mastication at elevated temperatures. Examples of such materials are thio-β-naphthol and *o,o'*-dibenzamidodiphenyl disulphide. In the case of thio-β-naphthol the level of use is normally 0·6 pphr.

The use of peptizing agents for polychloroprenes was discussed in Chapter 10.

21.13 RECLAIMED RUBBER

Reclaimed or regenerated rubber is the product obtained by chemical and/or thermal treatment of waste rubber goods or scrap. It should not, however, be considered simply as a devulcanized rubber, since much main-chain breakdown as well as some cross-link scission probably occurs. The end-products of reclaiming may be considered as highly branched but of low molecular weight.

Various types of reclaim exist but the best known is whole tyre reclaim, which itself may be produced by a variety of processes. In these processes the chopped, and usually partly debeaded, tyres are passed over vibratory screens with metal being removed by powerful magnets. The remaining rubber is fed to cracking mills and then to series of fibre-removing operations involving screens, hammer mills and air flotation tables. In digester processes the rubber is then heated with oils and peptizing agents in a digester in the presence of steam and fabric-destroying agents such as sodium hydroxide, calcium chloride or zinc chloride. This process may take some 4–10 h at 190°C. In an alternative process the rubber, oil and chemicals are fed to a screw extruder where the material is broken down by mechanical working.

Original interest in reclaimed rubber largely arose through cost savings in applications which were not technically very demanding. In recent years the differential in price between SBR, produced under severe price competition, and reclaimed rubber has been small, and this has severely affected the market for reclaim. The material is, however, of some interest as a processing aid, but in giving a faster cure rate to the compound scorching problems may occur, whilst the finish and dimensional control that may be achieved in extruded and calendered parts may also be a disadvantage.

21.14 REODORANTS

For some applications the odour of rubber products is aesthetically unacceptable. In many cases the solution may require the use of a thermoplastics material instead of a rubber but in other instances the incorporation of a chemical which provides an alternative odour that dominates existing odours may be acceptable. Many essential oils and proprietary materials are available on the market but probably the material most commonly used is vanillin.

21.15 TACKIFIERS

The ability of uncured rubber compounds to cohere, often referred to as natural tack, is a property of value where products are made by assembly operations as in tyres and belting. Whilst natural tack is exhibited by natural rubber it is notably absent in many synthetics and it is necessary to incorporate a tackifier into the compound.

Resinous materials such as those discussed in Chapter 20 are used for this purpose.

One group of interest are the esters of rosin and rosin derivatives which have a much lower oxygen absorption value than unesterified rosin or its derivatives and thus have a less adverse effect on ageing. The esterifying alcohols may be monohydric such as methanol, dihydric such as ethylene glycol, trihydric such as glycerol, and even tetrahydric such as pentaerythritol. The higher the functionality of the alcohol, the less tacky the ester but the stronger the vulcanizate. The rosins used may be hydrogenated, dimerized or 'polymerized' or just unreacted wood rosin. The dimerized and polymerized forms tend to give less tacky compounds but stronger vulcanizates.

Selection tends to be a compromise and in the case of natural rubber the preferred material is a glycerol ester of a polymerized rosin with a melting point of 110–120°C and a low oxygen absorption of about 0·25% (compared with about 1·05% for the unesterified polymerized rosin). For SBR and chloroprene rubbers, a pentaerythritol ester of wood rosin may be specified.

Another group of materials of interest are the terpene polymers obtained by cationic polymerization of β-pinene and which have typical molecular weights of about 1200 and softening points around 115°C. They confer good ageing properties to the rubber, have good tack retention and are claimed to be more compatible than most rosin derivatives.

BIBLIOGRAPHY

Anon. (1962). *Symposium: Factice as an aid to productivity in the rubber industry.* National College of Rubber Technology, London.

Nourry, A. (ed.) (1962). *Reclaimed Rubber.* Maclaren, London.

REFERENCE

Parkes, Alexander (1855). British Patent 2359.

Chapter 22

Compound Design

22.1 INTRODUCTION

The properties of an article made of vulcanized rubber will depend on a number of factors, of which the following are particularly important:

1. The type and specific grade of polymer used;
2. the selection of additives and the quantities used;
3. mixing, shaping and vulcanization conditions;
4. presence of reinforcing fillers or inserts into the product;
5. dimensions of the article.

This chapter will concern itself only with the first two aspects but in real situations it is important that all five aspects should be considered when designing a compound.

One disadvantage about rubbery materials is a lack of understanding by engineers and designers of both their possibilities and their limitations. As a car owner, the author has in the past suffered from the fact that small quite inexpensive rubber components have failed, which has led to considerable expense with repairs, subsequent disillusionment and change in brand loyalty. Fortunately car manufacturers now more fully appreciate the critical role of rubber components but there continue to be instances where unnecessary failures occur, often in an attempt to cut costs but at cost to the reputation of the company's products.

It is thus essential that before any product is marketed its technical requirements must be carefully specified and tests made to ensure that the product meets these requirements. The contents of this chapter can only suggest 'starter formulations'. The specific requirements in a real case would probably require some modification. It may also be that some formulations will give properties more than adequate for a specific need, so that it might be possible to use a cheaper polymer and/or more lower-cost additives.

22.2 SELECTION OF RUBBER

Previous chapters of this book have introduced the reader to something approaching 50 types of elastomer, most of which have many grades and a wide choice of suppliers. However, about 70% of tonnage consumption is taken up by NR and SBR and over 80% by NR, SBR and BR. Nine rubbers claim over 99% of the market (excluding polyurethanes and thermoplastic rubbers). Thus for many applications the choice is comparatively simple, the more sophisticated materials being used in the more critical applications.

Because of its low price structure, the first rubber the technologist is likely to consider is SBR, which when black-reinforced has good mechanical properties, particularly abrasion resistance. Where strong light-coloured stocks, gum stocks, compounds with good dynamic mechanical properties and compounds of high resilience are required, then NR is likely to be preferred. Both materials may be modified, with some property overlap, by blending with polybutadiene or a polyoctenamer.

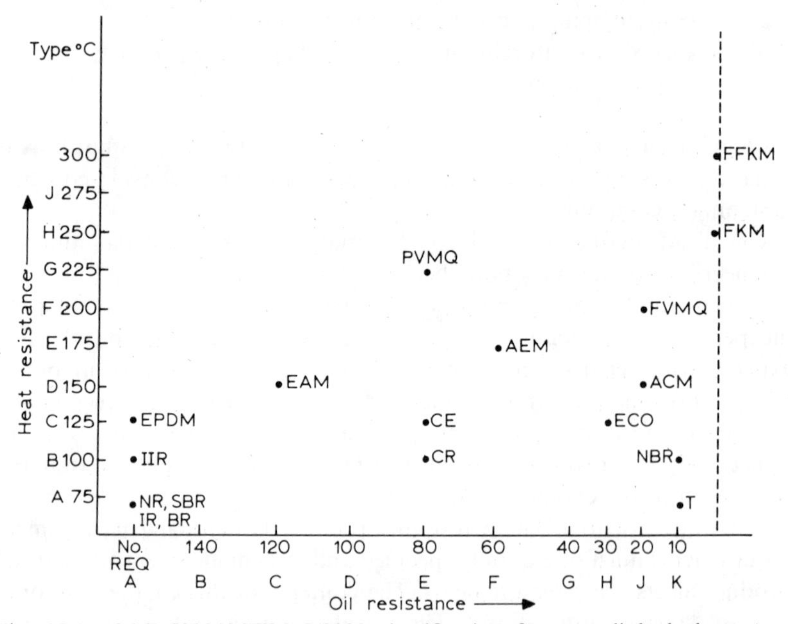

FIG. 22.1. SAE J200/ASTM D2000 classification for cross-linked elastomers. Whilst specific points are shown for clarity there will be a spread for each rubber, the exact position depending on the grade of polymer and compounding formula used. (See also Appendix D.)

NR, SBR, BR and the polyoctenamer all have limited resistance to heat, to UV light, to ozone and to oxidation. Where the material is to be used in anaerobic conditions (in the absence of air) such materials may be surprisingly durable and they have proved well equal to the task. Such an example is the use of NR for bridge bearings, an application which has proved to be highly successful in recent years.

However, where such diene rubbers are unsuitable for use under such conditions, a less unsaturated hydrocarbon rubber may be preferred. Thus EPDM has become very widely used as a 'general-purpose' non-tyre rubber, with IIR and its derivatives CIIR and BIIR being preferred for some purposes (e.g. tyre inner liners because of the low air permeability, and garden pool liners because of the high splice strength). Even with the unsaturated rubber, maximum continuous service temperatures are limited (about 120°C for IIR and 145°C for EPDM).

All the above hydrocarbon rubbers have limited resistance to hydrocarbon oils, so frequently encountered in engineering applications. Where the bulk of the rubber part is at some distance from the swelling liquid this may not be too serious a problem, because of the low rates of diffusion of viscous oils into a polymer. However, where this is not the case, a non-hydrocarbon rubber will be required. The selection of the polymer will then be made from those which meet both swelling and heat resistance criteria. For many applications there may be a low-temperature requirement. Figure 22.1 gives some comparative data on heat and swelling properties, whilst Table 22.1 provides a more comprehensive summary of the major elastomers.

22.3 STARTER FORMULATIONS FOR SELECTED APPLICATIONS

In the following sections some starter formulations for selected applications are given to illustrate some of the factors to be considered in compound design. It must, however, be stressed that in a real situation the operational requirements may be different from those assumed here. In addition, the mention of an ingredient does not imply freedom from toxic hazards, which should be checked with the supplier.

22.3.1 Bridge Bearing

Choice of polymer will be somewhat dependent on the location of the bridge, since this will determine the operating temperatures. In England,

TABLE 22.1
Summary of properties of principal rubbery materials

Property / ASTM D1418 class	NR	SBR	BR	IR	EPDM	IIR	CR	NBR
ASTM D2000 rating	AA	AA/BA	AA/BA	AA	CA	BA, AA	BC, BE	BF, BG, BK, CM
Specific gravity	0·92	0·94	0·93	0·93	0·86	0·92	1·23	1·00
Tensile strength (gum), MPa (psi)	>20·7 (3000)	<6·9 (1000)	<6·9 (1000)	>17·2 (2500)	<6·9 (1000)	>10·3 (1500)	>20·7 (3000)	<6·9 (1000)
Tensile strength (black-loaded), MPa (psi)	>20·7 (3000)	>13·8 (2000)	>20·7 (3000)	>20·7 (3000)	>20·7 (3000)	>13·8 (2000)	>20·7 (3000)	>13·8 (2000)
Tear resistance	Good–v. good	Fair	Fair	Good–v. good	Fair–good	Good	Good	Fair
Abrasion resistance	Excellent	Good–excellent	V. good	V. good	Good–excellent	Good	Excellent	Good
Compression set	Good	Good	Good	V. good	Good	Fair	Fair–good	Good
Rebound (cold)	Excellent	Good	Excellent	Excellent	V. good	Poor	V. good	Good
Rebound (hot)	Excellent	Good	Excellent	Excellent	Excellent	Good	V. good	Good
Dielectric strength	Excellent	Excellent	V. good	V. good	Excellent	Excellent	Good	Poor
Electrical insulation	Good–excellent	Good–excellent	V. good	V. good	Excellent	Good–excellent	Fair–good	Poor
Environmental resistance								
Oxidation	Good	Fair	Fair	Fair	Excellent	Excellent	V. good–excel.	Good
Ozone	Poor–fair	Poor	Poor	Poor	Outstanding	Excellent	V. good	Fair
Sunlight	Poor	Poor	Poor	Poor	Outstanding	V. good	V. good	Poor
Heat ageing (upper lim. cont. serv.), °C	85	90	90	85	145	120	95	120
Low-temperature flexibility	Excellent	V. good	Excellent	Excellent	Excellent	Good	Good	Fair–good
Fluid resistance								
Gas permeability	Fair	Fair	Fair	Fair	Fair	V. low	Low	V. low
Acid resistance (dil.)	Good	Good	Good	Good	Excellent	Excellent	Excellent	Good
Acid resistance (conc.)	Good	Good	Good	Good	Excellent	V. good	Good	Good
Swelling								
In aliphatic hydrocarbons	Poor	Poor	Poor	Poor	Poor	Poor	Fair–good	Excellent
In aromatic hydrocarbons	Poor	Poor	Poor	Poor	Poor	Poor	Fair	Good
In oxygenated solvents	Fair–good	Good	Good	Fair–good	Good	Good	Poor	Poor

(continued)

TABLE 22.1—*contd.*

Property / *ASTM D1418 class*	*ACM*	*ECO/CO*	*AEM*	*CSM*	*FKM*	*PVMQ*	*FVMQ*	*YBPO*
ASTM D2000 rating	DJ	CM	EF	CE	HK	GE	FJ	Not assigned
Specific gravity	1·10	1·27–1·36	1·08–1·12	1·12–1·28	1·85	1·06–1·42 (compound)	1·41 (compound)	1·17–1·25
Tensile strength (gum), MPa (psi)	—	< 6·9 (1000)	—	> 17·2 (2500)	> 12·4 (1800)	—	—	25–45(3500–6500)
Tensile strength (black-loaded), MPa (psi)	> 13·8 (2000)	17·2 (2000)	> 17·2 (2500)	> 20·7 (3000)	> 13·8 (2000)	5–10 (725–1450)	> 6·9 (1000)	—
Tear resistance	Fair	Fair–good	Good	Fair	Fair	Poor	Poor	Outstanding
Abrasion resistance	Fair	Fair–good	Good	Excellent	Good	Poor	Poor	Outstanding
Compression set	V. good	Poor	Good	Fair	Fair–good	Poor–fair	Poor–fair	Fair–poor
Rebound (cold)	Low	Good	Poor	Fair–good	Fair–good	Poor	Poor	Fair–v. good
Rebound (hot)	—	Good	Fair	Good	Good	—	—	Good–excellent
Dielectric strength	Good	Good	Good	V. good–excel.	Good	Good	Good	Fair–good
Electrical insulation	Fair	Good	Fair–good	Good	Fair–good	Excellent	V. good	Fair–good
Environmental resistance								
Oxidation	Good	Good	Excellent	Excellent	Outstanding	Excellent	Excellent	Excellent
Ozone	Excellent	Excellent	Outstanding	Outstanding	Outstanding	Excellent	Excellent	Excellent
Sunlight	Good	Good	Outstanding	Outstanding	Outstanding	Excellent	Excellent	V. good
Heat ageing (upper lim. cont. serv.), °C	177	135	170	135	205	235	230	100–110
Low-temperature flexibility	Fair–v. good	Good–v. good	Good	Good	Fair–good	Excellent	Excellent	Excellent
Fluid resistance								
Gas permeability	Fair	Low	V. low	V. low	V. low	High	High	Low
Acid resistance (dil.)	Fair	Fair–good	Good	Excellent	Excellent	Excellent	Excellent	Excellent
Acid resistance (conc.)	Poor	Fair	Poor	V. good	Excellent	Fair	Good	V. good
Swelling								
In aliphatic hydrocarbons	Excellent	Excellent	Good	Good	Excellent	Poor	Excellent	Excellent
In aromatic hydrocarbons	Fair–poor	Good	Fair	Fair	Excellent	Poor	Good–excellent	Good
In oxygenated solvents	Poor	Poor	Poor	Poor	Poor	Fair	Good–excellent	Fair

operating temperatures are in the range −15°C to +45°C. The rubber may also be in contact with water-borne salt from the road and subject to a variety of weather conditions. The rubber may also be exposed to hydrocarbon oils, oxygen and ozone.

These latter requirements might suggest that natural rubber would be precluded from selection. However, a bridge bearing consists of a series of parallel metal plates separated by layers of rubber; hence only the edges of the rubber layers are exposed to ozone, oxygen and oils. Because diffusion rates are so low, the effect of such agencies is only at the edges and the bulk of the rubber is unaffected. (In practice the bridge bearing may be 'wrapped' in a layer of rubber which protects the steel plates. Although this wrapping layer may be affected, it is of little importance as this layer has no significant 'mechanical' role.)

In principle several rubbers might be expected to be satisfactory for this application. Experience has shown that both natural rubber (NR) and polychloroprene rubber (CR) have performed satisfactorily and in this example the former polymer will be used.

There is no obvious reason to use anything other than an accelerated sulphur vulcanization system. The main decision to make is whether or not to use a conventional or an EV system. Whilst the EV systems have better ageing resistance, they are more expensive, tend to crystallize more rapidly at low temperatures and have relatively poor adhesion to metals. Thus a conventional system using a delayed-action accelerator such as CBS is indicated.

The application demands neither high strength nor high damping capacity so that a semi-reinforcing black such as the N660 (GPF) type should be adequate, the amount to be used depending on the modulus requirement of the rubber to give the desired shear stiffness. A typical loading level would be 50 pphr.

Since the rubber will be exposed to ozone and oxygen, both antiozonant and antioxidant are required. The function of the antiozonant may be improved by the presence of some microcrystalline wax.

A typical formulation would thus be:

		pphr
Rubber	Natural rubber	100
Vulcanizing system	Sulphur	2·5
	Zinc oxide	5
	Stearic acid	2
	CBS	0·6

and the second letter the class according to oil resistance. These are detailed in Appendix D.

To meet the first two requirements the rubber must have a designation in which both letters are equal to or higher (K being 'higher' than A) than in the specification. Designations for a number of elastomers were given in Table 22.1 and Fig. 22.1.

Inspection shows that there are several rubbers that meet this requirement, including some polyurethanes, fluororubbers and epichlorhydrin rubbers. For most applications the cheapest rubber would probably be a nitrile rubber and this will be chosen here. (It may be noted that polyether–ester thermoplastic rubbers do not come under the specification since they are not 'thermoset' elastomers but would probably meet both heat and swelling requirements. The ability to injection-mould without the need for vulcanization could in some circumstances make it an economical alternative.)

Since colour is unimportant there is no problem in using carbon black and hence there should be no great problem in meeting the tensile strength requirement. The general formulation of the compound will then be constructed using manufacturers' recommendations, particularly with respect to the curing system. Aspects of this, in the case of nitrile rubbers, have also been discussed in Chapter 9 of this book.

The remaining problem is to match both specific gravity and hardness simultaneously. If we assume that we have three variables—polymer, filler and plasticizer—then the use of triangular diagrams may be very helpful.

First take one blank diagram and let each corner of the triangle represent a composition that is 100% polymer, 100% filler or 100% plasticizer respectively. If we assume that the specific gravity of the NBR used is 1·00, that of the carbon black 1·80 and the plasticizer (a phthalate–phosphate blend) 1·1, then these densities may be inserted at these points. By interpolation the densities at other positions on the diagram may be calculated; some are shown. After only a few calculations it will be possible to draw a line going through compositions of equal density, the accuracy of the line depending on the number of points calculated. (In this calculation it has been assumed that the other ingredients form such a small part of the total that they may be ignored. It would be more accurate to replace the density of the polymer with the density of a blend of polymer/curing system/protective system/other additives.)

An experimental design should then be devised to produce a vulcanizate of the correct hardness as well as specific gravity. In one procedure, another triangular diagram is drawn up and a series of compounds are devised and prepared. Such compounds would have a specific gravity close to 1·2 and by experience would be expected to be fairly close to the expected hardness. The hardness of the compounds is then measured and marked on the diagrams. By interpolation, lines of equal hardness are plotted. The two diagrams are then compared and the point of intersection between the line passing through all compositions with a specific gravity of 1·2 and the line passing through compositions with a hardness of 60 will give the desired composition. A second, somewhat simpler, procedure is to make up three or four compounds with compositions on the line joining compositions with a specific gravity of 1·2 and to measure the hardness of these compounds. These figures are inserted along the line and by inspection and interpolation the compound giving the correct hardness is indicated. This compound is then prepared and tested to ensure that it meets the specification. This latter process is used in Fig. 22.2.

If three properties require a specific value (as opposed to being 'less than' or 'greater than') the position is much more complicated. An extra variable will be required (perhaps using both an inert and a reinforcing filler). In this case the compounds will be represented by a tetrahedron and the possible ones possessing a specific value of a property will be represented by a plane

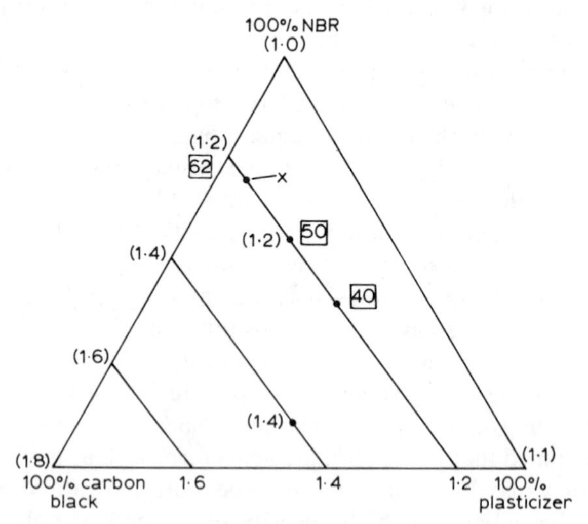

FIG. 22.2. The numbers in brackets are for specific gravities of various compositions obtained by calculation from the given data for polymer, carbon black and plasticizer blend. The numbers in square boxes are for hardness obtained from experimental compounds devised to have the correct density. The point X, obtained by interpolation, gives the predicted required composition.

rather than a line. The point where the three planes intersect will give the correct formulation. Similarly, to match four properties four-dimensional structures will be required, the point where three spaces meet providing the formulation. Such exercises would clearly require a computer program.

22.5 MASTERBATCHING TECHNIQUES

Much laboratory mixing is carried out on a two-roll mill, a process which without due care will lead to considerable batch-to-batch variation. It is important that the operator works as closely to a prescribed mixing schedule as possible but, even with care in both weighing and adding ingredients used in only small quantities, such as vulcanizing agents and accelerators, it is difficult to ensure that the correct amount of additive has been incorporated. Much of the error may be reduced if components such as the accelerators are made up as concentrates or masterbatches in the rubber, perhaps in the ratio 2:1 rubber/accelerator. The accelerator is then added as a masterbatch, reducing as appropriate the original amount of rubber because of the extra rubber being incorporated due to the masterbatch. Further reduction of scatter in the results may often be achieved by blending two test compounds to make additional ones. In the example given in Section 22.4, just two compounds of the correct specific gravity but markedly different composition could have been prepared. By blending these in various ratios (e.g. 1:1, 2:1, 1:2) a series of intermediate compounds could easily be prepared. Whilst the amount of error at this blending stage will be small, these compounds will be as good as the two original mixes and errors in these would give systematic errors of varying extents in the blends. By not blending there is likely to be greater scatter of the results but less hidden systematic error.

Blending is more useful when it is desired to study the general effect of varying compound composition rather than where it is important to know the absolute value of a property in order to develop a composition whose properties meet a required specification.

Appendix A

Specific Gravities of Some Common Additives

A.1 FILLERS AND OTHER MINERAL ADDITIVES (EXCLUDING COLOURANTS)

Aluminium silicate	2·10	Magnesium oxide	3·60
Barium sulphate (barytes)	4·50	Mica	2·95
Barium sulphate (blanc fixe)	4·25	Silica	1·95
Calcium silicate	2·10	Talc	2·80
Carbon black	1·81	Titanium dioxide (anatase)	3·90
China clay	2·50	Titanium dioxide (rutile)	4·20
Kieselguhr	2·20	Whiting (ground)	2·70
Litharge	9·30	Whiting (precipitated)	2·62
Lithopone (30% ZnS)	4·15	Zinc carbonate	3·30
Lithopone (50% ZnS)	4·06	Zinc oxide	5·57
Magnesium carbonate	2·21		

A.2 PLASTICIZERS, SOFTENERS AND RELATED ORGANIC ADDITIVES

Bitumen	1·04	Mineral oil (naphthenic)	0·932
Coumarone–indene resin	1·11	Mineral oil (paraffinic)	0·861
Dibutyl phthalate	1·045	Paraffin wax	0·90
Dibutyl sebacate	0·94	Petroleum jelly	0·90
Di(C_7–C_9) phthalate	0·995	Phenolic resin	1·27
Di(2-ethylhexyl) phthalate	0·985	Pine tar	1·08
Di-iso-octyl adipate	0·927	Polypropylene sebacate	1·060
Di-iso-octyl phthalate	0·984	Tritolyl phosphate	1·165
Dinonyl phthalate	0·970	Trixylyl phosphate	1·137
Factice	1·04	Zinc stearate	1·06
Mineral oil (aromatic)	1·018		

A.3 COMPONENTS OF CURING SYSTEMS

Accelerators	
CBS	1·30
DOTG	1·19
DPG	1·19
MBT	1·49
MBTS	1·54
TET	1·30
TMTD	1·42
TMTM	1·38
ZIX	1·54
ZMBT	1·64
Magnesia (light calcined)	3·60
Oleic acid	0·90
Stearic acid	0·85
Sulphur	2·05
Zinc oxide	5·57

A.4 MINERAL PIGMENTS

Antimony sulphide	3·30
Cadmium sulphide	4·40
Chromium oxide	5·21
Iron oxide (red)	5·14
Lead chromate	5·70
Ultramarine Blue	2·35
Vermilion	8·20

Appendix B

The Mooney Viscometer

In most chapters of this book reference has been made to Mooney viscometer values. This note is intended to provide a brief outline of the test and an explanation of the designations used when reporting results.

The viscometer was introduced by Mooney (1934) and consists essentially of a knurled disc which rotates slowly and continuously in one direction in a closed cavity which itself has a serrated or grooved surface (Fig. B.1).

In the usual test the rotation speed is $2 \pm 0{\cdot}02$ rpm. The space in the cavity is filled with rubber (which may be raw polymer, masticated polymer or compound) and the rubber allowed to warm up for a predetermined time before the rotor is started. A further specified interval then elapses before a reading is made of the torque required to rotate the shaft attached to the rotor. This is taken as a measure of the viscosity and is known as the Mooney viscosity.

Details of the test are given in ASTM D1646-87, ISO 289 and BS 1673 Part 3. Results would typically be expressed in the form

50-ML 1+4 (100°C)

which is interpreted as follows:

50-M indicates a Mooney viscosity of 50;

L indicates that a large rotor has been used (a small rotor may be used alternatively; it would be denoted by the letter S but this is seldom used);

1 indicates the warm-up time (in minutes) before the rotor is started;

4 is the time of rotation (in minutes) before the reading is taken;

100°C is the test temperature.

A 1-minute warm-up is used almost universally and a 4-minute rotation is most commonly chosen with SBR, BR, EPR, CR, NBR, ACM, EVA and FKM elastomers. For these rubbers the test temperature is also usually 100°C. In the case of butyl rubbers a longer rotation time (commonly 8 minutes) is required to reach a steady value. Some suppliers also use

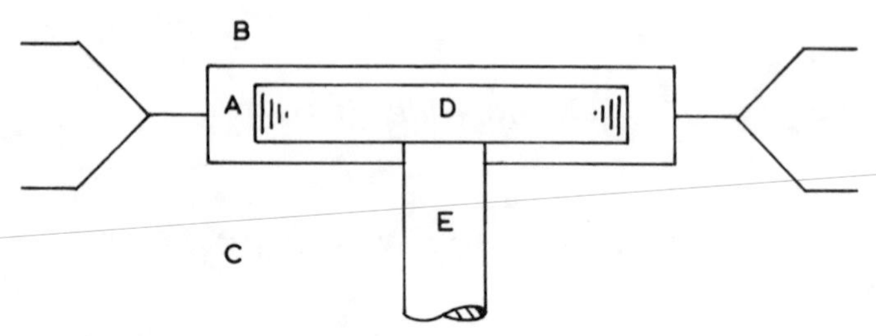

FIG. B.1. The Mooney viscometer. The chamber A is formed by two mating discs B and C. The serrated rotor D is driven via shaft E. The chamber surface is either serrated or grooved.

different test temperatures; in the case of higher-molecular-weight butyl rubber, 125°C is often specified. The International Institute of Synthetic Rubber Producers (IISRP) specifies a nominal Mooney viscosity for most of the Regular Institute Grades.

The Mooney viscosity gives some measure, *within a class of rubbers*, of the molecular size. Because of the very low shear rates involved, it gives little indication of the viscosities pertaining to high shear rate processes such as injection moulding.

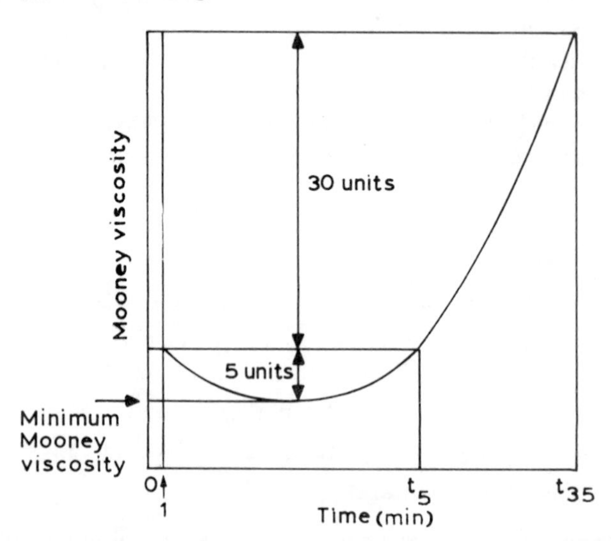

FIG. B.2. Use of Mooney viscometer for assessing scorch and cure rate. The scorch time is the time taken between start of warm-up and the point where the viscosity has risen five units above the minimum. The interval $(t_{35} - t_5)$ is known as the cure index, a low value indicating a high rate of cure.

The viscometer is also widely used to assess the tendency to scorch, and sometimes the rate of cure of a compound. In this case the machine is run at a temperature considered appropriate to the situation. For example, if during compounding and shaping the compound is subject to temperatures up to 120°C, the test could be run at that temperature. After a while curing will commence and it has been found that the time taken from the beginning of the warm-up period to that at which the Mooney value rises five units above its minimum value is a useful estimate of scorch behaviour and is known as the *scorch time* (Fig. B.2).

The time taken for the viscosity to rise from five units above the minimum to 35 units above is known as the *cure index* (sometimes, rather illogically, as the cure rate) at the test temperature. In practice, the test for cure would be more useful if carried out at the proposed vulcanization temperature. The Mooney equipment is somewhat limited for this purpose since when cure becomes well under way the rubber can no longer deform indefinitely and the rotor will start to slip. For this reason oscillating-disc rheometers and similar related equipment (discussed in Chapter 18) are more useful.

REFERENCE

Mooney, M. (1934). *Ind. Eng. Chem., Analyt. Edn*, **6**, 147.

Appendix C

A Short List of General Literature Sources

Bateman, L. (ed.) (1963). *The Chemistry and Physics of Rubber-Like Substances.* Applied Science, London.

Concerned largely with natural rubber, this may best be considered as a celebration and tribute to the work of the Malaysian Rubber Producers' Research Association and its predecessors.*

Blackley, D. C. (1966). *High Polymer Latices*, Vols 1 and 2. Applied Science, London.

In spite of its age this remains the most comprehensive review of this specialized topic.

Blackley, D. C. (1983). *Synthetic Rubbers—Their Chemistry and Technology.* Applied Science, London.

A scholarly work with greatest emphasis on the polymerization chemistry.

Blow, C. M. & Hepburn, C. (eds) (1982). *Rubber Technology and Manufacture*, 2nd edn. Published for the Plastics and Rubber Institute by Butterworth, London.

This is a comprehensive edited single-volume text giving broad coverage to the subject of rubber technology, excluding latex technology.

Brown, R. P. (1979). *The Physical Testing of Rubbers.* Applied Science, London.

The standard text on the subject.

Brydson, J. A. (1978). *Rubber Chemistry.* Applied Science, London.

This is an attempt to cover in 462 pages the chemistry of rubbery materials.

* It is understood that MRPRA are publishing a new book as part of their Golden Jubilee celebrations. It may be anticipated that this work will be of high quality.

Brydson, J. A. (1982). *Plastics Materials*, 4th edn. Butterworth, London.

In similar style to this book, but as the title suggests mainly concerned with plastics. There is some material on rubber and several chapters dealing with the relationship between structure and properties in a largely qualitative manner. A fifth edition is now in preparation.

Kennedy, J. P. & Törnqvist, E. G. M. (1968, 1969). *Polymer Chemistry of Synthetic Elastomers*, Parts 1 (1968) and 2 (1969). Interscience, New York.

Somewhat dated but excellent in its time. Chapter 2 of Part 1 (by Törnqvist) provides a fascinating account of the early history of synthetic rubbers.

Saltman, W. M. (ed.) (1977). *The Stereo Rubbers*. Wiley–Interscience, New York.

Mainly concerned with the polymerization chemistry of these rubbers.

Schidrowitz, P. & Dawson, T. R. (1952). *History of the Rubber Industry*. The Institution of the Rubber Industry, London.

An excellent source of information but see also Kennedy & Törnqvist (1968, 1969).

Whelan, A. & Lee, K. S. (eds) (1979, 1981, 1982, 1987). *Developments in Rubber Technology—1* (1979); *2* (1981); *3* (1982); *4* (1987). Applied Science, London.

Each volume consists of a set of reviews, primarily on materials aspects. These edited volumes have been frequently referred to in the text of this book.

In addition to books, the reader is also commended to *Rubber Chemistry and Technology*, a quarterly journal produced by the American Chemical Society containing both original articles and, periodically, excellent reviews. A further source is the *Encyclopedia of Polymer Science and Technology*, Wiley, New York. The multi-volumed first edition was produced in the 1960s whilst the somewhat less comprehensive second edition is now becoming available volume by volume and may be complete by the time this book is published.

Appendix D

Classification System for Vulcanized Rubbers

The American Society for Testing and Materials (ASTM) and the Society of Automotive Engineers (SAE) have issued a report entitled *Classification System for Rubber Materials for Automotive Applications* (SAE J200/ ASTM D2000); BS 5176 issued by the British Standards Institution is almost identical. The classifications do not at present apply to thermoplastic rubbers.

The classification provides basic classification codes based on test values obtained in standard laboratory tests. Each code is made up of two parts, the first part based on *basic requirements*, the second on *supplementary or suffix requirements*. These would provide, for a given specification, a *line call-out*.

The first part of the line call-out results from the basic requirements and consists of:

1. A letter M to indicate that SI units are being used (this may be followed or preceded, according to which specification is being used, by a number indicating supplementary requirements).
2. A letter indicating the type—assigned on the level of heat resistance required.
3. A letter indicating the class—assigned on the level of mineral oil resistance required.
4. A single figure indicating the hardness range.
5. A double figure indicating the tensile strength.

The *type* is based on changes of tensile strength of not more than 30%, an elongation change of −50% maximum and a hardness change of not more than 15 points. The temperatures at which the materials are tested for determining type are as shown in Table D.1

The *class* is based on the resistance of material to swelling in ASTM Oil No. 3 (or in the case of BS 5176, Oil No. 3 of BS 903). In test the immersion time shall be 70 h and the oil temperature shall be the type temperature or 150°C, whichever is the lower. Limits of swelling are listed in Table D.2.

TABLE D.1
Classification of rubber by type—assigned on the level of heat resistance

Type	*Test temperature (°C)*	*Type*	*Test temperature (°C)*
A	70	F	200
B	100	G	225
C	125	H	250
D	150	J	275
E	175		

TABLE D.2
Classification of rubber by class—assigned on the level of resistance to mineral oil

Class	*Maximum volume swell (%)*	*Class*	*Maximum volume swell (%)*
A	No requirement	F	60
B	140	G	40
C	120	H	30
D	100	J	20
E	80	K	10

The basic mechanical properties are indicated by a single figure for hardness followed by a double figure for tensile strength. Thus the set 810 indicates a Shore hardness of 80 ± 5 and a tensile strength of 10 MPa minimum. (The BS specification uses IRHD and with slightly different tolerance limits of $+5$ to -4.)

An example of an ASTM D2000 line call-out would be

M6 HK 810 A1-11 B38 EF31

This would indicate the use of SI units, an elastomer of Type H (resistant to 250°C) and of Class K (10% swell), a hardness of 80 ± 5 Shore and a minimum tensile strength of 10 MPa. The number 6 after M provides a key to supplementary (suffix) requirements which are detailed in the second part of the call-out (A1-11 B38 EF31). Interpretation of suffix requirements will require reference to details in the appropriate standards but it may be noted that in the sample call-out these refer to special heat resistance, compression set and fuel resistance requirements.

Index